The Classification of

LOWER ORGANISMS

By Herbert F. Copeland

It is usually accepted as a law of nature that all living things make up two kingdoms, plants and animals. As to nearly all creatures visible to the naked eye, this view is sound; but the microscope has revealed various groups which are not definitely either plants or animals, some of them being quite definitely neither.

During the nineteenth century, a series of scientists—the first was Bory de St. Vincent, and the best known was Haeckel—proposed placing the questionable organisms in a separate kingdom. These proposals failed to win acceptance. Their authors disagreed as to the groups to be referred to the proposed third kingdom. To the generality of biologists, it appeared that recognition of a third kingdom would create intolerable confusion.

The present work resumes the endeavor of Haeckel and his predecessors to establish a satisfactory system of kingdoms. Every effort has been made to avoid the deficiencies by which they fell short of success. Definite limits are given to the kingdoms of plants and animals. To escape the extreme heterogeneity of the "third kingdom" as proposed by some authors, the organisms excluded from plants and animals are organized as two kingdoms instead of one, *Mychota* and *Protoctista*. The resulting system is as follows.

Kingdom I. *Mychota*. Organisms without nuclei; the bacteria and blue-green algae.

Kingdom II. *Protoctista*. Nucleate organisms not of the characters of plants and animals; the protozoa, the red and brown algae, and the fungi.

Kingdom III. *Plantae*. Organisms in whose cells occur chloroplasts, being plastids of a bright green color, containing the pigments chlorophyll *a*, chlorophyll *b*, carotin, and xanthophyll, and no others; and which produce sucrose, true starch, and true cellulose.

Kingdom IV. *Animalia*. Multicellular organisms which pass during development through the stages called blastula and gastrula; typically predatory, and accordingly consisting of unwalled cells and attaining high complexity of structure and function.

Complete systems of the phyla, classes, and orders of Mychota and Protoctista are set forth, and the lower groups within these kingdoms are presented to a considerable extent. Scientific names are applied in strict conformity to the recognized principles of nomenclature, with detailed documented synonymy.

THE AUTHOR

Herbert F. Copeland is the son of E. B. Copeland, the distinguished botanist who, under authority of the Insular government and University, established the Philippine College of Agriculture. By teaching his classes that many of the so-called lower plants are not as a matter of nature plants at all, his father put before him the problem to which this book offers a solution.

He holds the academic degrees B.A. Wisconsin 1922; M.S. California 1925; Ph.D. Stanford 1931. Since 1927 he has been a teacher of biological sciences in Sacramento Junior College. He is the author of numerous technical papers, chiefly in the field of morphological botany. This is his first book. It represents forty years of study and reflection, of which more than the last ten have been spent in the actual composition.

The Classification of Lower Organisms

Ernst Heinrich Haeckel in 1874
From Bölsche (1906).
By permission of Macrae Smith Company.

The Classification

of

LOWER ORGANISMS

By

HERBERT FAULKNER COPELAND

PACIFIC BOOKS

PALO ALTO, CALIFORNIA

409
C782c

Published by

PACIFIC BOOKS

Palo Alto, California

Printed and bound in the United States of America

CONTENTS

CHAPTER   PAGE

I. INTRODUCTION 1
II. AN ESSAY ON NOMENCLATURE 6
III. KINGDOM MYCHOTA 12
    Phylum Archezoa 17
        Class 1. Schizophyta 18
            Order 1. Schizosporea 18
            Order 2. Actinomycetalea 24
            Order 3. Caulobacterialea 25
        Class 2. Myxoschizomycetes 27
            Order 1. Myxobactralea 27
            Order 2. Spirochaetalea 28
        Class 3. Archiplastidea 29
            Order 1. Rhodobacteria 31
            Order 2. Sphaerotilalea 33
            Order 3. Coccogonea 33
            Order 4. Gloiophycea 33
IV. KINGDOM PROTOCTISTA 37
V. PHYLUM RHODOPHYTA 40
        Class 1. Bangialea 41
            Order Bangiacea 41
        Class 2. Heterocarpea 44
            Order 1. Cryptospermea 47
            Order 2. Sphaerococcoidea 47
            Order 3. Gelidialea 49
            Order 4. Furcellariea 50
            Order 5. Coeloblastea 51
            Order 6. Floridea 51
VI. PHYLUM PHAEOPHYTA 53
        Class 1. Heterokonta 55
            Order 1. Ochromonadalea 57
            Order 2. Silicoflagellata 61
            Order 3. Vaucheriacea 63
            Order 4. Choanoflagellata 67
            Order 5. Hyphochytrialea 69
        Class 2. Bacillariacea 69
            Order 1. Disciformia 73
            Order 2. Diatomea 74
        Class 3. Oomycetes 76
            Order 1. Saprolegnina 77
            Order 2. Peronosporina 80
            Order 3. Lagenidialea 81
        Class 4. Melanophycea 82
            Order 1. Phaeozoosporea 86
            Order 2. Sphacelarialea 86
            Order 3. Dictyotea 86
            Order 4. Sporochnoidea 87

v

36662

CHAPTER   PAGE

        Order 5. Cutlerialea 88
        Order 6. Laminariea 89
        Order 7. Fucoidea 91

VII. PHYLUM PYRRHOPHYTA 94
    Class Mastigophora 95
        Order 1. Cryptomonadalea 96
        Order 2. Adiniferidea 98
        Order 3. Cystoflagellata 99
        Order 4. Cilioflagellata 102
        Order 5. Astoma 105

VIII. PHYLUM OPISTHOKONTA 110
    Class Archimycetes 111
        Order 1. Monoblepharidalea 111
        Order 2. Chytridinea 113

IX. PHYLUM INOPHYTA 119
    Class 1. Zygomycetes 121
        Order 1. Mucorina 121
        Order 2. Entomophthorinea 124
    Class 2. Ascomycetes 125
        Order 1. Endomycetalea 129
        Order 2. Mucedines 130
        Order 3. Perisporiacea 131
        Order 4. Phacidialea 133
        Order 5. Cupulata 134
        Order 6. Exoascalea 137
        Order 7. Sclerocarpa 137
        Order 8. Laboulbenialea 140
    Class 3. Hyphomycetes 140
        Order 1. Phomatalea 141
        Order 2. Melanconialea 141
        Order 3. Nematothecia 141
    Class 4. Basidiomycetes 142
        Order 1. Protobasidiomycetes 146
        Order 2. Hypodermia 147
        Order 3. Ustilaginea 149
        Order 4. Tremellina 149
        Order 5. Dacryomycetalea 150
        Order 6. Fungi 150
        Order 7. Dermatocarpa 152

X. PHYLUM PROTOPLASTA 157
    Class 1. Zoomastigoda 157
        Order 1. Rhizoflagellata 158
        Order 2. Polymastigida 163
        Order 3. Trichomonadina 166
        Order 4. Hypermastigina 168
    Class 2. Mycetozoa 171
        Order 1. Enteridiea 171
        Order 2. Exosporea 177

CHAPTER	PAGE
Order 3. Phytomyxida	177
Class 3. Rhizopoda	179
Order 1. Monosomatia	183
Order 2. Miliolidea	185
Order 3. Foraminifera	185
Order 4. Globigerinidea	187
Order 5. Nummulitinidea	188
Class 4. Heliozoa	189
Order 1. Radioflagellata	190
Order 2. Radiolaria	194
Order 3. Acantharia	195
Order 4. Monopylaria	198
Order 5. Phaeosphaeria	198
Class 5. Sarkodina	200
Order 1. Nuda	201
Order 2. Lampramoebae	205
XI. PHYLUM FUNGILLI	206
Class 1. Sporozoa	207
Order 1. Oligosporea	209
Order 2. Polysporea	211
Order 3. Gymnosporidiida	211
Order 4. Dolichocystida	214
Order 5. Schizogregarinida	215
Order 6. Monocystidea	215
Order 7. Polycystidea	216
Order 8. Haplosporidiidea	218
Class 2. Neosporidia	219
Order 1. Phaenocystes	219
Order 2. Actinomyxida	221
Order 3. Cryptocystes	222
XII. PHYLUM CILIOPHORA	223
Class 1. Infusoria	228
Order 1. Opalinalea	228
Order 2. Holotricha	229
Order 3. Heterotricha	230
Order 4. Hypotricha	233
Order 5. Stomatoda	233
Class 2. Tentaculifera	235
Order Suctoria	235
LIST OF NOMENCLATURAL NOVELTIES	237
BIBLIOGRAPHY	238
INDEX	271

LIST OF ILLUSTRATIONS

Portrait of Ernst Heinrich Haeckel . . . . Frontispiece

FIGURE   PAGE

1. Structure of cells of blue-green algae 13
2. Photographs of *Escherichia coli* 15
3. Caulobacterialea; Myxobactralea; *Cristispira Veneris* 26
4. Coccogonea; Gloiophycea 32
5. Bangialea 42
6. Nuclear phenomena in *Polysiphonia violacea* 45
7. Heterocarpea 48
8. Ochromonadalea 54
9. Ochromonadalea; Silicoflagellata 56
10. Vaucheriacea 64
11. Choanoflagellata 68
12. Hyphochytrialea 70
13. Bacillariacea 72
14. Oomycetes 78
15. Stages of nuclear division in Stypocaulon 84
16. Familiar kelps of Pacific North America 90
17. Microscopic reproductive structures of *Laminaria yezoensis* . . . 92
18. Cryptomonadalea 97
19. Cystoflagellata; Cilioflagellata 104
20. Astoma 106
21. Astoma 108
22. Monoblepharidalea 114
23. Chytridinea 116
24. Zygomycetes 122
25. Ascomycetes 132
26. Ascomycetes 136
27. *Mycosphaerella personata* 138
28. Basidiomycetes 144
29. Fruits of Agaricacea 153
30. Rhizoflagellata 160
31. Polymastigida; Trichomonadina 164
32. Hypermastigina 170
33. Mycetozoa 176
34. *Ceratiomyxa fruticulosa* 178
35. Life cycle of "*Tretomphalus*," i. e., *Discorbis* or *Cymbalopora* . . . 180
36. Shells of Rhizopoda 184
37. Radioflagellata 192
38. Radiolaria; Acantharia; Monopylaria; Phaeosphaeria 196
39. *Chaos Protheus* 200
40. Sarkodina 204
41. Life cycle of *Goussia Schubergi* 208
42. Life cycle of *Plasmodium; Babesia bigemina* 212
43. Life cycle of *Myxoceros Blennius* 220
44. Infusoria, order Hypotricha 232
45. *Tokophrya Lemnarum* 234

Chapter I

INTRODUCTION

The purpose of this work is to persuade the community of biologists that the accepted primary classification of living things as two kingdoms, plants and animals, should be abandoned; that the kingdoms of plants and animals are to be given definite limits, and that the organisms excluded from them are to be organized as two other kingdoms. The names of the additional kingdoms, as fixed by generally accepted principles of nomenclature, appear to be respectively Mychota and Protoctista.

These ideas originated, so far as I am concerned, in the instruction of Edwin Bingham Copeland, my father, who, when I was scarcely of high school age, admitted me to his college course in elementary botany. He thought it right to teach freshmen the fundamental principles of classification. These include the following:

The kinds of organisms constitute a system of groups; the groups and the system exist in nature, and are to be discovered by man, not devised or constructed. The system is of a definite and peculiar pattern. By every feature of this pattern, we are inductively convinced that the kinds of organisms, the groups, and the system are products of evolution. It is this system that is properly designated the natural system or the natural classification of organisms. It is only by metaphor or ellipsis that these terms can be applied to systems formulated by men and published in books.

Men have developed a classification of organisms which may be called the taxonomic system. Its function—the purpose for which men have constructed it—is to serve as an index to all that is known about organisms. This system is subject to certain conventions which experience has shown to be expedient. Among natural groups, there are intergradations; taxonomic groups are conceived as sharply limited. Natural groups are not of definite grades; taxonomic groups are assigned to grades. When we say that Pisces and Filicineae are classes, we are expressing a fact of human convenience, not a fact of nature. The names assigned to groups are obviously conventional.

Since the taxonomic system represents knowledge, and since knowledge is advancing, this system is inherently subject to change. It is the right and duty of every person who thinks that the taxonomic system can be improved to propose to change it. A salutary convention requires that proposals in taxonomy be unequivocal: one proposes a change by publishing it as in effect; it comes actually into effect in the degree that the generality of students of classification accept it. The changes which are accepted are those which appear to make the taxonomic system, within its conventions, a better representation of the natural system. Different presentations of the taxonomic system are related to the natural system as pictures of a tree, by artists of different degrees of skill or of different schools, are related to the actual tree; the taxonomic system is a conventionalized representation of the natural system so far as the natural system is known.

These statements are intended to make several points. First, as a personal matter, advancement of knowledge of natural classification, and corresponding improvement of the taxonomic system, have been my purpose during the greater part of a normal lifetime. Secondly, I have pursued this purpose, and continue to pursue it, under the guidance of principles which all students of classification will accept (perhaps with variations in the words in which they are stated). In the third place, I have tried to answer the question which scientists other than students of classification, and likewise the laity, are always asking us: why can one not leave accepted classification undisturbed? One proposes changes in order to express what one supposes to be improved knowledge of the kinds of organisms which belong together as facts of nature. If here I place bacteria in a different kingdom from plants, and Infusoria in a different kingdom from animals, it is because I believe that everyone will have a better understanding of each of these four groups if he does not think of any two of them as belonging to the same kingdom.

The course of evolution believed to have produced those features of the natural system to which the present work gives taxonomic expression is next to be described.

Life originated on this earth, by natural processes, under conditions other than those of the present, once only. These are the opinions of Oparin (1938)[1], and appear sound, although some of the details which he suggested may not be. When the crust of the earth first became cool, it was covered by an atmosphere of ammonia, water vapor, and methane, and by an ocean containing the gases in the atmosphere above it and minerals dissolved from the crust. This is to state the hypotheses that organic matter in the form of methane is older than life; and that whereas conditions on the face of the earth tend now to cause oxidation, they tended originally to cause reduction. In a medium of the nature of the supposed primitive ocean, spontanous chemical changes will occur and produce organic compounds of considerable complexity: this has repeatedly been demonstrated by experiment. To convert a solution of ammonia, methane, and minerals into protoplasm, Oparin postulates a very long series of changes, producing successively more complicated compounds and mixtures, and requiring perhaps hundreds of millions of years. The changes are conceived as accidents; they are supposed to have been probable accidents, like throwing a seven at dice, not events which could only very rarely occur by accident, like throwing twenty sevens in succession. By supposing that some of these processes used up the materials necessary for them, Oparin provides an explanation of the single origin of life: we are confident that all life is of one origin, because all protoplasm is of the same general nature, and all life consists of essentially the same processes. The course of events described would have yielded, as the original form of life, anaerobic saprophytes; this is in harmony with the fact that anaerobic energesis is in a sense the basic metabolic process. The original organisms would scarcely have possessed nuclei: Oparin's theories indicate, as the most primitive form of life which has been able to survive, the anaerobic bacteria. The anaerobic bacteria are indeed very far removed from any lifeless things; their protoplasm and their metabolism are fundamentally the same as ours.

Life requires energy. Under anaerobic conditions, an organism can obtain energy by converting sugars to alcohol, but it can not use alcohol as a source of energy. This example means that anaerobic energesis yields energy in strictly limited quantity and produces incompletely oxidized compounds. So long as all life was anaerobic, it was engaged in converting the organic matter upon which it depended into forms which it could not use; life under these conditions, at least if they persisted for any great period of time, was surely very sluggish. A further series of changes in the metabolic system, occurring accidentally in certain organisms and preserved by natural selection, brought photosynthesis into existence. The purple bacteria are believed to represent stages in the evolution of photosynthesis, which exists in its fully developed form, involving the release of elemental oxygen, in the blue-green algae. Once photo-

---

[1] Dates in parentheses are references to works which have been consulted and listed in the bibliography.

synthesis was established in certain organisms, aerobic energesis became possible both to these and to others. This made possible a manner of life more vigorously active than before. The inconsiderable groups of autotrophic bacteria—the organisms which live by oxidizing inorganic matter—appear to be secondary developments dependent upon the existence of photosynthesis.

The organisms whose origin has been suggested thus far—the ordinary bacteria, anaerobic and aerobic, the autotrophic bacteria, the purple bacteria, and the blue-green algae—are relatively simple in structure and function; all consist of minute physiologically independent cells. The first step in the evolution of more complex organisms was the evolution of the nucleus.

Morphologically, the nucleus is a part of a protoplast which is set apart by a membrane and which originates ordinarily by division of a pre-existent nucleus in the manner called mitosis. In this process, a definite number of definite chromosomes appear and undergo equal division. The nucleus exercises control over the protoplast in which it lies. Its controlling action depends upon the chromosomes which go into it, and mitosis has the effect that all nuclei which are derived from one original nucleus strictly by normal processes of mitosis are identical in the controlling effects which they exert. Thus the nucleus serves for the precise transmission of a complicated heredity. Beside mitosis, there are two other processes—two only—meiosis and karyogamy, by which nuclei may produce other normal and enduringly viable nuclei. In a sequence of generations of individuals sexually produced, these processes occur alternately, each one at one point in each cycle of sexual reproduction. Mendelian heredity is produced by changes, in the sets of chromosomes (or parts of chromosomes) in individual nuclei, which occur during meiosis and karyogamy. The role of the nucleus in sexual reproduction is one of its essential characters: the nucleus is related to sexual reproduction, including Mendelian heredity, as structure to function.

The existence of organisms without nuclei shows that the nucleus evolved after life did: it did not evolve at the same time as protoplasm. The essential uniformity of the nucleus and of its association with sexual reproduction shows that these things evolved only once, and together. There are a very few organisms, as *Porphyridium* and *Prasiola,* in which the presence or absence of nuclei is not certain; there is accordingly scant evidence for speculation as to the manner of this evolution. As to the time, we know only that microfossils representing nucleate organisms occur in the uppermost strata of the Proterozoic era.

By making possible the precise transmission of a complicated heredity, the nucleus has made possible the development of complexities of structure and function exceeding by far anything occurring in non-nucleate organisms. It appears that as soon as the nucleus was in existence, organisms provided with it entered upon evolution in many characters and gave rise to many distinguishable groups. Among these groups, those which consist respectively of the typical plants and the typical animals are the greatest. There is, however, neither any *a priori* reason, nor any evidence from nature, for a belief that all groups of nucleate organisms must naturally belong to one or the other of these two. Several other groups, in general much less considerable than these, are thoroughly distinct and appear equally ancient.

E. B. Copeland understood the history of life very much as it has just been presented. In his teaching, he treated the bacteria and blue-green algae as standing entirely apart both from plants and from animals, and pointed out several other groups which are not as a matter of nature either plants or animals. It was his opinion that these groups should be treated as a series of minor kingdoms; he excused himself from the attempt to formulate a definite and comprehensive system. This teaching was the original stimulus which has led to the present work. I bear witness that E. B. Copeland taught these things in 1914; he did not publish them until he had ceased to teach (1927).

In the year 1926, when the teaching of elementary botany was first fully my own responsibility, I came to the conclusion that the establishment of several kingdoms of nucleate organisms in addition to plants and animals is not feasible; that all of these organisms are to be treated as one kingdom. This is one of the few points of originality which I claim for my work. It is true that the kingdom thus described is not very different from the third kingdom which various early authors proposed and which Haeckel (1866) named Protista. Haeckel, however, in his varied presentations of the kingdom Protista, included always the bacteria. By setting apart the bacteria and blue-green algae as yet another kingdom, one meets, at least in part, the objection to the "third kingdom" that it is heterogeneous beyond what can be tolerated.

It has been necessary to meet also the objection that the "third kingdom" substitutes, for an acknowledgedly vague boundary between plants and animals, two vague boundaries: it has been necessary to recognize characters by which sharp definition can be given to plants and animals. It is my contention that these characters have long been known. The kingdom of plants, as the taxonomic representation of a natural group, is to be defined by the system of chloroplast pigments described by Willstätter and Stoll (1913), and also by the production of certain carbohydrates which occur only sporadically elsewhere. The kingdom of animals is defined by embryonic development through the stages called blastula and gastrula, as pointed out by Haeckel (1872). It is believed that no organisms exhibit both of these sets of characters; the "third kingdom" includes the nucleate organisms which exhibit neither. The kingdoms of plants and animals as here defined are essentially those which are traditionally and popularly accepted. They include all the creatures which Linnaeus listed as plants and animals, with the exceptions of forms of which he knew little, and which he listed superficially at the ends of his treatments of the respective kingdoms.

Of course, the definitions are not warranted to describe the kingdoms without exception. For one thing, each is supposed to have come into existence by evolution through a line of organisms which exhibited its characters imperfectly. For another, evolution can erase what it has created; it is proper to include in a group organisms which have by degeneration lost its formal characters. These things are true of all taxonomic groups.

In due form, then, the system of kingdoms here maintained is as follows:

Kingdom I. Mychota. Organisms without nuclei; the bacteria and blue-green algae.

Kingdom II. Protoctista. Nucleate organisms not of the characters of plants and animals; the protozoa, the red and brown algae, and the fungi.

Kingdom III. Plantae. Organisms in whose cells occur chloroplasts, being plastids of a bright green color, containing the pigments chlorophyll *a*, chlorophyll *b*, carotin, and xanthophyll, and no others; and which produce sucrose, true starch, and true cellulose.

Kingdom IV. Animalia. Multicellular organisms which pass during development through the stages called blastula and gastrula; typically predatory, and accordingly consisting of unwalled cells and attaining high complexity of structure and function.

This system has twice been given brief publication (1938, 1947). I am glad to say that Barkley (1939, 1949) and Rothmaler (1948) maintain a system of kingdoms which differs from this in a single significant detail.

Assuming that this system is tenable as a matter of reason, it will nevertheless not be accepted among taxonomists unless they have some knowledge of what it means in detail. No person is called upon to recognize the kingdoms Mychota and Protoctista until systems of their subordinate groups are available. The bulk of the present work consists of such systems. Complete systems of divisions or phyla, classes, and orders are presented. Groups of lower rank are presented in part, as examples. As a matter of facility, the groups of lower rank are presented more fully in the smaller or better known groups than in the larger or more obscure.

The preparation of this work has taken more than ten years. In the course of it I have received much help. Among those who have answered queries, or who have in various drafts scrutinized the whole work or parts of it for faults of every degree of significance, are Dr. G. M. Smith of Stanford University; Dr. A. S. Campbell of St. Mary's College; Dr. Herbert Graham, formerly of Mills College; Dr. Lee Bonar, Dr. G. L. Papenfuss, and Dr. H. L. Mason of the University of California at Berkeley; Dr. E. R. Noble of the University of California at Santa Barbara; and Dr. H. C. Day of Sacramento Junior College. The counsel of E. B. Copeland has not been withheld. It is a matter of grief that two distinguished zoologists of the University of California, Dr. S. F. Light and Dr. Harold Kirby, have passed away during the long course of this work; as have two colleagues who were my closest friends, Dr. H. J. Child and Dr. C. C. Wright.

The portrait of Haeckel which is my frontispiece is used by permission of Macrae Smith Company, Philadelphia. Two figures of *Chrysocapsa* are used by permission of the Cambridge University Press. Numerous figures have been taken from the *Archiv für Protistenkunde* with the gracious permission of Prof. Dr. Max Hartmann.

We do well to realize our indebtedness to libraries and librarians. To a great extent, this work has been made possible by the unstinted hospitality of the Biology Library of the University of California at Berkeley.

Two statements appear regularly in prefaces; they are of truths which are strongly impressed upon authors. In the first place, those who have given help have made the work better; the author alone is responsible for deficiencies. The foregoing list of good friends and good scholars does not claim them as proponents of the thesis of this work.

In the second place, the work is not offered as perfect or nearly so. The scholar in a strictly limited field may become master of the available knowledge. One who attempts studies in a broad field realizes that he is dealing with many subjects of which others know far more than he; that he has not wrung dry the existing literature; that some of the problems which puzzle him will be solved if he will wait a little longer. His colleagues have a right to raise these matters as criticisms. But surely, it is not desired that studies in broad fields be never attempted or indefinitely delayed.

A matter which is particularly likely to arouse criticism is that of the names which are here applied to the groups. The principles according to which this has been done are set forth in the following chapter. I beg my colleagues, in dealing with this chapter and with the names subsequently applied, not to imagine that I have acted without grave thought. I have decided, that as in classification, so also in nomenclature, I should set before the community of biologists an experiment in the application of principles; among which principles there are surely some whose strict application will be to the good of our science.

Chapter II
AN ESSAY ON NOMENCLATURE

Whoever sets forth a system of groups finds himself under the necessity of making responsible decisions as to names. The kingdoms have received more names than one (Table 1), and so have nearly all of the major groups within them: it has here been necessary to decide as to the validity and application of the names Flagellata and Mastigophora, Rhodophyceae and Florideae, Rhizopoda and Sarcodina, and many others.

TABLE 1. Names Applied by Various Authors to the Kingdoms
of Systems of Four Kingdoms

Authors {	Haeckel, 1894	Copeland, 1938, and Barkley, 1939	Rothmaler, 1948	Copeland, 1947 and here
Kingdoms {	I Protophyta	Monera	Anucleobionta	Mychota
	II Protozoa	Protista	Protobionta	Protoctista
	III Metaphyta	Plantae	Cormobionta	Plantae
	IV Metazoa	Animalia	Gastrobionta	Animalia

In dealing with plants, with animals, or with bacteria, it is necessary to observe the codes of nomenclature enacted by international congresses for the respective groups: the botanical code (Fournier, 1867; Lanjouw, 1952), with amendments enacted in 1954; the zoological code of 1889 as amended in 1948 and 1953 (issue of an edition incorporating the amendments is expected; Hemming, 1954); and the bacteriological code (Buchanan et al., 1948). Breach of the appropriate code renders an author liable to the penalty of having his work treated as nullity.

The existence of three sets of rules for one thing, and the continual amendment of the older codes, are evidence of imperfection. It will not be purely destructive to point out certain anomalies in the codes as they stand.

The zoological code pretends to overrule the principles of grammar in treating specific epithets as names. It is true that some of these words are names: the *Catus* in *Felis Catus* is a name of the cat, and the *Mays* in *Zea Mays* is a name of maize. But the great majority are adjectives; the *sapiens* in *Homo sapiens* is not by itself a designation of man, and the *vulgare* in *Hordeum vulgare* is not a name of barley. It is a further offense against grammar that the code prescribes, as the names of all families of animals, adjectives in the feminine. Applied originally to families of birds, *Aves*, these names were unobjectionable; but the names of the kingdom and of the overwhelming majority of its subordinate groups are neuter.

The botanical code as published with its appendages makes a book of more than two hundred pages. A statement of principles, in which the last clause provides for exceptions, occupies two pages. The definite rules and recommendations occupy about thirty-five pages; one who studies them critically will find that they prescribe more than one procedure not warranted by principle. A list of names maintained or rejected irrespective of principle occupies about seventy pages. These things mean that current botanical nomenclature is only within limits a matter of rule; it is to a considerable extent governed by enactments of the nature of *ex post facto* laws and bills of attainder.

The bacteriological code is for the most part a condensation of an earlier edition of the botanical code. It includes the odd feature that the name of a genus of bacteria is to be changed if it had previously been used either among plants or among Protozoa. Since there is an earlier *Phytomonas* among flagellates, bacteriologists have given a new name to the bacterium *Phytomonas*. The avoidance of homonyms which they desire will not, however, be attained: no zoologist will allow a new name for the flagellate *Klebsiella* on account of an earlier *Klebsiella* among bacteria.

The grounds upon which these things are treated as wrong are provided by a passage in the botanical laws of 1867 which is believed to define the legitimate authority of congresses and codes:

"Les règles de la nomenclature ne peuvent être ni arbitraires ni imposées. Elles doivent être bassées sur des motifs assez clairs et assez forts pour que chacun les comprenne et soit disposé à les accepter."

It is implied by this statement that principles, appealing to the reason and found sound by the trial of experience, were in existence when it was written; and this is the truth. By this statement, the legitimate powers of congresses are those of courts of common law, which avoid the explicit making of law, but discover the law, interpret it, and apply it. Congresses and codes may legitimately (a) state explicitly corollaries of the principles when they are not obvious; and (b) determine arbitrarily matters which are necessarily determined arbitrarily, not being within the range of principle. One would not in theory deny a power (c) to validate breaches of principle when these are of an expedience verging on necessity; but its use by botanical congresses to produce a roll of exceptions of twice the bulk of the text of the code leads one to doubt the expedience of this admission. It has been through failure to recognize the legitimate limits of their powers—through a conception that their powers are sovereign or plenary—that international congresses have come to enact codes conflicting with each other and giving incomplete satisfaction in themselves.

Under these circumstances, a nomenclature of superior legitimacy can be applied in groups treated as removed from the jurisdiction of the codes. Not without diffidence, this assumption is extended to the bacteria; it will be agreed that the nomenclatural practice applied to the bacteria must be the same as that which is applied to the blue-green algae.

Here one attempts a brief formulation of those principles, appealing to reason and proven sound in practice, to which all nomenclature must conform.

1. Scientific names are words of the Latin language. They are not "of Latin form" or "construed as Latin"; they are Latin. This is to treat Latin as a living language and scientific names as subject to the rules of its grammar. They are not code-designations, nor words of any language or none, as chemical names are.

2. The name of a group of the kind called a genus is a proper noun in the singular. Linnaeus replaced all generic names which were adjectives; all of us his successors should do likewise.

3. The names of groups of genera are proper nouns, or adjectives used as proper nouns, in the plural.

The foregoing principles are of pre-Linnaean origin; beginning with his first significant work (1735), Linnaeus took them for granted. For the principle next to be stated, authority is the practice of Linnaeus in later works (1753 and subsequently):

4. The name of a species consists of the name of the genus to which it belongs followed by one epithet, ordinarily an adjective, occasionally a noun in apposition or in the genitive.

A fifth principle represents Linnaean practices as subsequently modified:

5. Named taxonomic groups are necessarily of certain fixed ranks called categories, i.e., lists. There are seven principal categories, specified as follows. Every individual organism belongs to a group conceived as the single kind and called a species. Every species belongs to a genus; every genus to a family; every family to an order; every order to a class; every class to a division or phylum; every division or phylum to a kingdom. These conventions have the effect that the groups of each principal category embrace the entire range of the kinds of organisms.

The categories of genera and species come down from classic antiquity. Linnaeus originated orders; he originated classes in the sense of named definite groups; and it appears that he is responsible for kingdoms: the writer knows of no earlier authority for the traditional three kingdoms of nature. The category next below that of kingdoms has been variously called; originally it was *embranchements* (Cuvier, 1812). The history of the category of families is somewhat involved. It originated in the work of Adanson (1763); in the following year, Linnaeus (1764) treated the groups which Adanson had called families as natural orders. Botanists for a long time held that families and orders are the same thing. Zoological practice gradually made families a separate category. Authority for the list of seven principal categories as given is Agassiz (1857).

Nothing prevents the assignment of groups to categories other than these, to subclasses, tribes, and the like. These may be called subordinate categories. The groups of any subordinate category embrace only fragments of the range of kinds of organisms.

The work of Linnaeus was largely innovation, and he did not have the face to declare binding the generally accepted rule of priority. Definite authority for the rule is de Candolle (1813). As currently applied, it may be stated as follows:

6. The valid name of a group is its oldest published name, conforming to the rules, and not previously applied in the same kingdom.

As corollaries of the rule of priority, when groups are combined, the oldest name of any of them must be applied to the whole, and when a group is divided, its name must be retained for one of the parts. The part to which the original name is to be applied is determined by the method of types, formulated by Strickland and his associates (1843):

7. When a group is divided, its name must be applied to the portion which includes whatever part of it the original author would have regarded as typical. The part thus specified is the nomenclatural type of the group.

In the application of these principles to the naming of the groups of Mychota and Protoctista, the following practices appear expedient.

A name is applied by publication in such fashion that the community of biologists may reasonably be held responsible for knowing of its existence and recognizing the entity to which it is to be applied. This means that it is to be printed in a technical book or journal and defined in a language for which the generality of biologists will not require an interpreter, namely Latin, English, French, or German. Any regulation more detailed than this is an excuse for breaches of priority. Definition is not necessarily by description: nearly all of the Linnaean genera of plants were established by the listing of species in the *Species Plantarum*.

When two or more groups published in the same work at the same time are to be combined, their names are of equal priority. The choice of one of their names by the first author who combines them is binding.

A type as specified in the original publication of a group, or as implied by the inclusion of a single subordinate group, is unchangeable. Linnaeus and his immediate successors had no conception of the device of types, and it is practically impossible to be certain of the elements which they would have regarded as typical in some of their groups. It remains necessary that the type system be applied to these groups. In some of them, it may be expedient that international authority, proceeding with due caution, declare types arbitrarily. An individual scholar will do better to call what he supposes to be the type of a group by a different term, namely standard (Sprague, 1926) : the standard of a group is a supposed type which remains open to debate. The framers of codes have undertaken to make binding the choice of a type by the first author who divides a group. On various occasions, however, this action has been demonstrably mistaken.

Certain venerable names, as Vermes and Algae as used by Linnaeus, were applied to altogether miscellaneous collections of organisms among which the selection of a standard would be purely arbitrary. Such names are called *nomina confusa,* and are to be abandoned.

It follows from the principle of the binomial nomenclature of species that no genus is named until one or more of its species are designated by binomial names. It follows also that in works in which the nomenclature of species is not definitely binomial no names are of any standing. Hence, the point of time from which priority is effective is that of the introduction of binomial nomenclature, namely 1753. The enactment of other starting points for the nomenclature of particular groups is pretended law which is not law, like the pretended laws of American states which attempt to regulate interstate commerce under the appearance of doing something else.

The original spelling of names, so far as it is tolerable Latin, is not to be changed. Errors of gender or number, obvious mistakes of spelling, and misprints, are to be corrected. Good Latin is written without diacritical marks: a German *Umlaut* in a name as published is corrected by inserting an *e*; accents, cedilles, and other barbarisms are dropped. The codes err in prescribing changes in spelling beyond those which are here admitted. If they should establish uniformity in the future, it would be at the expense of divergence from the most respected works of the past.

Specific epithets are capitalized if they are (1) names in the nominative, in apposition with the generic names; (2) names of persons, places, or organisms in the genitive; (3) adjectives derived from names of persons.

Transfer of groups from one kingdom to another does not warrant any meddling with names. When a group is transferred from one kingdom to another, its valid name in the former—its oldest name not previously used in the kingdom in which it was originally published—has priority from the date of its original publication.

Names of groups higher than genera are in the plural. Some are proper nouns; the remainder are adjectives used as proper nouns, agreeing in gender with the names of the kingdoms in which they are included; either expressing characters of the groups which they designate, or consisting of generic names modified by terminations signifying "resembling" or "of the group of." Plurals of generic names are not tenable (de Candolle, 1813) : *Ericae* means the species of the genus *Erica*; it does not mean, and can not be used to designate, the genus together with its allies. Names consisting of words other than generic names modified by terminations signifying "resembling" or "of the group of" are not tenable, because they are nonsense: the name Coniferinae, applied by Engler to a class, is an adjective with an additional adjectival termination superimposed.

A name once applied in any principal category may not be transferred to another, unless it be of a form barred in the former and prescribed in the latter. The main clause of this statement is a consequence of the rule of priority. The exception is a concession to the practice of using names with uniform endings in certain categories.

Names of groups not of principal categories do not have priority as against names applied in principal categories. This practice, which denies to names in subordinate categories the full sanction of priority, is justified by the fact that groups in these categories are of concern only to specialists in the groups in which they occur; one is not in reason responsible for being aware of their names in groups outside of ones own specialty.

Almost all families of plants have had names with the uniform ending -*aceae* from the point of time at which the category of families was distinguished from that of orders. Such names were applied to algae, liverworts, and mosses by Rabenhorst (1863) and to higher plants by Braun (in Ascherson, 1864). They are adjectives in the feminine, agreeing with the name of the kingdom Plantae. It is altogether expedient that names of this form be held obligatory throughout the kingdom of plants. A uniform termination for names of families of animals has been in use for many years, but these names are not equally positively sound both grammatically and by priority. There has been a strong tendency to apply uniform terminations to the names of groups of other categories. So far as concerns groups of subordinate categories— suborders, subfamilies, and so forth—this practice appears expedient; these groups being of concern only to experts in the groups in which they occur, it is as well that their designations be of the nature of code designations rather than names. In attempting to put this practice into effect, some zoologists have made the mistake of applying the same adjective in different genders to different groups; they have not realized that Amoebida is the same word as Amoebidae. Meanwhile, uniform terminations for names of phyla, classes, and orders, beside involving wholesale violation of priority, is something of an insult to the intelligence.

The terminations of ordinal names in -*ales* and of family names in -*aceae*, currently in use among the Mychota, are here changed to -*alea* and -*acea* to agree with the neuter name of the kingdom. A change of the gender of an adjective does not create a new word, and the original authorities for the names will stand. Accordingly:

The name of an order of Mychota, if based on that of a genus, must bear the termination -*alea*. Names of this form are valid in no other category of this kingdom, and may be reapplied to orders. They have priority and authority by publication explicitly as orders. Such names do not supersede older ordinal names not based on names of genera.

The name of a family of Mychota is formed of the stem of a generic name (not necessarily a valid name, but never a later homonym) by adding the termination -*acea*. Names of this form are not valid in any other category, and may be reapplied to families. They have priority and authority by publication explicitly as families.

The names of families of Protoctista, unlike those of Mychota, of plants, and of animals, do not have by priority prevalently a uniform termination. Many of the oldest were first named in -*ina*. Those of flagellates and myxomycetes have double sets of names, respectively in -*aceae* and -*idae*, in current use. It is not expedient to impose uniform terminations on the names of these groups, at least not in the present work. Accordingly:

Each group of Protoctista is called by its oldest name of tenable form in the correct category, barring any previously used in other principal categories, irrespective of termination. All names which are adjectives are used in the neuter, but ascribed to the original authors.

The practices described have resulted in the use of many names which will seem strange, producing lists which are undeniably heterogeneous. A friendly critic notes as an example of these things the list of classes, Heterokonta, Bacillariacea, Oomycetes, and Melanophycea, on page 55. It will be realized that the three among these names which are adjectives must be in the feminine if the groups are construed as Plantae, neuter if Protoctista. Taking this fact into account, these are actually the first names, not previously used in other principal categories, applied to these groups as classes. What other names could one use? Everyone will know what groups are intended. Would any person understand them better if new names had been created by applying a uniform termination to the old roots?

Enough about nomenclature. We should begin to deal with organisms.

Chapter III
KINGDOM MYCHOTA

Kingdom I. MYCHOTA Enderlein

Stamm *Moneres* Haeckel Gen. Morph. 2: xxii (1866), in part.
SCHIZOPHYTAE Cohn in Beitr. Biol. Pfl. 1, Heft 3: 201 (1875).
Class SCHIZOPHYTA or *Protophyta* McNab in Jour. of Bot. 15: 340 (1877); not section *Protophyta* nor cohors *Protophyta* Endlicher (1836).
Kingdoms *Protophyta* and *Protozoa* Haeckel Syst. Phylog. 1: 90 (1894), in part; not *Protophyta* Endlicher nor class *Protozoa* Goldfuss (1818).
Subdivision *Schizophyta* Engler in Engler and Prantl Nat. Pflanzenfam. I Teil, Abt. la: iii (1900).
Division *Schizophyta* Wettstein Handb. Syst. Bot. 1: 56 (1901).
Phylum *Protophyta* Schaffner in Ohio Naturalist 9: 446 (1909), in part.
Kingdom MYCHOTA Enderlein Bakt.-Cyclog. 236 (1925).
Kingdom *Monera* Copeland f. in Quart. Rev. Biol. 13: 385 (1938).
Kingdom *Anucleobionta* Rothmaler in Biol. Zentralbl. 67: 248 (1948).
Organisms without nuclei.

The common name of Mychota in general is bacteria, but those which contain chlorophyll together with other pigments which make the green color impure are called blue-green algae.

The cells of Mychota are always separate or physiologically independent: multicellular bodies with distinct tissues do not occur. The cells are of various shapes; most often they are cylindrical, being of diameters from a fraction of one micron to a few microns, rarely more. Except in the groups of myxobacteria and spirochaets, they are walled; the thickness of the walls is of the order of 0.02μ (Knasyi, 1944). The walls may contain cellulose, but consist chiefly of pectates, compounds of slightly oxidized polysaccharides with sulfate, calcium, and magnesium (Kylin, 1943). These compounds are readily rendered gelatinous by hydration or hydrolysis, and the cells are often imbedded in gelatinous layers called sheaths or capsules.

In describing the Mychota as lacking nuclei, one commits himself to one side of a controversy of many years duration. Because of the greater size of the cells of the blue-green algae, the facts are more easily ascertained in this group than in the proper bacteria.

The cells of blue-green algae (Gardner, 1906; Swellengrebel, 1910; Haupt, 1923) are divided into outer and inner parts which are not sharply distinct. Pigments occur in a dissolved or colloidal condition in the outer part, which contains also granules of stored food. The granules are not carbohydrate, although a form of glycogen distinct from that of higher organisms has been extracted (Gardner; Kylin, 1943). The inner part contains rods and granules, some of which stain like chromatin, while others ("red granules of Bütschli") are stained red by methylene blue. Cell division is by constriction. Olive (1904) interpreted the inner part of the cell as a nucleus continually in process of mitosis, and accordingly without a membrane. It is true that in series of disk-shaped cells one may recognize series of corresponding granules. Where the cells are more elongate, the rods and granules of the interior are divided at random. Haupt expressed the impropriety of calling any part of these cells a nucleus.

Recent studies of typical bacteria by conventional microtechnical methods (Robinow, 1942, 1949; Tulasne and Vendrely, 1947) and by the electron microscope (Hillier, Mudd, and Smith, 1949) have made it possible to recognize the essential identity of the structure of their cells with those of the blue-green algae. The protoplast consists of outer and inner parts. The outer part, considered as a substance, may be called ectoplasm (Knasyi, 1930), and the inner, considered as a body, may be called the central body (Bütschli, 1890). The ectoplasm is very thin, occupying usually less than one fifth of the radius of the cell. The spiral bands which have often been seen

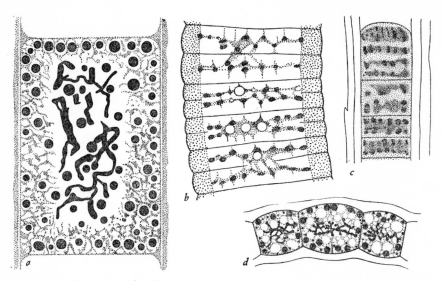

Fig. 1.—Structure of cells of blue-green algae. **a,** *Symploca Muscorum* after Gardner (1906). **b,** *Oscillatoria Princeps* after Olive (1904). **c,** *Lyngbya* sp. from a slide prepared by Dr. P. Maheshwari, x 1,000. **d,** *Anabaena circinnalis* after Haupt (1923) x 2,000.

in cells of bacteria, and which Swellengrebel (1906) mistook for a nucleus, are thickenings of the ectoplasm. Specific stains for nucleoprotein (chromatin), as Feulgen or Giemsa, usually color uniformly the entire central body. If the cells are exposed to hydrochloric acid, a part of the nucleoprotein, containing ribonucleic acid, dissolves. The remainder, containing desoxyribonucleic acid, persists in the form, basically, of a single fairly large granule in each cell. In rod-shaped bacteria, this granule appears usually to divide by constriction before the cell begins to divide, and may redivide, so that the cell may contain two dumb-bell shaped bodies. De Lamater and Hunter (1951) succeeded in a partial de-staining of the dumb-bell shaped bodies and interpreted them as dividing nuclei containing centrosomes and definite numbers of chromosomes; typical chromosomes, however, are never as small as the bodies they describe, and are not imbedded in bodies of nucleoprotein from which they can be distinguished only by the most refined technique. Enderlein (1916) observed in rod-shaped bacteria series of granules of which some at least are identical with the dumb-bell shaped bodies. He named these granules mychits. It might be held that the mychit is a chromosome, and the central body of bacteria a nucleus of a single chromosome, if it were not true that the blue-green algae contain comparable bodies of variable form and indefinite number.

Many bacteria swim by means of flagella. The diameter of the flagella, as revealed by the electron microscope, is of the order of 0.02μ. Their positions and lengths were made known, before the invention of the electron microscope, by the technique of Loeffler (1889), which consists essentially of depositing upon them a heavy layer of tannic acid. By the absence or presence and arrangement of flagella, bacteria are classified as of four types: atrichous, without flagella; monotrichous, with one flagellum at one end; lophotrichous, with a tuft of flagella at one end; peritrichous, with flagella on the sides.

Myxobacteria, spirochaets, and such blue-green algae as are sheathless filaments, are capable of bending movements (some spirochaets, observed with the electron microscope, are found also to have flagella at the ends of the cells). Spirochaets swim vigorously; in myxobacteria and blue-green algae, the bending movements are a matter of slow writhing. Filaments and cells of blue-green algae are capable also of a moderately rapid gliding movement. The mechanism of this movement has been the subject of much speculation, reviewed by Burkholder (1934), but remains uncertain. The appearance of the movement is as though it were caused by local secretion of substances affecting surface tension.

The normal reproduction of Mychota is by constriction of the cells, each into two equal daughter cells; whence the various names in *schizo-* (Greek σχίζω, to split). Henrici (1928) studied the changes undergone by bacteria during multiplication. As the cells become numerous, decreasing the food supply and producing substances harmful to themselves, they begin to attain greater length before dividing. Subsequently there is a gradual transition to enlarged and distorted forms called involution forms, which divide irregularly, cutting off minute fragments. These observations suggest the idea that the involution forms are the true normal forms of bacteria, the so-called normal forms being a temporary stage adapted to rapid multiplication under favorable conditions.

In many rod-shaped bacteria, when conditions cease to be ideal, the protoplasts produce within themselves walled bodies of dehydrated protoplasm called spores (endospores). In general, each cell produces only one spore. No experiment has definitely shown how long these spores can remain alive; it is surely a matter of centuries, doubtfully of millenia.

Löhnis and Smith (1916, 1923) observed of *Azotobacter* that numbers of protoplasts might escape from their walls and unite in a common mass, which they named the symplasm. The existence of this stage has never been confirmed by other authorities. If the symplasm exists, it is a device for achieving the effect which nucleate organisms attain by sexual reproduction, that is, combination of the heredity of different lines of ancestry.

That Mychota can actually combine characters from different lines of ancestry was first demonstrated beyond question by Tatum and Lederberg (1947). They mixed cultures of pairs of varieties of *Escherichia coli,* differing in two or more physiological characters, and isolated from the mixtures races having characters derived from both components. Further work, reviewed by Lederberg and Tatum (1953), has abundantly demonstrated phenomena analogous to typical sexual reproduction.

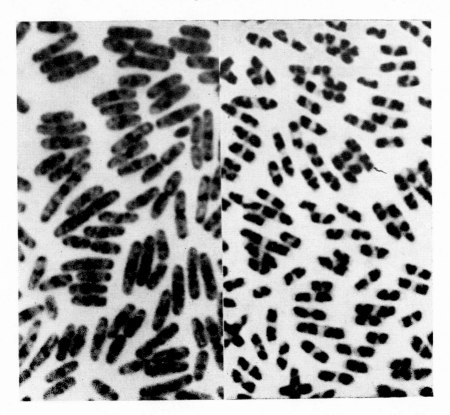

Fig. 2.—Photographs of *Escherichia coli* by Dr. C. F. Robinow, reproduced by Hillier, Mudd, and Smith (1949); left, stained to show the ectoplasm, in which there are thickenings which tend to be spiral; right, stained to show the large repeatedly dividing granule in the central body. About x 2,000. By courtesy of Dr. Robinow and of the Society of American Bacteriologists.

The metabolic systems of the Mychota are remarkably diverse. The most superficial list of physiological types would include the following: (a) anaerobic parasites and saprophytes; (b) facultatively aerobic parasites and saprophytes; (c) the vinegar bacteria, being apparently the only known organisms which, while requiring organic matter, are incapable of anaerobic energesis; (d) the autotrophic bacteria, the only organisms which maintain life by oxidation of inorganic matter; (e) organisms living by incomplete photosynthesis; and (f) organisms capable of typical photosynthesis.

Geologically, the Mychota are ancient. Iron deposits and certain other formations believed to have been produced by them occur in Archeozoic rocks estimated as more than a billion years old.

More than five thousand names have been applied to species of bacteria, but in the attempt to distinguish them, only about fifteen hundred are enumerated (Bergey's Manual, 6th ed., 1948). The species of blue-green algae are probably fewer than one thousand.

The classification of this group is inescapably highly tentative. The morphology is simple and not highly varied; the physiological characters likewise appear simple, but are highly varied, including many which are not known in other groups. The antiquity of the Mychota makes it probable that many groups which appear to belong together consist actually of parallel developments. The undoubted antiquity of the apparent main groups would lead one to place them in the category of divisions or phyla; but it is not expedient to make many divisions of a group of 2500 species: this would produce too many divisions of a single class or classes of a single order. The kingdom is accordingly treated as a single phylum, and its main divisions as classes.

Phylum **ARCHEZOA** Haeckel

Phyla *Archephyta* and Archezoa Haeckel Syst. Phylog. 1:90 (1894); not Phylum *Archephyta* Haeckel (1866).

Phylum *Myxophyceae* Bessey in Univ. Nebraska Studies 7: 279 (1907).

Phyla *Dimychota* and *Monomychota* Enderlein Bakt.-Cyclog. 236 (1925).

Bacteriophyta and *Cyanophyta* Steinecke (1931).

Stämme *Cyanophyta* and *Schizomycophyta* Pascher in Beih. bot. Centralbl. 48, Abt. 2: 330 (1931).

Divisions *Cyanophyta* and *Schizomycetae* Stanier and van Niel in Jour. Bact. 42: 464 (1941).

Characters of the kingdom.

Archezoa is Haeckel's name, at the point cited, for the bacteria. The name had been applied otherwise by Perty (1852), but not in a principal category. It will not be considered inappropriate, if it be remembered that the meaning of *zoe* is as much *life* as *animal*.

The conventional division of the group into two classes, bacteria and blue-green algae, is not perfectly natural. All of the recognized blue-green algae belong together; but the recognized bacteria are a wide miscellany, some of them belonging with the blue-green algae. Here three classes are recognized.

1. Cells without internal pigment, heterotrophic
   or living by chemosynthesis; not usually producing filaments with prominent sheaths.

2. Cells with firm walls, non-motile or
motile by means of flagella . . . . .Class 1. SCHIZOPHYTA.
2. Cells with thin walls or none, motile by
means of changes of shape, also some-
times by flagella . . . . .Class 2. MYXOSCHIZOMYCETES.
1. Cells mostly with internal pigment, living by
photosynthesis or chemosynthesis, exceptionally heterotrophic; often producing filaments
with prominent sheaths .Class 3. ARCHIPLASTIDEA.

Class 1. SCHIZOPHYTA (Cohn) McNab

Schizomycetes Nägeli ex Caspary in Bot. Zeit. 15: 760 (1857).
Class SCHIZOPHYTA or *Protophyta* McNab in Jour. of Bot. 15: 340 (1877).
Class *Schizomycetes* Winter in Rabenhorst Kryptog.-Fl. Deutschland 1, Abt. 1:
33 (1879).
Class *Schizomycetae* Schaffner in Ohio Naturalist 9: 447 (1909).
Classes *Holocyclomorpha* and *Hemicyclomorpha* Enderlein Bakt.-Cyclog. 236
(1925).
Dependent or chemosynthetic Mychota, with walled cells, without photosynthetic
pigments and not producing sheathed filaments.
This class includes as orders the typical bacteria and two minor groups.
1. Cells solitary or loosely gathered into clusters
or filaments, spherical, rod-shaped, or spiral,
not differentiated along the axis . . . . Order 1. SCHIZOSPOREA.
1. Consisting of branched filaments not divided
into cells . . . . .Order 2. ACTINOMYCETALEA.
1. Cells attached by stalks, the attached and
free ends differentiated . . . . .Order 3. CAULOBACTERIALEA.

Order 1. **Schizosporea** [Schizosporeae] Cohn in Hedwigia 11: 17 (1872).
Order *Schizomycetes* (Nägeli) McNab in Jour. of Bot. 15: 340 (1877).
Order *Eubacteria* Schröter 1886.
Order *Haplobacteriacei* Fischer in Jahrb. wiss. Bot. 27: 139 (1895).
Orders *Cephalotrichinae* and *Peritrichinae* Orla-Jensen in Centralbl. Bkt. Abt.
2,22: 334, 344 (1909).
Order *Eubacteriales* Buchanan in Jour. Bact. 2: 162 (1917).
Mychota whose cells in the typical condition are without internal pigment, walled,
of the form of rods, spheres, or spirals, not differentiated along the axis. As this is a
numerous group, likely with advancing knowledge to require division, it will be well
to provide it with a nomenclatural standard, and to suggest as such Cohn's principal
discovery among bacteria, namely *Bacillus subtilis*.
These are the typical bacteria. As originally described by Leeuwenhoeck (1677),
they were taken to be a few kinds of "animacules" distinguished only by extremely
small size. Only after many years were they shown to be numerous and varied, and
highly important as causes of diseases and of other natural phenomena.
The natural classification of the typical bacteria has been hard to discern. The
characters by which groups can be distinguished include forms of cells and of clusters
of cells; absence or presence and arrangement of flagella; non-formation or formation

of endospores; metabolic products; and the peculiar character called Gram reaction.

The method of staining invented by Gram, 1884, consists of staining successively with gentian violet and iodine. It gives an intense blue-black color. From some bacteria, this color is washed out by alcohol; others retain it; the former are said to be Gram negative, the latter Gram positive. In practice one applies successively gentian violet, iodine, alcohol, and safranine, the last being a red dye whose function is to make the Gram negative bacteria visible. The substance stained by gentian violet plus iodine is believed to be lipoid, such as occurs in all cells. The Gram positive quality is believed to consist in a relatively low isoelectric point, a capacity, that is, to combine with anions in a relatively acid medium. This quality lies in the ectoplasm of the cells and disappears in aging cultures.

The classification given in Bergey's Manual (1923, 1925, 1930, 1934, 1939, 1948) is accepted (at least among Americans) as standard. The following system of thirteen families is a moderate rearrangement of the Bergeyan system, with certain ideas or names from Enderlein (1917, 1925), Buchanan (1925), Pribram (1929) and Stanier and van Niel (1941).

1. Gram positive, with exceptions many of which are intracellular parasites; atrichous or peritrichous.
　　2. Spheres dividing in more planes than one.
　　　　3. Gram positive.................. Family 1. MICROCOCCACEA.
　　　　3. Gram negative; intracellular pathogens in animals.................. Family 2. NEISSERIACEA.
　　2. Rods, or spheres dividing in one plane.
　　　　3. Not producing endospores.
　　　　　　4. Atrichous.
　　　　　　　　5. Not intracellular parasites.. Family 3. CORYNEBACTERIACEA.
　　　　　　　　5. Intracellular parasites..... Family 4. RICKETTSIACEA.
　　　　　　4. Peritrichous................. Family 5. KURTHIACEA .
　　　　3. Producing endospores............. Family 6. BACILLACEA.
1. Gram negative.
　　2. Atrichous or peritrichous, requiring comparatively complicated organic food.
　　　　3. Not plant pathogens.
　　　　　　4. Not fixing nitrogen.
　　　　　　　　5. Capable of growth on ordinary media........... Family 7. ACHROMOBACTERIACEA.
　　　　　　　　5. Requiring special media; minute atrichous pathogens. Family 8. PASTEURELLACEA.
　　　　　　4. Fixing nitrogen.............. Family 10. AZOTOBACTERIACEA.
　　　　3. Plant pathogens................. Family 9. RHIZOBIACEA.
　　2. Atrichous, monotrichous, or lophotrichous; the atrichous representatives, and many others, can survive with organic foods simpler than carbohydrates, or with none.
　　　　3. Mostly requiring at least carbohydrates....................... Family 11. SPIRILLACEA.

3. Not requiring carbohydrates.
    4. Oxidizing alcohol to acetic
    acid, and acetic acid to CO_2
    and H_2O.....................Family 12. ACETOBACTERIACEA.
    4. Not as above; many examples
    strictly autotrophic............ Family 13. NITROBACTERIACEA.

Family 1. **Micrococcacea** [Micrococcaceae] Pribram in Jour. Bact. 18: 370, 385 (1929). Family *Coccaceae* Zopf 1884; but the genus *Coccus* is a scale insect. Gram positive spheres producing packets or irregular masses. *Micrococcus*, saprophytic or parasitic, producing irregular masses of cells; the pathogenic species have been treated as a separate genus *Staphylococcus*. *Sarcina*, saprophytic or commensal spheres producing packets.

Family 2. **Neisseriacea** [Neisseriaceae] Prévot ex Bergey et al. Manual ed. 5: 278 (1938). Family *Neisseriacées* Prévot in Ann. Sci. Nat. Bot. sér. 10, 15: 119 (1933). Obligate parasites, the Gram negative spherical cells occurring chiefly in pairs within leucocytes in the lesions of disease. *Neisseria gonorrhoeae*, the gonococcus; *N. Weichselbaumii* Trevisan (*N. intracellularis, N. meningitidis*, Auctt.), the meningococcus.

Family 3. **Corynebacteriacea** [Corynebacteriaceae] Lehmann and Neumann 1907. Family *Corynebacteriidae* Enderlein in Sitzber. Gess. naturf. Freunde Berlin (1917): 314. Family *Lactobacillaceae* Winslow et al. in Jour. Bact. 2: 561 (1917). Family *Lactobacteriaceae* Orla-Jensen 1921. Family *Leptotrichaceae* Pribram in Jour. Bact. 18: 372 (1929), not family *Leptotrichacei* Schröter 1886. Gram positive rods, or spheres dividing in one plane and producing chains, non-motile.

Streptococcus, spheres in chains; saprophytes in milk, involved in the making of butter and cheese; and commensals and serious pathogens causing, for example, abscesses, septicemia, erysipelas, and pneumonia.

Diplococcus, spheres usually in pairs, encapsulated. *D. pneumoniae* occurs in many immunologically distinct races which are the usual causes of pneumonia.

Lactobacillus, rods, microaerophilic, producing lactic acid. In milk, involved in the making of butter and cheese; in the oral cavity, being the usual agent of dental caries (Rosebury, Linton, and Buchbinder, 1929); common in sewage.

Leptotrichia, rods which become exceptionally long before dividing. Oral cavity of man and beasts.

Corynebacterium, rods, becoming club-shaped, staining in a banded pattern. The type species is the agent of diphtheria, *C. diphtheriae;* the genus includes also many harmless commensals important only as making diagnosis difficult. The cells divide in an exceptional fashion, by breaking violently from one side to the other near one end; the cut-off end swings around beside the main body and proceeds to grow. Repeated division in this manner produces clusters of parallel cells (Park, Williams, and Krumweide, 1924).

Family 4. **Rickettsiacea** [Rickettsiaceae] Pinkerton 1936. Families *Bartonellaceae* Gieszszykiewicz 1939 and *Chlamydozoaceae* Moshkovsky 1945. Minute obligate intracellular parasites of varied form, commonly Gram negative but with Gram positive granules.

There have been many observations of bodies of the characters stated, but a satisfactory classification of them is not yet possible. Howard Taylor Ricketts showed that Rocky Mountain spotted fever is transmitted by the tick *Dermocentor,* and observed, in the cells of diseased tissues, minute irregularly staining bodies; in 1910, in the course of further studies of the disease, he contracted it and died. Stanislas Prowazek, called into the Austrian military medical service in 1914, began to study typhus, which is transmitted by lice; observed similar intracellular bodies; contracted typhus, and died in February, 1915 (Hartmann, 1915). The cause of Rocky Mountain spotted fever is *Rickettsia Rickettsii,* and that of typhus is *R. Prowazekii.* Several other species are known. By serological methods, Anigstein (1927) showed that *R. Melophagi* is closely related to *Corynebacterium.*

In cases of the disease of the west slope of the Andes called verruga peruana, Oroya Fever, or Carrion's disease, there occur intracellular bodies named *Bartonella bacilliformis.* Noguchi and others (1928) completed the demonstration that the disease is transmitted by biting flies of the genus *Phlebotomus.* Good authority has construed *Bartonella* as a sporozoan.

Students of flagellates, Sarkodina, and Infusoria have occasionally observed in the cytoplasm or nuclei of these organisms minute bodies multiplying to form considerable masses. These parasites have generally been construed as chytrids, but have little in common with proper chytrids. The genus *Caryococcus* Dangeard includes at least a part of them.

Family 5. **Kurthiacea,** fam. nov. Gram positive peritrichous rods, not producing endospores. *Kurthia,* harmless; *Listeria* Pirie ex Murray in Bergey's Manual 6th ed. 408 (1948), pathogenic in sheep and man.

Family 6. **Bacillacea** [Bacillacei] Fischer in Jahrb. wiss. Bot. 27: 139 (1895). The spore-forming rods, always Gram positive, mostly peritrichous, very numerous in species, common, and important.

Bacillus Cohn 1872, is one of the oldest generic names of rod-shaped bacteria which can be definitely applied: it can be definitely applied because the type species *B. subtilis* was so described as to be recognizable. The genus has been used to include rods in general or at random. Defined as aerobic spore-formers, as proposed by Buchanan, 1917, it is a thoroughly natural group. As treated in the fifth edition of Bergey's Manual, it included nearly 150 duly distinguished species; in the sixth edition, this number is cut to thirty-three. The great majority are saprophytic. Exceptions, important pathogens, are *B. anthracis;* and *B. alvei* and other species causing foulbrood of bees.

The anaerobic spore-formers constitute the genus *Clostridium.* The type species was discovered and named three times in different connections. As an anaerobe involved in the fermentations which give butter its flavor, it is *C. butyricum* Prazmowski. As an organisms whose cells contain granules staining like starch, it is *Bacillus Amylobacter* van Tieghem. It has the property of fixing nitrogen; discovered in this capacity by Winogradsky (1902) it was named *C. Pastorianum.* The species of *Clostridium,* as of *Bacillus,* are numerous. They are primarily saprophytic, but many species produce powerful toxins and are serious pathogens. Examples are *C. tetani; C. botulinum;* and *C. septicum* and a whole roll of other species, causing various forms of gangrene, occasion for the study and distinction of which was found during World War I.

Family 7. **Achromobacteriacea** [Achromobacteriaceae] Breed 1945. Family *Bacteriaceae* McNab in Jour. of Bot. 15: 340 (1877), based on a generic name which must be abandoned as a *nomen confusum.* Family *Enterobacteriaceae* Rahn 1937, not based on a generic name. Gram negative rods which lack the dictinctive characters of the families subsequently to be treated.

The nine genera listed first occur normally in animals, mostly in the gut and mostly as commensals; exceptions are important pathogens. Most of them produce acid, and many of them produce gas, from sugar. These genera are the traditional colon-typhoid-dysentery group.

Escherichia coli, the colon bacillus, and *Aerobacter aerogenes,* the gas bacillus, are common commensals which produce acid and gas from dextrose and lactose. The standard method of testing waters for contamination is essentially a test for the presence of these organisms.

Klebsiella also produces acid and gas from sugars. It inhabits the respiratory tract. The cells are heavily capsulated and non-motile. The type species *K. pneumoniae* is an important pathogen, the pneumobacillus of Friedländer.

Proteus vulgaris (this is at least the third genus to bear the name *Proteus,* but the first in this kingdom) produces acid and gas from dextrose but not lactose, and liquefies gelatine. It is usually isolated from spoiled meat.

Salmonella is distinguished from *Proteus* by non-liquefaction of gelatine. Many of its species are harmless commensals; others cause paratyphoid fevers. Immunological study of cultures of *Salmonella* from cases of disease and from waters have resulted in the distinction of fully 150 races, mostly unnamed and identifiable only by immunological reactions. *Eberthella* includes motile rods producing acid but not gas from sugars, and belonging to the same immunological system as the various races of *Salmonella. Eberthella typhi* causes typhoid fever.

Shigella is distinguished from *Eberthella* by non-motility. The Shiga bacillus, *S. dystenteriae,* is the cause of dystentery.

Bacteroides is a numerous group of acid-producing gut bacteria, motile or non-motile, generally harmless, distinguished from the foregoing as strictly anaerobic.

Alcaligenes fecalis, an apparently harmless organism isolated from intestinal contents, does not produce acid from sugars; grown in milk, it produces an alkaline reaction.

Numerous races of bacteria which have been isolated from soil and are capable of attacking cellulose are assigned to the genus *Cellulomonas.* Bacteria which produce an extracellular red pigment are *Serratia* (one of the oldest generic names for bacteria); those which produce yellow pigment are *Flavobacterium;* those which produce blue, black, or violet growths are *Chromobacterium.* Cultures which lack the distinctive characters of all of the above named genera (most such cultures have been isolated from water) are called *Achromobacter.*

Family 8. **Pasteurellacea** nom. nov. Family *Parvobacteriaceae* Rahn; there is no corresponding generic name. Minute non-motile Gram negative rods, pathogenic, requiring special media for cultivation. *Pasteurella avicida* is the cause of chicken cholera, upon which Pasteur made important studies. Of greater direct importance to man is *Pasteurella pestis,* the cause of plague. *Hemophilus* includes the agents of whooping cough, soft chancre, and conjunctivitis. *Brucella* includes the organisms which cause Malta fever, undulant fever, Bang's disease, contagious abortion. *Pfeifferella mallei* is the cause of glanders.

Family 9. **Rhizobiacea** [Rhizobiaceae] Conn in Jour. Bact. 36: 321 (1938). Gram negative rods, atrichous or peritrichous, parasites on plants. Cultured in the presence of sugars, these organisms produce acid; they are evident allies of the colon group.

Erwinia commemorates Erwin F. Smith, the discoverer of many bacteria pathogenic to plants. Typical species cause blights, wilts, or dry necroses. The discovery by Burrill, 1882, of *Erwinia amylovora,* the cause of the fire blight of pears, should have prevented the formulation of a theory, once entertained, that all bacteria require neutral media, and are accordingly incapable of causing diseases of plants. The species of *Pectobacterium*, as *P. carotovorum*, cause rots. Those of *Agrobacterium* cause galls; *A. tumefaciens* causes crown gall of many plants.

Rhizobium includes the species which produce little galls ("nodules") on the roots of plants and which benefit their hosts by fixing nitrogen. The best known hosts of *Rhizobium* are plants of the family Leguminosae; the relationship between Leguminosae and *Rhizobium* is a classic example of symbiosis. There are several or many species of *Rhizobium*, scarcely distinguishable morphologically, but living on different groups of legumes. The race which was first recognized and isolated, *R. Leguminosarum* Frank 1890 (*Schinzia Leguminosarum* Frank 1879; *Bacillus Radicicola* Beijerinck 1888) is that which attacks plants of the pea tribe. Bewley and Hutchinson (1920) accounted for the variety of forms which *Rhizobium* can assume. In the roots of plants it occurs as involution forms. In culture, it is a peritrichous rod, but the flagella are often reduced to one, and it has been confused with the monotrichous bacteria (Conn and Wolfe, 1938).

Family 10. **Azotobacteriacea** [Azotobacteriaceae] Bergey, Breed, and Murray in Bergey's Manual 5th ed., preprint, v and 71 (1938). These are the organisms which were originally isolated by Beijerinck (1901) by inoculating with garden soil shallow layers of a nitrogen-free nutrient solution containing mannite. The commonest species, *Azotobacter Chroococcum*, is usually seen as ellipsoid cells, as much as 4µ thick and 7µ long, solitary, with peritrichous flagella, or forming non-motile clusters imbedded in a heavy capsule. Beijerinck observed the occurrence of globular involution forms as much as 15µ in diameter. Löhnis and Smith (1916) made a thorough study of variations in form, and reported a remarkable variety of other stages, including the symplasm.

The Pasteurellacea and Rhizobiacea are apparently reasonably close allies of the Achromobacteriacea. The Azotobacteriacea stand somewhat apart. The remaining families of the present order are more definitely distinct, being marked by monotrichous or lophotrichous flagella.

Family 11. **Spirillacea** [Spirillaceae] Migula 1894. Family *Pseudomonadaceae* Winslow et al. in Jour. Bact. 2: 555 (1917). Rods and spirals, Gram negative, monotrichous or lophotrichous; not producing much acetic acid, and mostly heterotrophic.

Pseudomonas is a numerous genus of rods which may or may not produce a fluorescent pigment soluble in water; they do not produce a yellow pigment which is insoluble in water. The original species, *P. aeruginosa,* was isolated from pus, in which it produces a blue-green discoloration; it is by itself weakly if at all pathogenic. Other species have been isolated from fresh and salt waters and brines; the bacteria which produce phosphorescence on salt fish are of this genus. Many further species are pathogenic to plants, producing chiefly leaf spots.

Phytomonas Bergey et al. 1923 (*Xanthomonas* Dowson 1948) includes numerous plant pathogens which in culture produce an insoluble yellow pigment; among them are the causes of cabbage rot, walnut blight, and leaf spots on many plants.

Pacinia Trevisan 1885 includes monotrichous curved rods. The type species *P. cholerae-asiaticae* is the cause of Asiatic cholera. Among numerous other species the majority are harmless saprophytes in waters. Recent authorities have treated the cholera organism as the type of the genus *Vibrio* Müller (1773); their action is an intolerable falsification of the usage of a full century preceding the discovery of the cholera organism.

Spirillum includes the typical spirals, lophotrichous, a small number of species of harmless saprophytes in foul waters.

Thiospira includes large lophotrichous spirals, colorless, containing granules of sulfur. They are believed to live by chemosynthesis.

Family 12. **Acetobacteriacea** [Acetobacteriaceae] Bergey, Breed, and Murray 1938. As gross objects, growths of *Acetobacter aceti* Beijerinck have been known since prehistoric times. With included yeasts they constitute mother of vinegar (the old names *Mycoderma mesentericum* Persoon, *Ulvina aceti* Kützing, and *Umbina aceti* Nägeli designated the combination of bacteria and yeasts, and it seems proper to reject them). Free-swimming cells with polar flagella have been observed; ordinarily the cells appear as rods in chains, heavily encapsulated, or as involution forms. The organic food required by *Acetobacter* is alternatively alcohol, which is oxidized to acetic acid, or acetic acid, which is oxidized to carbon dioxide and water. These processes are strictly aerobic: to make vinegar, one exposes wine to air; to preserve it, one seals the vessels.

Family 13. **Nitrobacteriacea** [Nitrobacteriaceae] Buchanan in Jour. Bact. 2: 349 (1917). Organisms oxidizing the simplest organic compounds; or facultatively capable of chemosynthesis; or living strictly by chemosynthesis and strictly aerobic: mostly Gram negative monotrichous or atrichous rods.

Methanomonas is capable of oxidizing methane; *Carboxidomonas* of oxidizing carbon monoxide; *Hydrogenomonas,* of oxidizing elemental hydrogen. *Thiobacillus* includes organisms which oxidize hydrogen sulfide or elemental sulfur.

Winogradsky had discovered chemosynthesis in the course of studies of *Beggiatoa* and other sulfur-oxidizing organisms before he undertook to isolate bacteria which cause nitrification, that is, the natural production of nitrates in soil and waters. He achieved success (1890) by inoculating, with soil or sewage, media which contained salts of ammonia but no food; he saw the nitrifying organisms first as minute motile rods which he named *Nitromonas.* Further study and the use of solid media showed that nitrification takes place in two stages and is the work of several kinds of organisms. Winogradsky distinguished *Nitrosomonas europaea* and *N. javanensis,* monotrichous rods from different regions as indicated, oxidizing ammonia to nitrites; *Nitrosococcus,* non-motile spheres from South Amerca, effecting the same oxidation as *Nitrosomonas;* and *Nitrobacter,* non-motile rods oxidizing nitrites to nitrates. Subsequent authors have validated Winogradsky's names by creating the combinations *Nitrosococcus nitrosus* and *Nitrobacter Winogradskyi.* Subsequently, Winogradsky discovered yet other bacteria capable of the same oxidations.

The presence of nitrifying bacteria is necessary for the normal growth of most crops. So active are the nitrifying bacteria that no more than traces of ammonia and nitrites are found in normal soils, and so avidly do plants absorb nitrates that these accumulate only in fallow fields.

Order 2. **Actinomycetalea** [Actinomycetales] Buchanan in Jour. Bact. 2: 162 (1917).

Organisms which consist typically of slender filaments not divided into cells, but which are capable of producing conidia, that is, minute spherical or elongate bodies cut off by constriction from the ends of the filaments, or of breaking up into cells of the form of regular or irregular rods. Non-motile; Gram positive or Gram negative; often of the staining character called acid fast.

The order may be treated as a single family.

Family **Mycobacteriacea** [Mycobacteriaceae] Chester 1907. Family *Actinomyce-taceae* Buchanan in Jour. Bact. 3: 403 (1918). Family *Streptomycetaceae* Waksman and Henrici 1943. Characters of the order. Three genera require discussion.

Streptomyces Waksman and Henrici 1943. The original name of this genus is *Streptothrix* Cohn (1875); there is an older genus *Streptothrix* among plants, and the numerous species of the present genus have generally been included in *Actinomyces*. Cultures are readily isolated from air or soil. They appear as slowly growing colonies which may at first be of various colors and have shiny surfaces. Their texture is tough; a blunt needle will more often tear a colony from the medium than penetrate it. As the colonies grow, they become truncate; the exposed surfaces become white and powdery; pigments, black, brown, red, or yellow, in various races, are produced, and discolor the medium. The toughness of the colonies is a consequence of their structure, of myriad crooked branching filaments about 1µ in diameter, without joints; the white and powdery surface is produced by myriad conidia released in basipetal succession. The cultures are of an odor which may be described as that of earth under the first rain after drouth: undoubtedly, this familiar odor is that of *Streptomyces* in the soil. Drechsler (1919), from careful study of several species of *Streptomyces*, concluded that they are fungi; their filaments are, however, much finer than those of fungi, and no definite nuclei have been seen.

Certain species of *Streptomyces* cause a scabbiness of potatoes. Except for this, the genus was for a long time regarded as quite unimportant. When the capacity of the fungus *Penicillium notatum* to inhibit the growth of bacteria had been observed, and had led to the discovery of the drug penicillin, Waksman, the leading authority on the classification of Actinomycetalea, sought comparable drugs produced by *Streptomyces,* and had the great success of discovering streptomycin.

Actinomyces Bovis Harz 1877 is one of several species of the same general nature as *Streptothrix* which are pathogenic to animals. It causes lumpy jaw of cattle.

Mycobacterium Lehmann and Neumann 1896 is typified by *M. tuberculosis,* the agent of one of the most important diseases of man, supposed originally to have attacked cattle, and to have spread around the world with European cattle. It is a chronic disease, destroying the tissues slowly and producing a nugatory sort of immunity which makes it possible to test for the disease, but does not check it. The cells are recognized in sputum and in diseased tissues by the acid fast reaction: the dye carbol fuchsin must be applied hot in order to color them; once it has done so, it does not wash out in acid alcohol. It is cultivated with difficulty. The growth is dry, powdery, wrinkled, with an odor described as sickening-sweet. It consists of branching filaments which break up readily into rod-shaped or irregular fragments.

Lesions of leprosy contain acid fast organisms named *Mycobacterium leprae.* Gay (1935) has discussed the results of attempts to cultivate this species. They have yielded either "diphtheroid" cells or a "streptothrix." He concludes that most of the reports are of the same organism reacting variously to various conditions.

Order 3. **Caulobacterialea** [Caulobacteriales] Henrici and Johnson in Jour. Bact. 29: 4 (1935).

Aquatic bacteria, the cells of most examples secreting gelatinous matter in such a manner as to produce stalks. Henrici and Johnson provided a system of four families, five genera, and nine species. Stanier and van Niel (1941) rejected the group as artificial, placing some of the genera among Eubacteria and leaving others unplaced. The order may be maintained for the accommodation of the latter and divided into two families.

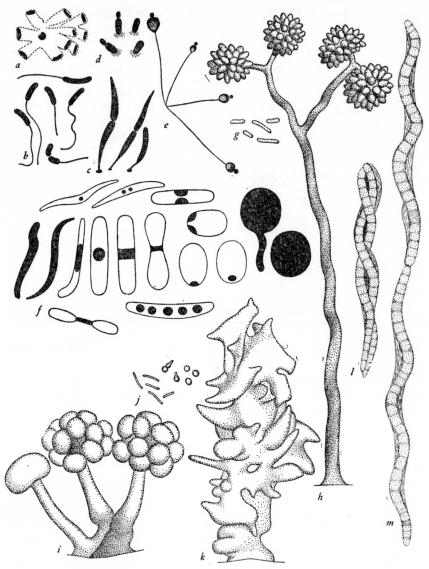

Fig. 3—**a-e,** CAULOBACTERIALEA after Henrici and Johnson (1935) x 2,000: **a,** *Nevskia* sp.; **b,** *Caulobacter vibrioides*; **c,** *Caulobacter* sp.; **d,** *Pasteuria* sp.; **e,** *Blastocaulis* sp. **f,** Various stages of *Cytophaga Hutchinsonii* (*Spirochaeta cytophaga*) after Hutchinson and Clayton (1919). **g-k,** MYXOBACTRALEA after Thaxter (1892), the cells x 1,000, in the fruits x 200. **g, h,** Cells and fruit of *Chondromyces crocatus;* **i,** fruit of *C. aurantiacus;* **j,k,** vegetative cells and spores, and fruit, of *Myxococcus coralloi des.* **l, m,** Dividing cells of *Cristispira Veneris* after Dobell (1911) x 2,000.

Family 1. **Leptotrichacea** [Leptotrichacei] Schröter 1886. The cells not elongated in the direction of the axis of the stalk.

Didymohelix ferruginea (Ehrenberg) Griffith (first named, and usually listed, under *Gallionella*, which is a misspelling of the name of a genus of diatoms) occurs in waters containing iron. Older authors described it as consisting of paired filaments, less than 1μ in diameter, colored bright yellow with imbedded iron oxide, and coiled about each other. In fact, the supposed paired filaments are the margins of a single twisted band, which is not itself an organism but the stalk secreted by a terminal cell. *Spirophyllum* Ellis is either the same species or a closely related larger one.

Leptothrix Kützing Phyc. Gen. 198 (1843) was inadequately described; the species which was first named, and which is accepted as the type, was *L. ochracea*. It is believed that this name properly designates the masses of ochraceous matter seen in iron springs. Under the microscope, this matter is seen to consist of fine yellow filaments, straight and unbranched. Ellis (1916) described them as consisting of a cylinder of protoplasm, not divided into cells, enclosed in a sheath. Almost surely, these structures, generally recognized as of the same nature as *Didymohelix*, are likewise stalks secreted by minute terminal cells.

Siderocapsa Molisch and *Sideromonas* Cholodny, described as minute spheres or rods imbedded in capsules colored by ferric oxide and attached to plants in waters containing iron, are perhaps to be interpreted as stalkless members of the present group.

Nevskia Famintzin, forming minute gelatinous colonies floating on water, does not accumulate iron.

Family 2. **Caulobacteriacea** [Caulobacteriaceae] Henrici and Johnson 1. c. (1935). The cells elongated in the direction of the long axes of the stalks. *Caulobacter, Pasteuria,* and *Blastocaulis,* colorless saprophytes in waters or parasites in aquatic animacules.

Class 2. MYXOSCHIZOMYCETES Schaffner

Class MYXOSCHIZOMYCETAE Schaffner in Ohio Naturalist 9: 447 (1909).
Class *Polyyangidae* Jahn Beitr. bot. Protistol. 1: 65 (1924).
Class *Spirochaetae* Stanier and van Niel in Jour. Bact. 42: 459 (1941).
Parasitic or saprophytic Mychota, the elongate cells with thin walls or none, capable of bending movements and sluggishly or actively motile. In many examples there is a resting stage: the cell contracts generally, so as to diminish the surface, and deposits a definite wall. The structure so produced is a spore of the type called an arthrospore or chlamydospore.

The two orders Myxobactralea and Spirochaetalea have not previously been combined to form a separate class. A certain species which Hutchinson and Clayton (1919) described as a spirochaet, *Spirochaeta cytophaga,* has subsequently been found to be a myxobacterium. The hint of relationship thus conveyed is confirmed by the whole character of both groups, as may be seen from the discussions of them by Stanier and van Niel (1941) and Knasyi (1944).

Order 1. **Myxobactralea** [Myxobactrales] Clements Gen. Fung. 8 (1909).
Order MYXOBACTERIACEAE Thaxter in Bot. Gaz. 17: 389 (1892).
Order *Myxobacteriales* Buchanan in Jour. Bact. 2: 163 (1917).
The cells not definitely of spiral form, sluggishly motile. In typical examples, the

cells occur in swarms imbedded in slime; the entire mass moves concertedly, and is eventually converted into macroscopically visible fruiting bodies.

The group was first recognized by Thaxter. He took note that the fruiting bodies of *Chondromyces* had already been described by Berkeley and Curtis as those of a gasteromycete, and learned subsequently that *Polyangium* Link, also described as of the puffball group, is an older name for his *Myxobacter*. The swarms of cells live in air on damp substrata (commonly the feces of various kinds of animals), moving across them and digesting and absorbing food as they proceed. Labratory culture is fairly easy. As a reaction, apparently, to exhaustion of the available food, the cells change into chlamydospores; the masses of spores held together by dried slime are called cysts. These may be borne on simple or branched stalks built up from the slime as a preliminary to the formation of the cysts and spores. The group is of essentially no economic importance.

The accepted classification is that of Jahn (1924); to the four families which he recognized, one more has been prefixed for the accommodation of the genus *Cytophaga*.

Family 1. **Cytophagacea** [Cytophagacae] Stanier 1940. The chlamydospores formed sporadically by individual cells, not in cysts. *Cytophaga Hutchinsonii* Winogradsky (*Spirochaeta cytophaga* Hutchinson and Clayton) is one of several species discovered as active fermenters of cellulose. The slenderly spindle-shaped cells are sluggishly motile, and produce ellipsoid chlamydospores resembling yeasts.

Family 2. **Archangiacea** [Archangiacae] Jahn op. cit. 66. Spores elongate in irregularly extensive masses, not in cysts. *Archangium, Stelangium*.

Family 3. **Sorangiacea** [Sorangiaceae] Jahn op. cit. 73. Spores elongate, the cysts angular, in masses, not stalked. *Sorangium*.

Family 4. **Myxobacteriacea** [Myxobacteriaceae] (Thaxter) E. F. Smith 1905. Family *Polyangiaceae* Jahn op. cit. 75. Spores elongate, in distinct rounded cysts, clustered or solitary, sessile or borne on simple or branched stalks. *Polyangium* Link 1795 (*Myxobacter* Thaxter 1892), *Stelangium, Melitangium, Podangium, Chondromyces*.

Family 5. **Myxococcacea** [Myxococcaceae] Jahn op. cit. 83. Spores spherical; cysts indefinite or definite. *Myxococcus, Chondrococcus, Angiococcus*.

Order 2. **Spirochaetalea** [Spirochaetales] Buchanan in Jour. Bact. 2: 163 (1917). Cells solitary, spiral in shape, actively motile.

The first known species of this group was *Spirochaeta plicatilis*, observed in foul waters by Ehrenberg (1838). The next was the species now known as *Borrelia recurrentis* (Lebert) Bergey et al., observed in the blood of relapsing fever patients by Obermeier, 1873.

During the last years of the nineteenth century, many attempts to identify the agent of syphilis by standard bacteriological methods were unsuccessful. The German government directed Schaudinn and Hoffmann to continue this work. Fritz Schaudinn, 1871-1906 (Stokes, 1931), had attained distinction as a student of pathogenic protozoa. Within a few weeks, by the microscopic examination of lesions, he attained success where the bacteriologists had failed, and discovered *Treponema pallidum* (Schaudinn and Hoffmann, 1905).

Spirochaets were first cultivated by Noguchi; few others have been successful in this difficult practice. It requires a medium of aseptic, not sterilized, animal material, under more or less anaerobic conditions. Each species requires its peculiar variant of the conditions, to which it is quite sensitive.

Spirochaeta plicatilis and other saprophytic species, together with certain species parasitic in mollusks, are fairly large. The species which are parasitic or commensal in other animals may be extremely small. It is chiefly by study of the larger species that the structure is known. The internal structure is septate. Dobell (1911) found in *Cristispira,* at the margin of each septum, a whorl of granules staining like chromatin, and interpreted these granules collectively as a nucleus. Noguchi (in Jordan and Falk, 1928) saw in the interior of the smaller species no chambered structure, but a lengthwise rod. This has been interpreted as a nucleus, as a locomotor or skeletal structure, or as an artifact. The electron microscope has shown actual flagella at the ends of cells of *Treponema pallidum.* Reproduction is normally by transverse division into two. During division, the daughter cells may coil about one another, giving a false appearance of lengthwise division. Gross (1913) observed that *Cristispira* is capable of breaking up into cylindrical *Stäbchen* corresponding to the chambers.

The discovery of *Treponema* by an eminent protozoologist; the character of spirochaetal diseases, several of which are spread by biting insects, and produce only that nugatory immunity which makes diagnosis possible but does not check the disease; and the supposed lengthwise division of the cells; led to the hypothesis that the spirochaets are protozoa. Dobell was surely correct in dismissing this hypothesis, insisting that the spirochaets are neither protozoa nor typical bacteria, but a group *sui generis.*

The larger and smaller spirochaets are reasonably treated as separate families.

Family 1. **Spirochaetacea** [Spirochaetaceae] Swellengrebel 1907. The cells comparatively large, 80-500μ long. *Spirochaeta, Saprospira, Cristispira.*

Family 2. **Treponematacea** [Treponemataceae] Robinson in Bergey Man. 6th ed. (1948). Family *Treponemidae* Schaudinn 1905. The cells 4-15μ long.

Treponema Schaudinn. The cells comparatively loosely coiled. *T. pallidum,* the agent of syphillis. *T. pertenue,* the agent of yaws. *T. macrodentium* and *T. microdentium,* harmless commensals in the mouth.

Borrelia Swellengrebel is doubtfully distinct from the foregoing; Noguchi reduced it. *B. recurrentis* and other species cause relapsing fevers. *B. Vincenti* causes Vincent's angina (trench mouth). The fusiform cells always found associated with it and supposed to be ordinary bacteria of a genus *Fusiformis* or *Fusobacterium* may be its chlamydospores.

Leptospira Noguchi. The cells tightly coiled. *L. icterohaemorrhagiae* is the agent of infectious jaundice. *L. icteroides,* isolated by Noguchi in South America, supposedly from cases of yellow fever, is perhaps the same thing: it is now known that yellow fever is caused by a virus. It was in pursuing in Africa his study of yellow fever that Noguchi lost his life by this disease (Flexner, 1929; Eckstein, 1931).

Class 3. **ARCHIPLASTIDEA** Bessey

Myxophykea Wallroth 1853.

Myxophyceae Stizenberger 1860.

Division (of Class *Algen*) *Phycochromaceae* and order Gloiophyceae Rabenhorst Krytog.-Fl. Sachsen 1: 56 (1863).

Cyanophyceae Sachs Lehrb. Bot. ed. 4: 248 (1874).

Order *Cyanophyceae* or *Phycochromaceae* McNab in Jour. of Bot. 15: 340 (1877).

Schizophyceae Cohn 1879, not suborder *Schizophyceae* Rabenhorst Deutschland's Kryptog.-Fl. 2, Abt. 2: 16 (1847).

Order *Schizophyceae* Schenck in Strasburger et al. Lehrb. Bot. 1894.
Class *Schizophyceae* Engler in Engler and Prantl Nat. Pflanzenfam. I Teil, Abt.
   1a: iii (1900).
Class Archiplastideae Bessey in Univ. Nebraska Studies 7: 279 (1907).
Class *Cyanophyceae* Schaffner in Ohio Naturalist 9: 446 (1909).
Class *Myxophyceae* G. M. Smith (1918).
Subclass *Myxophyceae* Setchell and Gardner in Univ. California Publ. Bot. 8,
   part 1: 3 (1919).
Cyanophyta Steinke (1931).
Stamm *Cyanophyta* Pascher in Beih. Bot. Centralbl. 48, Abt. 2: 330 (1931).

Mychota most of which live by phytosynthesis of primitive or typical character, many of them, and most of the saprophytic and chemosynthetic organisms included with them, being of the form of sheathed filaments.

This is primarily the group of the blue-green algae. Blue-green algae are familiar things as forming dark scums in water and on wet surfaces. Rabenhorst (1863) appears first to have recognized them as a group definitely distinct from green algae; he named most of the recognized families. Revisions by Thuret (1875), Bornet and Flahault (1886-1888), and Gomont (1892) failed to provide a satisfactory system of the group; Kirchner's revision (in Engler and Prantl, 1898) is the accepted system.

One of the important contributions of Cohn was his suggestion that the bacteria and blue-green algae belong together. He emphasized this view by mingling the genera of the two groups in two new groups, "tribes," named in effect slime-formers and thread-formers (1875). In this he went too far; but some of the arrangements which he suggested appear natural. *Beggiatoa,* the type of order Thiobacteria of Migula, appears to be a variant of the common blue-green alga *Oscillatoria,* differing from it in living by chemosynthesis. Most of the so-called iron bacteria, family Chlamydobacteriaceae of Migula, fall readily into scattered places among the blue-green algae. Only the genus *Sphaerotilus* remains at loose ends. It is credibly reported to produce cells swimming by means of flagella; no proper blue-green algae do this.

A variety of purple bacteria—bacteria, that is, which contain a red pigment—have been discovered from time to time. Engelmann (1888) observed that they swim toward the light, and convinced himself that they live by photosynthesis. Van Niel confirmed this, and showed that photosynthesis is in this group of a peculiar character; it requires the presence of reducing agents and does not release oxygen. This type of photosynthesis appears, in fact, to represent a stage of the evolution of typical photosynthesis; the group in which it occurs appears to represent the ancestry of the typical blue-green algae. The poorly known green bacteria appear to belong with the purple bacteria.

Various members of this class have been proved capable of fixing nitrogen (Sisler and ZoBell, 1951; Williams and Burris, 1952).

Four orders may be distinguished:

1. Possessing a red ("purple") intracellular
   pigment, or a green pigment not masked by
   others. . . . . Order 1. Rhodobacteria.
1. With green pigment masked by others, or
   colorless.
   2. Producing cells with flagella; non-pig-
      mented sheathed filaments not accumu-
      lating ferrugineous matter. . . . . Order 2. Sphaerotilalea.

2. Never producing cells with flagella.
  3. Cells dividing in more planes than one, growing to full size before re-dividing; unicellular or colonial, not filamentous.........................Order 3. Coccogonea.
  3. Cells dividing in one plane, and accordingly producing filaments; exceptional examples reproducing by budding (unequal division) or by repeated division into minute spores................................ Order 4. Gloiophycea.

Order 1. **Rhodobacteria** Molisch Purpurbakterien 27 (1907).

Rods, spheres, and spirals, solitary or colonial, with red or green pigment, performing in the presence of light and reducing substances a sort of photosynthesis in which no oxygen is released.

These organisms have generally been included in Thiobacteria, but do not include *Beggiatoa*, the type of that order. Molisch divided them into two families, Thiorhodaceae, aerobic, accumulating granules of sulfur, and Athiorhodaceae, microaerophilic or anaerobic, not accumulating granules of sulfur. The green bacteria are to be placed as a third family. The names originally applied to the families are not tenable.

Family 1. **Chromatiacea** (Migula) *nomen familiare novum*. Subfamily CHROMATIACEAE Migula. Family *Rhodobacteriaceae* Migula; Family *Thiorhodaceae* Molisch; the family does not include genera with corresponding names. Purple bacteria, aerobic, accumulating granules of sulfur. *Chromatium* Perty includes the organism of foul waters which was originally named *Monas Okenii*. It is a plump rod, often bent, sometimes exceeding 10μ in length, monotrichous or lophotrichous. There are a dozen other genera, rods, spheres, and spirals (*Thiospirillum*, which belongs here, is to be distinguished alike from *Spirillum, Thiospira*, and *Rhodospirillum*), solitary or colonial, motile or non-motile. Most of them were discovered by Winogradsky.

Family 2. **Rhodobacillacea** nom. nov. Family *Athiorhodaceae* Molisch. Molisch named in this family a genus *Rhodobacterium*, but the name Rhodobacteriaceae had already been applied by Migula to the preceding family. Purple bacteria, anaerobic, not accumulating granules of sulfur. Molisch discovered all known members of the present family. The method of culture was to place a mass of organic matter, for example an egg, in the bottom of a cylinder of water (the original account specified water of the River Moldau), cover the surface with oil, place in a north window, and wait several weeks. This method yielded organisms which were assigned to seven genera. Those of spiral form are *Rhodospirillum*. All others are by van Niel treated as a single genus, which may be called *Rhodobacillus* Molisch (*Rhodopseudomonas* van Niel).

Family 3. **Chlorobiacea** nom. nov. Family *Chlorobacteriaceae* Geitler and Pascher ex van Niel in Bergey's Manual ed. 6: 869 (1948). Geitler and Pascher (in Pascher Süsswasserfl. Deutschland, 1925) did not place this group in a definite category and name it unequivocally: they called it Cyanochloridinae or Chlorobacteriaceae. Minute spherical or elongate cells with a green pigment different from typical chlorophyll, anaerobic, non-motile, producing irregular or regular gelatinous colonies. *Chlorobium, Pelodictyon, Clathrochloris*, with a half a dozen known species. Certain

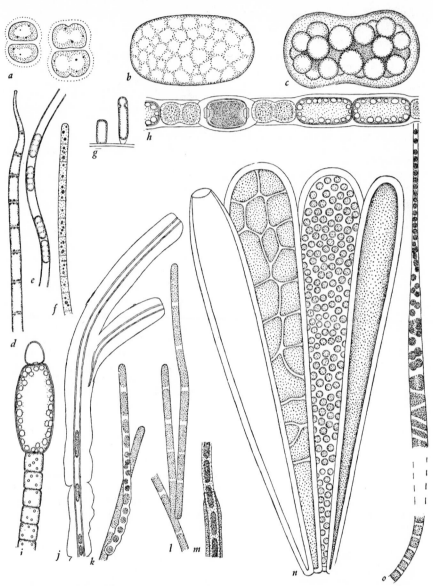

FIG. 4.—COCCOGONEA: **a**, *Chroococcus* sp.; **b**, **c**, *Achromatium oxaliferum*. GLOIO-
PHYCEA: **d**, *Oscillatoria splendida*; **e**, *Phormidium* sp.; **f**, *Beggiatoa* sp.; **g**, *Chamaesiphon incrustans*; **h**, *Anabaena inaequalis*; **i**, *Cylidrospermum majus*; **j**, *Chlamydothrix ochracea*; **k**, **l**, **m**, *Clonothrix fusca* after Kolk (1938); **n**, *Dermocarpa protea*
after Setchell and Gardner (1919); **o**, *Crenothrix polyspora* after Kolk (1938).   All
x 1,000.

organisms of this group, occurring in symbiotic combinations with larger bacteria or with protozoa, have been named as additional genera; one of these is *Chlorobacterium Lauterborn*, but the name is a later homonym.

Order 2. **Sphaerotilalea** nom. nov.
> Order *Desmobacteriales* Pribram in Jour. Bact. 18: 376 (1929); there is no corresponding generic name.

Cells colorless, elongate, in sheathed filaments which branch freely in the manner called "false": the cells divide strictly in one plane; those at a distance from the tip may so multiply as to break the continuity of the series by pushing a growing point laterally out of the sheath. The cells may escape from the filaments, become lophotrichous, and function as swarm spores. There is a single family:

Family **Sphaerotilacea** [Sphaerotilaceae] Pribram l. c. There is probably only one species, *Sphaerotilus natans* Kützing (*Cladothrix dichotoma* Cohn). It is found as minute gelatinous colonies floating on stagnant water; cells 2-4μ in diameter.

Order 3. **Coccogonea** [Coccogoneae] (Thuret) Campbell Univ. Textb. Bot. 84 (1902).
> Tribe *Chroococcaceae* (*Coccogoneae*) Thuret in Ann. Sci. Nat. Bot. sér. 6, 1: 377 (1875).
> Subclass *Coccogoneae* Engler in Engler and Prantl Nat. Pflanzenfam. I Teil, Abt. la: iii (1900).
> Order *Coccogonales* Atkinson 1903.
> Orders *Chroococcales* and *Entophysalidales* Geitler in Pascher et al. Süsswasserfl. Deutschland 12: 52, 120 (1925).

Cells solitary or colonial, not filamentous, never flagellate; mostly of blue-green color and living by photosynthesis.

Kirchner (in Engler and Prantl, 1898) placed here two families, Chroococcaceae and Chamaesiphonaceae, but the second belongs to the following order. A proper second family includes the colorless organisms of genus *Achromatium*.

Family 1. **Chroococcacea** [Chroococcaceae] (Nägeli) Rabenhorst Kryptog.-Fl. Sachsen 1:69 (1863). Order *Chroococcaceae* Nägeli Gatt. einzell. Alg. 44 (1849). Unicellular or colonial blue-green algae. *Chroococcus, Gloeocapsa, Merismopedia, Coelosphaerium, Gomphosphaeria*, etc., occur as plankton or as masses on damp surfaces or the bottoms of bodies of water. Certain species occur as symbionts or parasites within the cells of the green algae *Glaucocystis* and *Gloeochaete*. The resulting bodies, having the color of blue-green algae with the structure of green algae, resisted classification until Geitler (1923) explained their nature.

Family 2. **Achromatiacea** [Achromatiaceae] Buchanan. Cells solitary, large, ellipsoidal, without flagella, non-pigmented; protoplasm alveolar, with or without granules of sulfur in the alveoli. Half a dozen species have been described; Bersa (1920) was probably correct in reducing all to the original one, *Achromatium oxaliferum* Schewiakoff. It occurs on mud under still waters rich in organic matter.

Order 4. **Gloiophycea** [Gloiophyceae] Rabenhorst Kryptog.-Fl. Sachsen 1: 56 (1863).
> Tribe *Nostochineae* (*Hormonogoneae*) Thuret in Ann. Sci. Nat. Bot. sér. 6, 1: 377 (1875).

Family *Hormogoneae* Bornet and Flahault in Ann. Sci. Nat. Bot. sér. 7, 3: 337 (1886).

Subclass *Hormogoneae* Engler in Engler and Prantl Nat. Pflanzenfam. I Teil, Abt. la: iii (1900).

Order *Hormogoneae* Campbell Univ. Textb. Bot. 84 (1902).

Order *Hormogonales* Atkinson 1905.

Blue-green algae whose cells divide predominantly in one plane, so that filaments are produced, together with related colorless organisms.

So far as cell division is strictly in one plane, any branching of the filaments is of the type called "false": it occurs by breaks in the continuity of the series of cells, followed by the outgrowth, beside the original series, of the newly formed tips. In some members of the group, however, the cells are not strictly confined to division in one plane, with the result that "true" branching is possible. There are a few apparently derived examples in which cell division takes place freely in all planes.

Many of these algae produce spores of the type called arthrospores by the direct conversion of normal cells into thick-walled resting spores. Many (almost but not quite exactly the same ones which produce arthrospores) produce peculiarly differentiated cells called heterocysts (the word means "different cells"). These are enlarged thick-walled cells with colorless contents; their most obvious function is to furnish breaking points for the filaments. They are believed to be variants of the arthrospores; they have been seen to germinate and give rise to normal filaments.

Ten families may be distinguished as follows:

1. Cells dividing strictly in one plane; branching none or of the "false" type.
   2. The filaments not branching nor tapering nor producing spores or heterocysts.
      3. Filaments elongate.
         4. Pigmented, blue-green . . . . Family 1. OSCILLATORIACEA.
         4. Colorless organisms accumulating sulfur . . . . Family 2. BEGGIATOACEA.
      3. Filaments reduced to single cells which reproduce by budding . . . . Family 3. CHAMAESIPHONACEA.
   2. Filaments branching or tapering, or producing spores or heterocysts, or showing several of these characters.
      3. Filaments not tapering.
         4. Filaments not branching . . . . Family 4. NOSTOCACEA.
         4. Filaments branching.
            5. Blue-green algae mostly producing heterocysts . . . . . . Family 5. SCYTONEMATACEA.
            5. Minute colorless filaments without heterocysts . . . . Family 6. CHLAMYDOTRICHACEA.
      3. Filaments tapering . . . . Family 7. RIVULARIACEA.
1. Cells dividing in more planes than one, usually after a preliminary filamentous phase.
   2. Pigmented, blue-green.
      3. Producing extensive filaments with heterocysts . . . . Family 8. SIROSIPHONACEA.

3. Filaments more or less reduced, reproducing by minute spores (gonidia) formed by repeated division
in all planes......................Family 9. PLEUROCAPSACEA.
2. Colorless; filamentous, reproducing by
gonidia.............................Family 10. CRENOTRICHACEA.

Family 1. **Oscillatoriacea** [Oscillatoriaceae] Harvey 1858. Blue-green algae consisting of unbranched filaments, not tapering, without spores or heterocysts; mostly
actively motile by mechanisms as yet unknown. In the commonest genus, *Oscillatoria,*
the filaments are straight and lack sheaths. *Lyngbya* and *Phormidium* produce
sheathed filaments, in the latter genus very slender. *Microcoleus* and *Hydrocoleum*
have more than one filament in each sheath. In *Arthrospira* and *Spirulina* the filaments are coiled; those of *Spirulina* are not visibly septate, and are said to be unicellular.

Family 2. **Beggiatoacea** [Beggiatoaceae] Migula 1895. *Beggiatoa* Trevisan includes
slender colorless filaments, actively writhing, containing granules of sulfur, found
in foul waters and sulfur springs. The species were originally included in *Oscillatoria.*
Winogradsky (1887) showed that they live by chemosynthesis, and discovered the
related genera *Thiothrix* and *Thioploca.* From the time of these discoveries, these
organisms were construed as bacteria of an order Thiobacteria. Under the current
hypothesis that chemosynthesis is a derived character, we are free to believe that the
position originally assigned to the species of *Beggiatoa* was the natural one.

Family 3. **Chamaesiphonacea** [Chamaesiphonaceae] Borzi 1882. Order *Chamaesi-
phonales* Smith Freshw. Alg. 74 (1933). The only genus is *Chamaesiphon,* minute
organisms epiphytic on freshwater plants. The ellipsoid cells are attached at one
end and are enclosed in tenuous sheaths. They reproduce by transverse division, which
cuts loose small cells from the free ends. By the time two or three such cells are
produced, the sheath is ruptured at the free end, and the small cells drift away to
reproduce the organism elsewhere.

Family 4. **Nostocacea** [Nostocaceae] (Nägeli) Rabenhorst Kryptog.-Fl. Sachsen
1: 95 (1863). Order *Nostocaceae* Nägeli 1847. Of this family the most familiar genus
is *Nostoc,* seen as gelatinous bodies, usually globular, green, blue-green, yellow, or
brown, of sizes from barely visible to the naked eye up to 10 cm. or more in diameter,
in fresh water or on damp earth. Under the microscope, these bodies or colonies are
seen to consist of myriad crooked and tangled filaments of bead-like cells imbedded
in a gelatinous matrix. Heterocysts are always, and spores usually, present.

If in water one finds filaments of much the same structure as those of *Nostoc,*
but comparatively short, straight, and free or at least not in definite colonies, these
represent the genus *Anabaena.* Filaments floating on water, with cylindrical spores
not confined to the ends of the filaments, are *Aphanizomenon.* Filaments each with
one heterocyst and one spore at one end are *Cylindrospermum.*

Family 5. **Scytonematacea** [Scytonemataceae] Rabenhorst op. cit. 106. Members
of this family produce heavily sheathed filaments like those of *Lyngbya,* with the
difference that heterocysts are usually present. The multiplication of the cells of a
filament may produce the result that the cell next to a heterocyst is driven out of line
and forced obliquely through the sheath. With further growth, the file of cells ending
in one which was forced out of line may appear to be the main axis of a system of
branches, while the original summit of the filament appears to be a lateral branch.
The description of "false" branching thus given applies particularly to *Tolypothrix.*

In *Scytonema,* the pressure of multiplying cells causes waves of the filament to break laterally through the sheath and produce branches in pairs. *Plectonema* branches like *Tolypothrix* but has no heterocysts.

Family 6. **Chlamydotrichacea** [Chlamydotrichaceae] Pribram in Jour. Bact. 18: 377 (1929). Aquatic organisms consisting of colorless cylindrical cells in sheathed filaments, without heterocysts but exhibiting false branching, the sheaths of young filaments thin and colorless, those of older ones thick and yellow to brown. *Chlamydothrix ochracea* Migula was intended as a new name for *Leptothrix ochracea* Kützing, but the entity to which it is believed to apply is totally different from the one to which the latter name was applied above. *Chlamydothrix* is a filament of definite cells about 1μ in diameter. The only other definitely characterized species of this family is *Clonothrix fusca* Roze, the cells about 2μ in diameter, those near the tips of the filaments dividing repeatedly (always in one plane) to produce spherical non-motile gonidia (Kolk, 1938).

Family 7. **Rivulariacea** [Rivulariaceae] Rabenhorst op. cit. 101. The filaments include heterocysts and exhibit the false branching of *Tolypothrix;* the outgrowth of the filament below each heterocyst gives to the original terminal part the appearance of a branch of which the heterocyst is the basal cell. The ends of the filaments become attenuate and colorless. In *Calothrix* the filaments are mostly solitary; in other genera they remain together in gelatinous colonies. *Rivularia* is without spores; in *Gloeotrichia* there is a large cylindrical spore next to each heterocyst.

Family 8. **Sirosiphonacea** [Sirosiphonaceae] Rabenhorst op. cit. 114. Family *Stigonemataceae* Kirchner 1898. This family takes its name from the ancient generic name *Sirosiphon* Kützing 1843, which turned out to be identical with *Stigonema* Agardh 1824. The cells divide at first in one plane and produce filaments. Presently they exhibit a capacity to divide in other planes, and may produce true branches or multiseriate filaments or both. Heterocysts and spores are generally produced.

Family 9. **Pleurocapsacea** [Pleurocapsaceae] Geitler in Pascher et al. Süsswasser-Fl. Deutschland 12: 124 (1925). This group was formerly included in Chamaesiphonacea, but it appears probable that *Chamaesiphon* is related to *Oscillatoria,* and the present group to *Stigonema.* Most of the Pleurocapsacea are marine, epiphytic on seaweeds. Their apparently typical behavior, as exemplified by *Hyella* and *Radaisia,* consists of the production of branching filaments whose terminal cells become enlarged, after which their contents undergo division in many planes to produce numerous minute spores called gonidia. In *Pleurocapsa* and *Xenococcus* there is no filamentous phase; the gonidium gives rise to a cluster of cells all of which produce gonidia. In *Dermocarpa* the gonidium gives rise to a single vegetative cell which divides only to produce gonidia.

Family 10. **Crenotrichacea** [Crenotrichaceae] Hansgirg. This family includes the single known species *Crenothrix polyspora* Cohn, one of the traditional iron bacteria. There is every appearance that it is a colorless variant of the Pleurocapsacea. A germinating gonidium gives rise to an unbranched filament of cells, about 2μ in diameter, in a sheath which is at first thin and colorless, later becoming thicker and discolored by ferric oxide. Some cells may burst from the free end of the sheath as macrogonidia. Others may begin to divide lengthwise. These may at first grow before re-dividing, and may swell the sheath to a fusiform or trumpet-like shape. By further division they produce numerous microgonidia, which may sift out of the sheath or be released by its decay.

Such are the Mychota, the organisms which may properly be characterized as lacking nuclei.

Chapter IV

KINGDOM PROTOCTISTA

Kingdom II. **PROTOCTISTA** Hogg

Regne Psychodiaire, Psychodiés, Bory de Saint Vincent Dict. Class Hist. Nat. 8: 246 (1825), 14: 329 (1828).

Kingdom *Protozoa* Owen Palaeontology 5 (1860), not class *Protozoa* Goldfuss (1818).

Regnum Primigenium seu PROTOCTISTA Hogg in Edinburgh New Philos. Jour. n.s. 12: 223 (1860).

Kingdom *Acrita* or *Protozoa* Owen Palaeontology ed 2: 6 (1861).

Kingdom *Primalia* Wilson and Cassin in Proc. Acad. Nat. Sci. Philadelphia 1863: 117 (1864).

Kingdom *Protista* Haeckel Gen. Morph. 2: xix (1866).

Kingdom *Protobionta* Rothmaler in Biol. Centralbl. 67: 243 (1948).

Nucleate organisms other than Plantae and Animalia: the marine algae and the fungi and protozoa. *Amiba diffluens* may be construed as the standard.

The name Protista, of Haeckel, is the most familiar among those which have been applied to the kingdom here to be discussed, but it is not the earliest. Among followers of Cuvier, the animal kingdom consisted necessarily of four branches. Presumably, it was this tradition that induced Owen to refer the Infusoria and Amorphozoa (sponges) to a separate kingdom, which he called Protozoa. A year later, Owen published an alternative name for this kingdom; but Hogg had already published modifications of two of Owen's names, Protoctista and Amorphoctista(κτίζω, to establish, create), for the reason that names in *-zoa* appeared inappropriate to groups excluded from the animal kingdom.

The limits here given to the kingdom Protoctista were proposed by the present author (1938, 1947). They have been accepted, with exception in a single significant point, by Barkley (1939, 1949) and Rothmaler (1948).

It is assumed that the evolutionary origin of the Protoctista consisted of the evolutionary origin of the nucleus, and that all nuclei are essentially the same thing. Kofoid (1923) insisted that enduringly viable nuclei originate among protozoa, as among plants and animals, regularly by mitosis, never by binary or multipe fragmentation, nor by aggregation of stainable granules. He did not recognize the nucleus as essentially a device for sexual reproduction. Several considerable groups of protozoa, however, which Kofoid listed as not known to reproduce sexually, have been found to do so. Here, then, it is maintained that all nuclei, in this kingdom as among plants and animals, are the same thing; and that the nucleus is essentially a device for sexual reproduction, that is, for processes of reproduction which involve always one act of meiosis and one of karyogamy, and which produce Mendelian heredity as an effect.

Photosynthesis is believed to have evolved only once. As it occurs both among non-nucleate and nucleate organisms, the nucleus is believed to have evolved in organisms living by this function. The closest approach between non-nucleate and nucleate organisms is believed to be between the blue-green algae and the primitive red algae (Smith, 1933; Tilden, 1933). Thus it appears that the original nucleate organisms were not capable of swimming by means of flagella. Flagella appear to have evolved in unicellular nucleate photosynthetic organisms as a device for dissemination (Bessey, 1905). The flagella of nucleate organisms are not homologous with those of bacteria; they are much larger and of much more complicated structure.

The origin of flagella was apparently associated with a simplification of the system of photosynthetic pigments, by the loss of chromoproteins, leaving systems of chlorophylls and carotinoids. The association of these two courses of evolution may have been merely coincidental; Tilden suggested the idea that the loss of chromoproteins may have been occasioned by increasing illumination of the waters of the face of the earth.

Organisms of the body type of solitary walled cells, having chlorophylls and carotinoid pigments but not chromoproteins, and producing flagellate reproductive cells, appear to have undergone radiating evolution, producing a wide variety of types of organisms, distinguished by different specific chlorophylls and carotinoids, different types of flagella, and different specific metabolic products. The types of flagella occurring in nucleate organisms are here particularly to be noted.

Loeffler (1889), in the original publication of the standard method of staining the flagella of bacteria, remarked that he had applied this method also to certain larger organisms. He found that the flagellum of *Monas* bears numerous lateral appendages, and that the cilia of a certain infusorian bear solitary terminal appendages. Loeffler's method is difficult, and has not been much used. Fischer (1894) used it and coined terms, *Flimmergeisseln* and *Peitschengeisseln,* designating structures of the respective types seen by Loeffler. Petersen (1929), having applied Loeffler's method to a reasonable variety of flagellates, introduced refinements of terminology. Flagella of the type of the larger flagellum of *Monas* (the organism bears also a minute simple flagellum) became *allseitswendige Flimmergeisseln*; those of *Euglena,* which bear a single file of appengages, became *einseitswendige Flimmergeisseln.*

Deflandre (1934) devised a different method for seeing the appendages on flagella, and substituted, for the Teutonisms just quoted, French terms based on Greek. These may be Anglicised as follows. (1) The acroneme flagellum bears a single terminal appendage. The flagellum without appendages is said to be simple; so far as it appears among nucleate organisms, it appears to be a variant of the acroneme type. (2) The pantoneme flagellum bears appendages on all sides. (3) The pantacroneme flagellum bears both terminal and lateral appendages. It is a rarity, known only in the collared monads, and may be supposed to be a variant of the pantoneme type. (4) The stichoneme flagellum bears a single file of appendages.

The point in which Barkley and Rothmaler take exception to the limits here given to kingdom Protoctista is this, that they include in this kingdom the green algae. In the present work, scant attention is given to organisms whose plastids are bright green, containing chlorophylls *a* and *b*, carotin, and xanthophyll, and no other pigments; whose motile stages have acroneme flagella, more than one (usually two), and equal; and which produce essentially pure cellulose, true starch, and sucrose. These organisms represent the undoubted evolutionary origin of the higher plants; a classification which attempts to represent nature includes them necessarily in the plant kingdom.

Rothmaler set up a system of only four phyla, being the red organisms, basically without flagella; those which are typically yellow to brown, having pantoneme flagella; those with acroneme flagella, including the green algae; and the euglenid group, which have stichoneme flagella. The non-pigmented Protoctista were distributed among these groups. The system appears unsound by the fact that large blocks of non-pigmented organisms are placed where only portions of them belong.

In the present work, a less symmetrical system of phyla is offered. Its basis is an ingenuous system of red algae, brown algae, fungi, and the four traditional groups of protozoa; this has been radically modified in view of the great accumulation of knowledge subsequent to the formulation of these groups. The phylum Pyrrhophyta as here limited is tentative; the phylum Protoplasta, marked only by negative characters, amounts to a dumping ground for groups whose relationships are altogether obscure.

1. Living by photosynthesis, which takes place in plastids containing red or blue chromoprotein pigments; never producing flagellate cells . . . . Phylum 1. RHODOPHYTA.
1. Without chromoprotein pigments.
   2. Typically living by photosynthesis, brown, yellow, or green in color.
      3. Producing flagellate cells each with one pantoneme or pantacroneme flagellum, often with additional acroneme flagella . . . . Phylum 2. PHAEOPHYTA.
      3. Producing flagellate cells whose flagella are never pantoneme or pantacroneme, often stichoneme . . . . .Phylum 3. PYRRHOPHYTA.
   2. Dependent; motile cells with acroneme flagella or cilia, or amoeboid, or none.
      3. Not producing cilia, i. e., structures of the nature of acroneme flagella, numerous and widely distributed on the surfaces of the cells.
         4. Cells walled in the vegetative condition.
            5. Producing motile cells with single posterior flagella; bodies mostly with tapering rhizoids . . . . .Phylum 4. OPISTHOKONTA.
            5. Producing no motile cells; bodies filamentous . . . . Phylum 5. INOPHYTA.
         4. Cells not walled in the vegetative condition.
            5. Mostly predatory, flagellate or amoeboid or with flagellate or amoeboid stages . . . . .Phylum 6. PROTOPLASTA.
            5. Parasitic in animals, producing flagellate cells only as rare exceptions . . . . .Phylum 7. FUNGILLI.
      3. With cilia . . . . .Phylum 8. CILIOPHORA.

Chapter V

PHYLUM RHODOPHYTA

Phylum 1. **RHODOPHYTA** Wettstein

Order *Floridées* Lamouroux in Ann. Mus. Hist. Nat. Paris 20: 115 (1813).

FLORIDEAE C. Agardh Synops. Alg. Scand. xiii (1817).

Order FLORIDEAE C. Agardh Syst. Alg. xxxiii (1824).

Division (of order *Algae*) Rhodospermeae Harvey in Mackay Fl. Hibern. 160 (1836).

Class HETEROCARPEAE Kützing Phyc. Gen. 369 (1843).

Class *Florideae* J. Agardh Sp. Alg. 1: v (1848).

Rhodophyceae Ruprecht in Middendorff Sibirische Reise 1, Part 2: 200 (1851).

Stamm Florideae Haeckel Gen. Morph. 2: xxxiv (1866).

Phylum RHODOPHYTA Wettstein Handb. syst. Bot. 1: 182 (1901).

Division *Rhodophyceae* Engler Syllab. ed 3: 18 (1903).

Phylum *Carpophyceae* Bessey in Univ. Nebraska Studies 7: 291 (1907).

Phylum *Rhodophycophyta* Papenfuss in Bull. Torrey Bot. Club 73: 218 (1946).

Definitely nucleate organisms (*Porphyridium* and *Prasiola* doubtfully so); with few exceptions living by photosynthetic processes involving red and blue pigments (phycocyanin and phycoerythrin) as well as green and yellow (chlorophylls *a* and *d* and carotinoids); not producing true starch, and producing cellulose only in small quantity, the cells walled chiefly with modified carbohydrates which tend to become gelatinous; never producing flagellum-bearing cells, but sometimes producing cells which move in water without the use of definite organelles.

Tilden (1933) and Smith (1933) are authority for placing the red algae next to the blue-green algae, thus suggesting the inference that they include the most primitive of nucleate organisms. The resemblances between blue-green and red algae are in the following points. Both groups possess, along with the chlorophylls and carotinoids usual in photosynthetic organisms, other pigments, both blue and red. To these pigments as found in both groups, the same names, phycocyanin and phycoerythrin, are applied; they are not, however, the same chemical species (Kylin, 1930). Neither group produces true starch; carbohydrate is stored as substances of the general nature of dextrin or glycogen (occuring in the red algae as solid granules called floridean starch). Both groups produce cellulose only in scant quantities (Miwa, 1940; Kylin, 1943); the cell walls consist chiefly of materials, of the general nature of carbohydrates, which tend to become gelatinous. They share the negative character of never producing flagella, and the positive one of producing cells which can move actively upon surfaces, without motor organelles, by a mechanism as yet unknown (Rosenvinge, 1927).

The phylum is divisible into two classes:

1. Cells of most examples each with one central plastid, without protoplasmic interconnections, in aggregates of indefinite extent or organized as filaments or thalli with intercalary growth; zygotes producing spores directly by division...Class 1. BANGIALEA.

1. Cells with protoplasmic interconnections, containing except in the lowest examples several parietal plastids, organized as filaments with apical growth, the filaments usually massed as thalloid bodies; zygotes giving rise to spores indirectly..............................Class 2. HETEROCARPEA.

Class 1. BANGIALEA (Engler) Wettstein

Subclass *Bangioideae* de Toni Sylloge Algarum 4: 4 (1897).
Subclass BANGIALES Engler in Engler and Prantl Nat. Pflanzenfam. I Teil, Abt. 2: ix (1897).
Class BANGIALES Wettstein Handb. syst. Bot. 1: 187 (1901).
Class *Bangioideae* and orders *Bangiales* and *Rhodochaetales* Bessey in Univ. Nebraska Studies 7: 291 (1907).
Class *Bangieae* Schaffner in Ohio Naturalist 9: 448 (1909).
Protoflorideae Rosenvinge in Mem. Acad. Roy. Sci. Lett. Danemark, sér. 7, Sciences 7: 55 (1909).
Abtheilung (of *Stamm* RHODOPHYTA) *Bangiineae* Pascher in Beih. bot. Centralbl. 48, Abt. 2: 328 (1931).
Subclass *Protoflorideae* Smith Freshw. Algae 120 (1933).

Red algae (exceptionally green or of other colors), the cells with solitary central plastids (exceptionally with multiple parietal plastids), lacking protoplasmic interconnections, in irregular colonial masses or forming filaments or thalli with intercalary growth; the zygote produced in sexual reproduction dividing to produce spores directly.

The group is of one order, five families, about fifteen genera; the number of known species is about eighty.

Order **Bangiacea** [Bangiaceae] Nägeli 1847.
Characters of the class.
1. Cells forming irregular aggregates.............Family 1. PORPHYRIDIACEA.
1. Cells forming filaments or thalli.
   2. Vegetative cells becoming spores without dividing...........................Family 2. RHODOCHAETACEA.
   2. Vegetative cells undergoing division to produce spores.
      3. Organisms red, purplish, etc..........Family 3. PORPHYREA.
      3. Organisms green...................Family 4. SCHIZOGONIACEA.
   2. Spores formed solitary in special cells.......Family 5. COMPSOPOGONACEA.

Family 1. **Porphyridiacea** [Porphyridiaceae] Kylin in Kungl. Fysiog. Sällsk. Förhandl. 7, no. 10: 4 (1937). Order *Porphyridiales* Kylin l. c. The only well known species is *Porphyridium cruentum* (C. Agardh) Nägeli (1849). It is widely distributed in damp climates, forming extensive red patches like blood on damp earth or stone. The spherical cells are reported as varying widely in diameter (5-24μ), and Geitler (1932) and Kylin (1937) have distinguished additional species.

Porphyridium has been classified among blue-green, red, and green algae. Lewis and Zirkle (1920) found in each cell a central red plastid, occupying most of its volume, and having rays extending to the cell membrane. Within the plastid there is

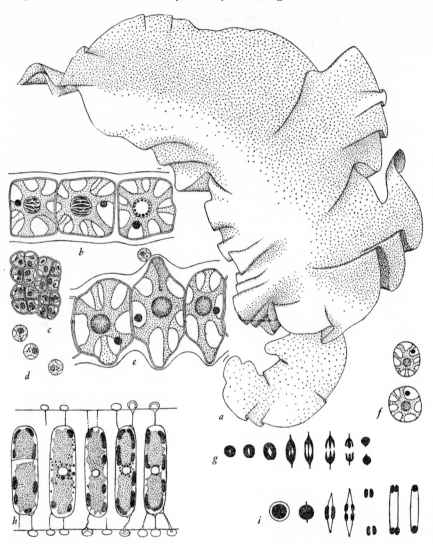

Fɪɢ. 5.—**a**, *Porphyra laciniata*, thallus x ½.   **b-g**, *Porphyra tenera* after Ishikawa (1921); **b**, cells; **c**, cell dividing to produce sperms; **d**, sperms; **e**, fertilization; **f**, "carpospores," i.e., cells produced by division of the zygote; **g**, stages of nuclear division x 2,000.   **h, i**, *Porphyra umbilicaris* after Dangeard (1927); **h**, fertilization; **i**, stages of nuclear division x 2,000.   All figures x 1,000 except as noted.

a moderately large stainable granule; outside the plastid, a single additional granule can usually be found. When a cell is to divide, the granules break up into considerable numbers of smaller ones, some of which become organized as a system of strands forming an irregular network on the surface of the plastid. The protoplast, the network, and the plastid undergo constriction; the processes by which the daughter cells return to the original structure were not clearly seen. Interpretation of these observations is difficult. It is possible that the granule outside of the plastid is a nucleus of the type of those which have been observed in *Bangia* and *Porphyra*.

Family 2. **Rhodochaetacea** [Rhodochaetaceae] Schmitz in Engler and Prantl Nat. Pflanzenfam. I Teil, Abt. 2: 317 (1896). Family *Goniotrichaceae* Smith Freshw. Algae 121 (1933). Branching filaments, sometimes becoming multiseriate by lengthwise division, the vegetative cells capable of escaping and functioning as spores. Sexual reproduction unknown. *Asterocystis,* uncommon, in fresh water; the remaining genera marine, epiphytic on other algae. *Goniotrichum. Rhodochaete* and *Goniotrichopsis,* the cells with numerous plastids.

Family 3. **Porphyrea** [Porphyreae] Kützing (1843). Family *Porphyraceae* Rabenhorst 1868. Family *Bangiaceae* (Nägeli) Schmitz (in Engler and Prantl, 1896). Filaments or thalli of a red or purple color; the cells, in producing spores, may release their protoplasts as wholes or may undergo division into many. Rosenvinge (1927) observed the active motion of these spores.

The most important genus is *Porphyra;* the individuals are thalli up to several centimeters in diameter, on rocks or other algae in ocean water along coasts. They are called purple lavers, *tsu'ai, amanori;* they are used as food, for making soup or in condiments, and are extensively cultivated in Japan (Tseng, 1944). *Bangia* is either freshwater or marine; in structure it differs from *Porphyra* in having filamentous bodies, uniseriate or pluriseriate.

During nuclear division in *Porphyra tenera* as described by Ishikawa (1921), polar appendages form at both ends of the nucleus, which becomes elongate and appears to consist of three strands. The strands break transversely, and each set of three fuses into a mass. Dangeard (1927), dealing with *Porphyra umbilicaris* and *Bangia fuscopurpurea,* observed nuclei 5μ in diameter, each consisting of a karyosome, that is, a mass of chromatin, lying in a clear space surrounded by a membrane. In mitosis, the membrane and the unstained matter disappear. Polar appendages grow out from the karyosome, and their tips become cut off as granules which may be regarded as centrosomes. The remainder of the karysome becomes organized as two masses, evidently chromosomes, connected to the centrosomes by fibers. Each chromosome divides into two; the daughter chromosomes move to the centrosomes and fuse with them to form karyosomes about which new membranes appear. This description represents a definite, if primitive, process of mitosis.

Sexual reproduction, here where we first encounter it, involves differentiated gametes. Naked sperms, indistinguishable from spores, move to the surface of other cells which function as eggs. A strand of protoplasm grows through the gelatinous wall of the egg from the sperm to the egg protoplast, and the protoplast of the sperm migrates through the passage thus formed. The zygote divides two or three times, producing spores. During the first two divisions, the two masses of chromatin which appear are somewhat different in appearance from the vegetative chromosomes (Dangeard, op. cit.); it may be supposed that these masses are tetrads and diads, and that the divisions are meiotic. Evidently, this is a life cycle of the primitive type, in which all cells except the zygotes are haploid.

Family 4. **Schizogoniacea** [Schizogoniaceae] Chodat. Family *Prasiolaceae* West. Family *Blastosporaceae* Wille. Filamentous or thallose algae, freshwater or marine, of the structure of Porphyrea, but of a green color; sexual reproduction unknown. Kylin (1930) found the pigmentation to be that of green algae rather than of red. Copeland (1955) was unable to discern nuclei. The sole genus *Prasiola* (*Schizogonium* represents a stage of development) is of about fifteen species. Setchell and Gardner (1920) and Ishikawa (1921) suggested the place in Bangiacea here given to this group.

Family 5. **Compsopogonacea** [Compsopogonaceae] Schmitz in Engler and Prantl Nat. Pflenzenfam. I Teil, Abt. 2: 318 (1896). Family *Erythrotrichiaceae* Smith Freshw. Algae 122 (1933). Filaments, unbranched or branched, uniseriate or pluriseriate, or thalli. Spore-formation is accomplished by the division of a vegetative cell, by an oblique wall, into two unequal cells; the protoplast of the smaller is released as a spore. Rosenvinge observed the spores of *Erythrotrichia carnea* to move as far as 140μ per minute. Sexual reproduction is much as in Porphyrea. *Erythrotrichia. Erythrocladia. Compsopogon,* in fresh water, the cells with numerous parietal plastids.

Class 2. **HETEROCARPEA** Kützing

Class HETEROCARPEAE Kützing Phyc. Gen. 369 (1843).
Class *Florideae* (C. Agardh) J. Agardh Sp. Alg. 1: v (1848).
Subclass *Florideae* Engler in Engler and Prantl Nat. Pflanzenfam. I Teil, Abt. 2: ix (1897).
Subclass *Eufiorideae* de Toni Sylloge Algarum 4: 4 (1897).
Abtheilung (of *Stamm Rhodophyta*) *Floridineae* Pascher in Beih. bot. Centralbl. 48, Abt. 2: 328 (1931).

As this is the type group of phylum Rhodophyta, most of the synonymy of that name applies to this one also.

Red algae whose bodies consist essentially of filaments growing apically, the cells with protoplasmic interconnections, the plastids (except in some of the lowest examples) of the form of multiple parietal disks; the filaments commonly compacted into cylindrical or thallose bodies; zygotes not dividing to form spores directly, producing spores by budding or indirectly by processes of growth of various degrees of complexity.

In undertaking to describe the varied, and often highly complicated, reproductive processes of the typical red algae, one notes that these organisms occur as haploid individuals, and that the majority occur as distinct male and female haploid individuals. Sperms (commonly called spermatia) are minute naked protoplasts released from small cells commonly occurring in patches on the surfaces of thalli. The egg is called a carpogonium (Schmitz, 1883). It is the terminal cell of a specialized filament, the carpogonial filament, and bears a filiform terminal extension, the trichogyne (Bornet and Thuret, 1867), whose function is to receive the sperms. The cell, often differentiated, from which the carpogonial filament grows, is the supporting cell (*Trugzelle*).

In the more primitive members of the class, the zygote gives rise by budding to a mass of cells called the cystocarp. The cells of the cystocarp release their protoplasts as spores called carpospores. These on germination produce haploid individuals like the original ones. The zygote nucleus is the only diploid nucleus in the life cycle; its first divisions are meiotic.

In more advanced examples, the first step of development after fertilization consists of the establishment of protoplasmic contact between the zygote and other cells. These may be adjacent cells, reached directly, or distant cells, reached by the outgrowth of connecting filaments from the zygotes. In the generality of the group, the cells with which contact is made give rise to cystocarps producing carpospores; in this situation, the cells in question are called auxiliary cells. In some examples, the connecting filaments, after making contact with cells called nurse cells, themselves give rise to the cystocarps. The carpospores, in all of these more advanced examples, give rise to diploid individuals. The diploid individuals are of the same vegetative

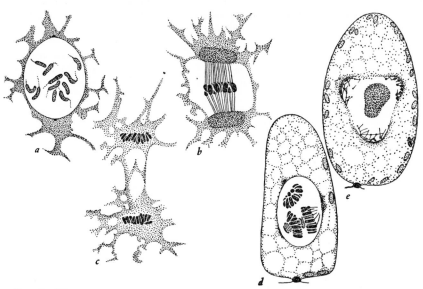

FIG. 6—Nuclear phenomena in *Polysiphonia violacea* after Yamanouchi (1906). **a, b, c,** Stages of mitosis. **d, e,** Stages of homeotypic division.

structure as the haploid individuals, but do not produce spermatia, carpogonia, or cystocarps. Certain cells, commonly scattered and imbedded in the body, produce sets of four spores which are accordingly called tetraspores; these give rise to haploid individuals.

This account means that these algae occur typically in somata of four types: male and female haploid individuals; cystocarps, being a preliminary, parasitic, multiplicative phase of the diploid stage (Janet named this stage the carposporophyte; Drew, 1954); and free-living diploid individuals, reproducing by tetraspores. The production of carpospores and tetraspores by different individuals of identical vegetative structure explains the oldest name applied to this class, namely Heterocarpea.

Understanding of the life cycle of typical Heterocarpea has been reached only by much labor and after a certain amount of confusion. The first significant observations were by Bornet and Thuret (1867). Schmitz (1883) showed that the zygote makes protoplasmic contact with other cells. He supposed that the contact of the zygote with an auxiliary cell is a second sexual fusion (*Copulation*) following upon proper fertilization. Oltmanns (1898) disproved this: he showed that the nuclei of auxiliary cells are inert, and that the nuclei of carpospores are derived entirely from zygote nuclei. Yamanouchi (1906) showed that the chromosome number of carposporic individuals of *Polysiphonia violacea* is 10, and that that of tetrasporic individuals is 20; and reported much more of the cytology. Centrosomes appear *de novo* during the earlier stages of mitosis, and fade out and disappear during the later stages. The mitotic spindle is formed, and the chromosomes take their place upon it, within an intact nuclear membrane, which fades out in later stages. In meiosis, which produces the nuclei of tetraspores, the tetrads and diads divide within the original nuclear membrane, which becomes tetrahedrally lobed, and then disappears except where the haploid groups of chromosomes lie against it, with the result that the membranes of the tetraspore nuclei are partly old and partly new.

There are some 2500 species of Heterocarpea, including comparatively few in fresh water, but the majority of the marine algae. Many of them are beautiful; their variety and beauty contribute to the pleasure which people find on coasts. Experienced naturalists can identify many genera by gross structure, but the systems of orders and families based on gross structure, such as those of Kützing (1843) and J. Agardh (1851-1863), have been found artificial and abandoned. A proper respect for the principles of nomenclature makes it necessary, however, to apply many of the names used in these systems. Schmitz applied his morphological studies to a classification of the typical red algae as four groups (1889); Engler (1897) made these groups definitely orders. Subsequent scholars have found this system sound in principle, but have found it necessary, on the basis of studies of additional examples (for example, by Kylin, 1923, 1924, 1925, 1928, 1930, 1932; Papenfuss, 1944; Sjöstedt, 1926; Svedelius, 1942) radically to rearrange the families and genera. At least four orders in addition to those of Engler have been proposed but reductions have decreased the number currently recognized to six.

The following key to the orders is a rather considerable modification of those published by Kylin (1932) and Smith (1944).

1. All free-living individuals haploid; tetra-
    spores not produced, or produced as carpospores. . Order 1. CRYPTOSPERMEA.
1. Free-living individuals of two types, the one
    producing gametes (the zygotes giving rise
    to carpospores), the other producing tetraspores.
       2. Without specialized auxiliary cells or
           nurse cells, the lower cells of the carpo-
           gonial filaments, or normal vegetative
           cells, serving as auxiliary cells. . . . . .Order 2. SPHAEROCOCCOIDEA.
       2. With specialized nurse cells, the carpo-
           spores produced from filaments which
           have made contact with these. . . . . Order 3. GELIDIALEA.
       2. With specialized auxiliary cells from
           which the carpogonia develop.
              3. The auxiliary cells being intercalary
                  cells in specialized filaments homol-
                  ogous with the carpogonial filaments. . . . Order 4. FURCELLARIEA.
              3. The auxiliary cells terminal in fila-
                  ments which grow from the support-
                  ing cells of the carpogonial fila-
                  ments before fertilization. . . . . Order 5. COELOBLASTEA.

3. The auxiliary cells originating after fertilization as branches of the supporting cells of the carpogonial filaments...........................Order 6. FLORIDEA.

Order 1. **Cryptospermea** [Cryptospermeae] Kützing Phyc. Gen. 321 (1843).
Order *Periblasteae* Kützing op. cit. 387, in part.
Orders *Helminthocladeae* J. Agardh Sp. Alg. 2: 410 (1851), *Chaetangieae* op. cit. 456 (1851), and *Wrangelieae* op. cit. 701 (1863).
Order *Batrachospermaceae* Rabenhorst Kryptog.-Fl. Sachsen 1: 278 (1863).
Nemalioninae Schmitz in Flora 72: 438 (1889).
Order *Nemalionales* Engler in Engler and Prantl Nat. Pflanzenfam. I Teil, Abt. 2: ix (1897).

Heterocarpea normally without diploid bodies, the carpogonium arising from the zygote or from an adjacent cell serving as an auxiliary cell, the carpospores producing haploid bodies like the original ones. Certain genera which are exceptional to these characters are noted below. *Batrachospermum* may be regarded as the standard genus.

In all recent literature, this order is called Nemalionales. Eight families are recognized. The forms consisting of mere filaments, *Acrochaetium, Rhodochorton,* and others, are family Acrochaetiacea [Acrochaetiaceae] Fritsch (Family *Chantransiaceae* Auctt., but *Chantransia* DC. as originally published included no members of this family; Papenfuss, 1945). In the remainder of the order, the filaments are differentiated, or, with or without differentiation, organized as bodies of definite form, simply cylindrical, branched, or flattened. Fresh-water examples (the only fresh-water Heterocarpea) include *Batrachospermum, Lemanea,* and *Thorea.* These organisms are not red, but bluish, green, or brown. Marine examples include *Nemalion* and *Cumagloia.*

In *Liagora tetrasporifera* and certain other species tetraspores are produced in the place of carpospores. Within this genus, then, there has been a change in the time of meiosis (which could be established, presumably, by a single mutation) from immediately after fertilization to the end of the cystocarp stage.

Galaxaura is a genus of tropical marine algae which are calcified, which is to say that they deposit much calcium carbonate in the tissues; they were originally classified as corals. They have distinct sexual and tetrasporic stages. Svedelius (1942) ascertained their life cycle. Carpospore-bearing filaments arise both from the zygote and from other cells, previously undifferentiated, which serve as auxiliary cells. The genus has the structure of the present order, and is to be placed here, in spite of exhibiting in unspecialized form the life cycle of the following orders.

Order 2. **Sphaerococcoidea** [Sphaeroccoideae] J. Agardh Sp. Alg. 2: 577 (1852).
Family *Gigartineae* Kützing (1843).
Orders *Gigartineae* and *Chaetangieae* J. Agardh op. cit. 229, 456 (1851).
Gigartininae Schmitz in Flora 72: 440 (1889).
Order *Gigartinales* Engler in Engler and Prantl Nat. Pflanzenfam. I Teil, Abt. 2: x (1897).
Order *Nemastomatales* Kylin in Kgl. Fysiog. Sällsk. Handl. n. f. 36, no. 9: 39 (1925).
Order *Sphaerococcales* Sjöstedt in Kgl. Fysiog. Sällsk. Handl. n. f. 37, no. 4: 75 (1926).

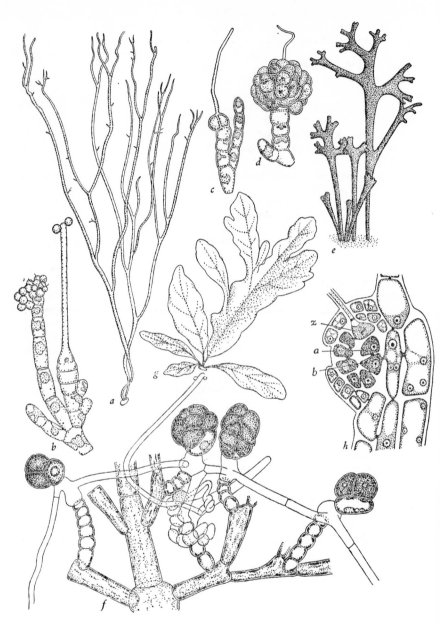

(Legend on bottom of page 49)

This order, in all recent literature called Gigartinales, is a numerous and varied one. The bodies are generally erect; they may be cylindrical or flattened, unbranched or branched. In some examples, *Haliarachnion, Rhodophyllis, Sebdenia,* the zygote sends out extensive filaments, which make contact with unspecialized cells scattered in the body. In other examples, the zygote makes contact with a lower cell of the carpogonial filament. In either case, the cells with which contact is made are auxiliary cells and give rise to cystocarps; these produce carpospores, and the carpospores produce tetrasporic individuals. Certain species of *Phyllophora, Gymnogongrus,* and *Ahnfeldtia* are exceptional in producing tetraspores in the place of carpospores; these species have no free-living tetrasporic generation. In these organisms, as contrasted with *Liagora tetrasporifera,* it is believed that this type of life cycle has been established by reduction of a longer one.

Kylin (1932) assigned twenty families to this order. *Gracilaria* is a minor source of agar agar. *Gigartina mammilosa* and *Chondrus crispus* (Irish moss or carageen) are well known as yielding a jelly, carageenin, resembling but distinct from agar agar (Tseng, 1945).

Various abnormal growths on red algae have been found to be parasitic red algae, almost always on hosts closely related to themselves (Setchell, 1914). To the present order belong *Gardneriella* and its host *Agardhiella; Plocamiocolax* and its host *Plocamium; Gracilariophila* and its host *Gracilaria* (Wilson, 1910).

Order 3. **Gelidialea** [Gelidiales] Kylin in Kgl. Svensk. Vetensk.-Akad. Handl. 63, no. 11: 132 (1923).
Family GELIDIEAE Kützing (1843).
Order *Gelidieae* J. Agardh Sp. Alg. 2: 464 (1851).
Heterocarpea in which the zygote sends out a single elongate filament which makes contact successively with several chains of nurse cells and gives rise to carpospores; bodies consisting of branched filaments, the ultimate tips of the lateral branches compacted into a firm layer covering a branching body, cylindrical or flattened; the surface adjacent to the masses of carpospores pushed out and punctured by pores through which the spores escape.

There is a single family Gelidiea [Gelidieae] Kützing (Family *Gelidiaceae* Schmitz and Hauptfleisch). Such economic importance as the red algae possess lies chiefly in

---

FIG. 7—**a**, Thallus of *Nemalion multifidum* x 1. **b, c, d**, production of sperms; beginning of production of carpospores; and cluster of carpospores of *Nemalion multifidum* after Bornet and Thuret (1867). **e**, Thallus of *Chondrus crispus* x 1. **f**, Reproduction of *Dudresnaya purpurifera* (order Furcellariea or Cryptonemiales) after Bornet and Thuret, *op. cit.* The trichogyne, whose free end with attached sperms is seen above, is irregularly twisted below; it leads to the egg (carpogonium); connecting filaments, growing from cells below the egg, make contact with auxiliary cells at the summits of specialized filaments; each auxiliary cell gives rise to a cluster of carpospores. **g**, Thallus of *Delesseria sinuosa* x 1. **h**, Longitudinal section of conceptacle of *Polysiphonia nigrescens* x 500, after Kylin (1923). The zygote *z* is the fourth and terminal cell of the carpogonial filament whose connection with the supporting cell *b* is not shown; the auxiliary cell *a* has grown from the supporting cell after fertilization.

this family, and particularly in the genus *Gelidium*. It is the chief source of agar agar. This is the principal material of the cell walls of *Gelidium*. It is a jelly consisting essentially of chains of galactose units, and has the property, that having been melted by heat, it does not again become solid until cooled to a much lower temperature. Algae containing it have long been used as foods in the orient. Brought into laboratory use by Koch, it has become a necessity in routine bacteriological work. The chief source is Japan.

Kylin construed this order as relatively primitive; but its reproductive processes, involving specialized nurse cells, appear less primitive than those of the Sphaerococcoidea. The production of elongate connecting filaments is shared with certain examples both of the preceding order and of the following, and the Gelidialea are probably derived by specialization from one or the other.

> Order 4. **Furcellariea** [Furcellarieae] Greville Alg. Brit. 66 (1830).
> Orders *Spongocarpeae* and *Gastrocarpeae* Greville op. cit. 68, 157 (1830).
> Order *Epiblasteae* Kützing Phyc. Gen. 382 (1843).
> Orders *Cryptonemeae, Dumontieae, Squamarieae,* and *Corallineae* J. Agardh Sp. Alg. 2: 155, 346, 385 (1851), 506 (1852).
> *Cryptoneminae* Schmitz in Flora 72: 452 (1889).
> Order *Cryptonemiales* Engler in Engler and Prantl Nat. Pflanzenfam. I Teil, Abt. 2: xi (1897).

The individuals are crustose or thallose, the thalli cylindrical or flattened, unbranched or branched. On the two or three types of individuals of each species, the reproductive structures may be scattered or clustered on the surfaces or gathered in specialized pits called conceptacles. The eggs are as usual the terminal cells of specialized filaments; other filaments, homologous with these but abortive, bear the auxiliary cells. After fertilization, the zygote may or may not establish connection with a lower cell of the same filaments. Under either circumstance, it sends out filaments which establish connection with the auxiliary cells, and these send out filaments which bear the carpospores. In less specialized examples, the filaments growing from the zygote may extend widely through the body; a single one, branching, may reach many auxiliary cells.

Kylin (1932) placed nine families here.

The family Corallinea [Corallineae] Kützing (family *Corallinaceae* Hauck) is one of the more specialized. The eggs, and subsequently the carpospores, are clustered in conceptacles. In each conceptacle the zygotes, the filaments from them, and the auxiliary cells, unite eventually in a single large multinucleate cell from whose margins grow the filaments which bear the carpospores. Members of this family have the property of accumulating and depositing calcareous material, and were originally classified as corals. In modern usage, the term coral means certain lower animals; but the coralline algae are associated with them in coral reefs, being indeed, according to Setchell (1926) and other authorities, responsible for the building of the reefs. Fossil coralline algae are known from the Ordovician.

The parasite *Callocolax* and its host *Callophyllis* belong to this order; *Coreocolax,* belonging to this order, attacks species of order Floridea.

The Furcellariea are a numerous group, rather unspecialized, varied almost to the extent of a miscellany. They are related to the Sphaerococcoidea, and are believed to represent the ancestry of the two following orders, and possibly also of the Gelidialea.

Order 5. **Coeloblastea** [Coeloblasteae] Kützing Phyc. Gen. 438 (1843).
Order *Rhodymenieae* J. Agardh Sp. Alg. 2: 337 (1851).
Rhodymeninae Schmitz in Flora 72: 442 (1889).
Order *Rhodymeniales* Engler in Engler and Prantl. Nat. Pfllanzenfam. I Teil,
Abt. 2: x (1897).

Heterocarpea producing auxiliary cells terminally on brief filaments which grow from the supporting cells of the carpogonial filaments before fertilization; cystocarps enclosed in cup- or vase-like pericarps; the thalli (cylindrical or flattened, branched or unbranched) usually hollow. *Champia* may be regarded as the standard genus.

In various red algae, the germinating carpospore or tetraspore gives rise to a globe of cells which grows to produce the thallus (Kylin, 1917). In the present group the sporeling is particularly blastula-like. Its upper layer of cells becomes a ring of apical cells, of definite number, distinguishing the group from others which grow by apical cells either of a single filament or of a fascicle of indefinite number. The apical cells are indeed homologous with the apical cells of filaments, but the cells derived from them are arranged in a three-dimensional pattern as in the tissues of higher organisms; it is only in the reproductive structures that the filamentous structure remains evident.

The order thus limited by Kylin (1932) is a specialized group including only the two families Rhodymeniacea [Rhodymeniaceae] Hauck and Champiea [Champieae] Kützing. The latter family is the more specialized; the hollow thalli are partitioned by transverse septa and the supporting cells produce usually just two auxiliary cells. In many examples of this family, after fertilization and the fusion of the zygote with the auxiliary cells, the latter proceed to unite with further neighboring cells to produce a massive coenocyte from which the brief carpospore-bearing filaments arise. The resulting structure is deceptively similar to that which occurs in the Corallinea. The parasite *Faucheocolax* and its host *Fauchea* belong to this order.

Order 6. **Floridea** [Florideae] C. Agardh Syst. Alg. xxxiii (1824).
Order *Floridées* Lamouroux in Ann. Mus. Hist. Nat. Paris 20: 115 (1813).
Section *Florideae* C. Agardh Synops. Alg. Scand. xiii (1817).
Orders *Trichoblasteae, Axonoblasteae,* and *Platynoblasteae* Kützing Phyc. Gen. 370, 413, 442 (1843).
Orders *Ceramieae, Spyridieae, Chondrieae,* and *Rhodomeleae* J. Agardh Sp. Alg. vol. 2 (1851-1863).
Ceramiales Oltmanns Morph. u. Biol. Alg. 1: 683 (1904).
Order *Ceramiales* Kylin in Kgl. Svensk. Vetensk.-Akad. Handl. 63, no. 11: 132 (1923).

The *Floridées* of Lamouroux included the whole group of red algae organized as four genera, *Chondrus* Stackhouse and the new genera *Claudea, Delesseria,* and *Gelidium.* Lamouroux listed first *Claudea* and *Delesseria,* belonging to the present order, to which the name is accordingly applied.

This order is characterized by specialized strict patterns in the development of the female reproductive structures. The carpogonial filament is always of four cells. The supporting cell initiates, in definite patterns, brief additional filaments. After fertilization, the supporting cell cuts off one more cell adjacent to the zygote, and this becomes the auxiliary cell. The spore-bearing structures developed from it are naked in the more primitive examples; in most, they are protected by pericarps, which, in some examples, begin to develop before fertilization.

There are four families, all numerous in species: Ceramiea (Harvey) Kützing,

Dasyea Kützing, Delesseriea Kützing, and Polysiphoniea Kützing (*Rhodomelaceae* Hauck). The Ceramiea are mostly filaments, uniseriate or becoming pluriseriate by lengthwise divisions. In many members of the other families the bodies are thallose, though consisting essentially of filaments produced in definite patterns. In many Delesseriea the branches of the thalli simulate leaves of higher plants.

Gonimophyllum is parasitic on *Botryoglossum;* both are Delesseriea. Various species of *Janczewskia,* a genus of Polysiphoniea, attack *Laurencia, Chondria,* and other members of the same family. This was the first genus of parasitic red algae to be recognized as such, by Solms-Laubach (1877).

Such are the red algae. The Bangialea appear to represent the transition between the organisms which lack nuclei and the generality of nucleate organisms. The Heterocarpea appear to be a specialized offshoot, leading to no other group.

Chapter VI
PHYLUM PHAEOPHYTA

Phylum 2. PHAEOPHYTA Wettstein

Fucoideae C. Agardh Synops Alg. Scand. ix (1817).

Orders Diatomeae and Fucoideae C. Agardh Syst. Alg. xii, xxxv (1824).

Stämme *Diatomea* and *Fucoideae* Haeckel Gen. Morph. 2: xxv, xxxv (1866).

Stämme *Zygophyta* in part and Phaeophyta Wettstein Handb. syst. Bot. 1: 71, 171 (1901).

Divisions *Zygophyceae* in part and *Phaeophyceae* Engler Syllab. ed. 3: 8, 15 (1903).

Chysophyta, with subordinate groups *Chrysophyceae, Bacillariales,* and *Heterokontae,* Pascher in Ber. deutschen bot. Gess. 32: 158 (1914).

Stamm *Chrysophyta* Pascher in Süsswasserfl. Deutschland 11: 17 (1925).

Phyla *Chrysophycophyta* and *Phaeophycophyta* Papenfuss in Bull. Torrey Bot. Club 73: 218 (1946).

Organisms typically living by photosynthesis, without chromoprotein pigments, the plastids containing chlorophylls *a* and *c*, carotin, and various xanthophylls. Lutein (the xanthophyll of typical plants) may be present but is usually exceeded in quantity by flavoxanthin, violoxanthin, isofucoxanthin, or fucoxanthin, particularly the last. The xanthophylls occur usually in quantity sufficient to give the organisms a yellow or brown color. True starch is not produced. Many examples contain granules of a white solid called leucosin, presumably a carbohydrate, which does not give a blue color with iodine. The cells are usually enclosed in walls consisting of cellulose together with larger quantities of other carbohydrates or oxidized or esterized carbohydrates. Silica or calcium carbonate may be deposited. Methanol extracts of the cells contain fucosterol, a sterol distinct from the sitosterol of typical plants. Flagellate cells are usually produced; these bear one pantoneme or pantacroneme flagellum, and usually, in addition, one acroneme or simple flagellum. Exceptional examples, non-pigmented or without flagellate stages, are rather numerous. The obvious standard genus of the phylum is *Fucus* L.

The chemical characters are stated on the authority chiefly of Carter, Heilbron, and Lythgoe (1939), Miwa (1940), and Tseng (1945). The character of the flagellation, positively known of rather few examples, is stated by authority of Petersen (1929), Vlk (1931, 1939), Couch (1938, 1941), Longest (1946), Manton (1952), and Ferris (1954).

These characters bind together an assemblage of organisms which is in some respects original here[1]. Engler (1897), West (1904), and Smith (1918, 1920) included the chrysomonad flagellates in the group of brown algae. Pascher (1914) combined as one group the chrysomonads, the diatoms, and the exceptional green algae called Heterokontae. Later (1927, 1930), he included also the colorless flagellates of family Monadina. He did not associate this group with the brown algae, and subsequent authors have in general followed him. Kylin (1933), however, considered the diatoms to be the closest allies of the brown algae, both groups being descended from the brown flagellates. Almost certainly, he was correct. Couch showed that the paired unlike flagella of the typical Oomycetes are respectively pantoneme and acroneme,

---

[1]Manton (1952) recognized this group, but omitted nomenclatural formalities.

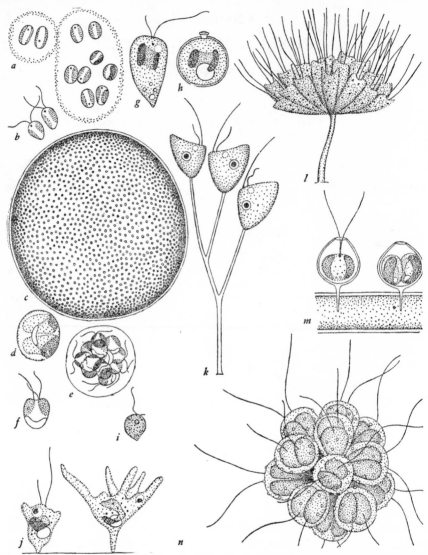

FIG. 8.—OCHROMONADALEA: **a**, **b**, *Chrysocapsa paludosa* after West (1904); **a**, a colony; **b**, zoospores. **c-f**, *Phaeocystis globosa* after Scherffel (1900); **c**, a colony x 50; **d**, a cell with two plastids, a mass of leucosin forming on a mound of protoplasm projecting into the central vacuole; **e**, production of zoospores; **f**, a zoospore. **g**, **h**, Cell and statospore of *Ochromonas granularis* after Doflein (1922). **i**, Cell of *Monas* sp. **j**. Two cells of *Brehmiella Chrysohydra* after Pascher (1928). **k**, A very young colony of *Dendromonas virgaria* after Stein (1878). **l**, Colony of *Cephalothamnium Cyclopum* after Stein, *op. cit.* **m**, Cells of *Epipyxis utriculus* after Stein, *op. cit.* **n**, Colony of *Synura Uvella.* x 1,000 except as noted.

and distinguished these fungi from practically all others by the presence of cellulose in their walls.

The phylum thus assembled may be organized as four classes.

1. Miscellaneous groups, mostly small and relatively unspecialized, of varied body type; not
   of the characters of the following groups...........Class 1. HETEROKONTA.
1. Comparatively numerous and specialized
   groups.
      2. Unicellular brown organisms with shells
        of silica consisting of two parts...............Class 2. BACILLARIACEA.
      2. Organisms of fungal or chytrid body
        types producing swimming spores with
        paired unlike flagella.......................Class 3. OOMYCETES.
      2. Filamentous and thallose brown algae..........Class 4. MELANOPHYCEA.

Class 1. HETEROKONTA Luther

Class *Flagellata* or *Mastigophora* Auctt., in part.

Class HETEROKONTAE Luther in Bihang Svensk. Vetensk.-Akad. Handl. 24, part 3, no. 13: 19 (1899).

Subclass *Chrysomonadineae* Engler in Engler and Prantl Nat. Pflanzenfam. I Teil, Abt. 1a: iv (1900).

Class *Silicoflagellatae* (Borgert) Lemmermann in Ber. deutschen bot. Gess. 19: 254 (1901).

Phylum *Siphonophyceae* and class *Vaucherioideae* Bessey in Univ. Nebraska Studies 7: 285, 286 (1907).

Chrysophyceae and *Heterokontae* Pascher in Ber. deutschen bot. Gess. 32: 158 (1914).

Divisions *Chrysophyceae* and *Heterokontae,* and classes *Chrysomonadineae, Rhizo-chrysidineae, Chrysocapsineae, Chrysosphaerineae, Chrysotrichineae, Hetero-chloridineae, Rhizochloridineae, Heterocapsineae, Heterococcineae, Hetero-trichineae,* and *Heterosiphoneae* Pascher in Beih. bot. Centralbl. 42, Abt. 2: 323, 324 (1931).

Classes *Ebriaceae, Silicoflagellata,* and *Coccolithophoridae* Deflandre, and *Chrys-omonadina* Hollande in Grassé Traité Zool. 1, fasc. 1: 407, 425, 438, 471 (1952).

Class *Phytomastigophorea* Hall Protozoology 117 (1953), in part.

Phaeophyta which lack the distinctive characters of the remaining three classes. Luther named the group on the occasion of his discovery of *Chlorosaccus,* and this genus may be regarded as the type.

The chrysomonad flagellates are the core of this class and of the first two among the five orders into which it is divided. In the classification of these two orders, three novelties will be noted.

(a) Pascher (1913) made of the chrysomonad flagellates three orders character-ized respectively by paired unequal flagella, paired equal flagella, and solitary flagella. Petersen (1929) found that the supposedly equal flagella of *Synura* are actually unlike, being respectively pantoneme and acroneme. Here, accordingly, Pascher's first two orders are combined.

(b) Pascher made separate classes or orders of groups related to the chrysomonad flagellates but of distinct body type, as palmelloid, chlorococcoid, filamentous, or

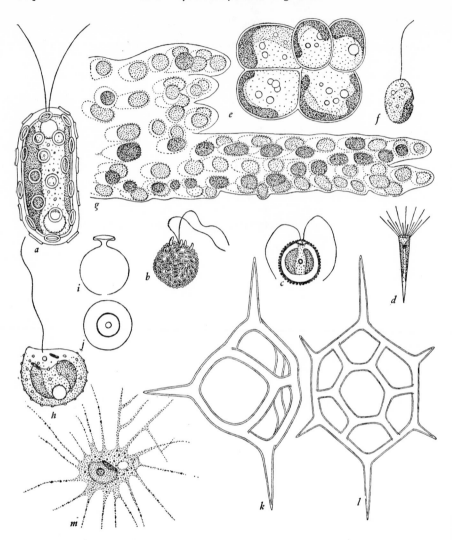

FIG. 9.—OCHROMONADALEA: **a,** *Mallomonas roseola,* based on Stein (1878) and Conrad (1926).   **b,** *Syracosphaera Quadricornu;* **c,** *Calyptosphaera insignis;* **d,** *Calciconus vitreus;* after Schiller (1925).   SILICOFLAGELLATA: **e, f,** Colony and zoospore of *Epichrysis* after Pascher (1925).   **g,** Part of the thallose growth of *Hydrurus foetidus.*   **h,** Cell, and **i, j,** statospores of *Chromulina Pascheri* after Hofeneder (1913).   **k, l,** Skeletons of *Dictyocha Fibula* and *Distephanus Speculum* from diatomaceous earth at Lompoc, California.   **m,** *Rhizochrysis Scherffeli* after Doflein (1916).   All x 1,000.

amoeboid. By Pascher's own principle of the repeated evolution of body types, these groups are surely artificial. Here most of them are broken up and their members distributed between the two chrysomonad orders according to whether the flagella of their motile stages are paired or single. It is not possible to divide by this character ameboid forms not known to produce flagellate stages; these are lumped in the second order.

(c) Since flagella appear to have evolved as a device for the dissemination of unicellular pigmented organisms, examples whose vegetative state is that of clusters of non-motile cells are placed in each order before those which are flagellate in the vegetative condition.

The two chrysomonad orders are particularly characterized by production of leucosin. They are further characterized by production of resting cells of a type called statospores. This occurs by the deposition within the protoplast of a globular shell impregnated with silica, punctured by a single pore, and often marked on the outer surface by warts, spines, or ridges, of definite pattern. The external protoplasm migrates through the pore to the interior of the shell, and the pore is then closed by deposition of a silicified plug.

The group which is treated as the third order of the present class includes the typical Heterokonta. Compared with typical green algae, these organisms give the impression of a markedly distinct class; placed next to the chrysomonads, they appear scarcely entitled to this rank. Their name is the oldest applicable to the present class, and is accordingly so applied. If it appear expedient to maintain the typical Heterokonta as a distinct class, the remainder of the present one will be called Chrysomonadinea [Chrysomonadineae] (Engler) Pascher.

Of including the choanoflagellates and anisochytrids in the present class as additional orders, one may say that it is not contrary to current knowledge.

1. Mostly pigmented; non-pigmented examples
   mostly producing motile cells with two
   flagella.
   2. Brown or colorless.
      3. Producing motile cells with two
         flagella (exceptionally more)......... Order 1. Ochromonadalea.
      3. Producing motile cells with one
         flagellum; or without known flagel-
         late stages....................... Order 2. Silicoflagellata.
   2. Green............................... Order 3. Vaucheriacea.
1. Non-pigmented, producing motile cells with
   one flagellum.
   2. Predatory, flagellate in the vegetative
      condition, each cell bearing a collar-like
      protoplasmic ridge.................... Order 4. Choanoflagellata.
   2. Parasitic or saprophytic, the vegetative
      cells non-motile, walled................ Order 5. Hyphochytrialea.

Order 1. **Ochromonadalea** [Ochromonadales] Pascher Süsswasserfl. Deutschland
   2: 10, 51 (1913).
Suborder *Monadina* Bütschli in Bronn Kl. u. Ord. Thierreichs 1: 810 (1884).
Order *Isochrysidales* Pascher op. cit. 10, 42.
Order *Syracosphaerinae* Schiller in Arch. Prot. 51: 108 (1925).

Orders *Heliolithae* and *Orthlithinae* Deflandre in Grassé Traité Zool. 1, fasc. 1: 452, 457 (1952).

Brown or colorless Heterokonta, the swimming cells of typical examples with two flagella which are respectively pantoneme and acroneme. In the exceptional family Trimastigida there are a pair of equal flagella and a third flagellum shorter or longer than these; the detailed structure of the flagella of this family is unknown.

Cells of pigmented types contain usually one or two lateral band-shaped plastids. Details of nuclear division are known chiefly by the observations of Doflein (1918, 1922) on *Ochromonas*. The flagella spring from a granule which may be identified as a blepharoplast, near which lies the nucleus. The blepharoplast is connected through two stainable strands (rhizoplasts) to two granules, recognizable as centrosomes, on the two sides of the nucleus. The spindle forms within the intact nuclear membrane with its poles at the centrosomes. The chromosome number appears to be about 4. The nuclear membrane presently disappears. At metaphase, the rhizoplasts are found to lead to separate blepharoplasts, each bearing two flagella. Sexual processes are scarcely known in this group. Schiller (1926) observed in *Dinobryon* the division of calls into two which are released to swim and conjugate in pairs.

This order is believed to represent the direct ancestry of the two following, and also of the typical brown algae.

1. Not filamentous.
  2. Flagellate stages with a pair of equal flagella and a third which is shorter or longer.............................. Family 1. TRIMASTIGIDA.
  2. Flagellate stages with two unequal flagella.
    3. Without calcareous structures attached to the cell walls.
      4. Cells not enclosed in loricae, i. e., open shells.
        5. Not flagellate in the vegetative condition........... Family 2. CHRYSOCAPSACEA.
        5. Flagellate in the vegetative condition, not forming free-swimming circular or globular colonies.......... Family 3. MONADINA.
        5. Free-swimming circular or globular colonies.......... Family 4. SYNCRYPTIDA.
      4. Cells enclosed in loricae........ Family 5. DINOBRYINA.
    3. With calcareous structures attached to the cell walls............ Family 6. HYMENOMONADACEA.
1. Filamentous..............................Family 7. PHAEOTHAMNIONACEA.

Family 1. **Trimastigida** [Trimastigidae] Kent Man. Inf. 1: 307 (1880). Family *Trimastigaceae* Senn in Engler and Prantl. Nat. Pflanzenfam. I Teil, Abt. la: 141 (1900). Family *Prymnesiidae* Hall Protozoology 127 (1953). Organisms producing swimming cells with a pair of equal flagella and a third flagellum longer or shorter than these. With a vegetative stage as globular non-motile colonies as large as pinheads, of pigmented cells; marine: *Phaeocystis*. Motile solitary cells, pigmented: *Prymnesium, Chrysochromulina; Platychrysis* with an amoeboid stage. Motile solitary cells, not pigmented: *Dallingeria, Trimastix, Macromastix.*

Family 2. **Chrysocapsacea** [Chrysocapsaceae] Pascher in Süsswasserfl. Deutschland 2: 85 (1913). Family *Chrysocapsidae* Poche in Arch. Prot. 30: 156 (1913). Nonmotile cells with brown plastids (usually two), imbedded in gelatinous matter and forming colonial aggregates, the protoplasts sometimes escaping as zoospores with two flagella. *Chrysocapsa* Pascher, in fresh water, the colonies few-celled. *Phaeosphaera* West and West, the colonies more extensive.

Family 3. **Monadina** Ehrenberg Infusionsthierchen 1 (1838). Family *Monades* Goldfuss (1818), the mere plural of a generic name. Family *Dendromonadina* Stein Org. Inf. 3, I Hälfte: x (1878). Family *Monadidae* Kent (1880). Family *Heteromonadina* Bütschli in Bronn Kl. u. Ord. Thierreichs 1: 815 (1884). Family *Chrysomonadaceae* Engler in Engler and Prantl Nat. Pflanzenfam. I Teil, Abt. 2: 570 (1897), not family CHRYSOMONADINA Stein. Family *Ochromonadaceae* Senn in Engler and Prantl Nat. Pflanzenfam. I Teil, Abt. la: 163 (1900). Family *Ochromonadidae* Doflein. Pigmented or colorless Ochromonadalea, flagellate in the vegetative condition, not forming circular or globular free-swimming colonies, nor loricate, nor bearing calcareous structures on the cell walls (these being the distinctions respectively of the three following families).

Ochromonas is considered to be in its normal condition when it occurs as solitary swimming cells; it occurs also as gelatinous colonies like those of *Chrysocapsa*. *Stylochrysalis* consists of *Ochromonas-like* cells attached by a stalk at the end away from the flagella. *Chrysodendron* is similar but colonial, the cells attached by branched stalks. *Brehmiella* Pascher (1928) may occur as free-swimming *Ochromonas*-like cells, or these may become attached by the end away from the flagella and develop a whorl of pseudopodia at the free end. Pseudopodia are a device for predatory nutrition, here occurring in an organism which is capable also of photosynthesis. *Heterochromonas* includes organisms of the structure of Ochromonas but without plastids, being presumably saprophytic, and containing only a pigmented speck by which it is supposed that the direction of light is perceived. The historical generic name *Monas* O. F. Müller, as restricted in application by scholars up to Ehrenberg and as applied ever since, designates totally non-pigmented cells, saprophytic or predatory, freeswimming like *Ochromonas* or attached like *Stylochrysalis* (*Physomonas* Kent designates cells of *Monas* in the attached condition). There are believed to be several species, but the group remains poorly known. It was in some member of it that Loeffler (1889) first observed the pantoneme character of flagella. *Dendromonas* consists of similar cells forming colonies like those of *Chrysodendron*. In *Cephalothamnium* Stein, *Monas*-like cells are gathered in capitate clusters on stout stalks. *Anthophysis* Bory is an organism which Leeuwenhoeck had described as a microscopic water plant: it consists of *Monas*-like cells at the ends of branching stalks colored yellow by deposits of iron. The comparatively unfamiliar original spellings of the two generic names just mentioned were restored by Kudo (1946). The name *Uvella* Bory appears to represent small clusters of cells of *Cephalothamnium* or *Anthophysis* which have broken loose to swim free.

Family 3. **Syncryptida** [Syncryptidae] Poche in Arch. Prot. 30: 156 (1913). Family *Isochrysidaceae* Pascher in Süsswasserfl. Deutschland 2: 43 (1913), not based on a generic name. Family *Isochrysidae* Calkins Biol. Prot. 262 (1926). Families *Synuraceae* and *Syncryptaceae* Smith Freshw. Algae (1933). *Ochromonas*-like cells forming circular or globular free-swimming colonies. Flagella markedly unequal, colonies circular: *Cyclonexis*; colonies globular: *Uroglena, Uroglenopsis*. Flagella apparently equal: *Syncrypta, Synura*.

Family 4. **Dinobryina** Ehrenberg Infusionsthierchen 122 (1838). Family *Dinobryaceae* Engler in Engler and Prantl Nat. Pflanzenfam. I Teil, Abt. 2: 570 (1897). Pigmented or colorless cells of the characters of *Ochromonas* or *Monas,* sheltered in loricae, that is, in transparent open shells, solitary or colonial. The pigmented examples have generally been referred to Ochromonadaceae (or whatever), the colorless to Monadidae (or whatever). Pigmented, solitary, flagella markedly unequal: *Epipyxis, Stylopyxis;* flagella apparently equal: *Chrysopyxis* Stein (*Derepyxis* Stokes). Pigmented, forming branching colonies: *Dinobryon, Hyalobryon. Poteriochromonas* Scherffel resembles *Stylopyxis,* but the protoplast can project pseudopodia from its lorica, thus supplementing photosynthesis by predatory nutrition. Non-pigmented, solitary, flagella markedly unequal: *Stokesiella;* flagella apparently equal: *Diplomita.* Non-pigmented cells in colonies quite of the character of those of *Dinobryon*: *Stylobryon.*

Family 5. **Hymenomonadacea** [Hymenomonadaceae] Senn in Engler and Prantl Nat. Pflanzenfam. I Teil, Abt. 1a: 159 (1900). Family *Coccolithophoridae* Lohman in Arch. Prot. 1: 127 (1902). Family *Hymenomonadidae* Doflein. Family *Coccolithidae* Poche in Arch. Prot. 30: 157 (1913). Order *Syracosphaerinae* and family *Pontosphaeraceae* Schiller in Arch. Prot. 51: 8 (1925). Families *Syracosphaeraceae, Halopappaceae, Deutschlandiaceae,* and *Coccolithaceae* Kamptner. Family *Thoracosphaeracee* Schiller in Rabenhorst Kryptog.-Fl. Deutschland ed. 2, 10, Abt. 2: 156 (1930). Families *Syracosphaeridae, Calcisolenidae, Thoracosphaeridae,* and *Braadrudosphaeridae* Deflandre in Grassé Traité Zool. 1, fasc. 1: 452, 457, 458 (1952). Family *Discoasteridae* Tan Sin Hok. Suborder *Coccolithina* Hall Protozoology 130 (1953). Solitary cells with one or two brown plastids, usually with two apparently equal flagella, having a thin cell wall from which project bodies of calcium carbonate (coccoliths) of definite form.

More than twenty genera and nearly 150 species have been described (Lohman; Schiller; Kamptner, 1940). Neither the number of species nor the variety of form appears to warrant making more than one family of the group. Nearly all examples are marine. In *Pontosphaera, Calyptosphaera,* and allied genera, the coccoliths are disks or hemispheres, sometimes umbonate and sometimes marked by one or more pits. In *Syracosphaera* the coccoliths, or a few of them near the insertion of the flagella, bear horn-like projections. In *Najadea, Halopappus,* and *Calciconus,* each cell bears a whorl or elongate bristles. Cells of *Calcisolenia* are fusiform, without flagella, with an armor of two layers of spiral bands of calcareous matter. In *Hymenomonas* and *Coccolithus* Swartz 1894 (*Coccosphaera* Wallich 1877, non Perty 1852; *Coccolithophora* Lohman 1902) the coccoliths are punctured and accordingly ringshaped; *Hymenomonas* differs from most of the group in occurring in fresh water. In *Discosphaera* and *Rhabdosphaera* the punctured calcareous bodies are drawn out to the form of tubes, spools, or trumpets.

These obscure organisms are not without importance. They occur in all oceans, being most abundant in gulfs, such as the Adriatic, where the salinity is diminished by rivers (Schiller, 1925). According to Bernard (1947) turbidity in the Mediterranean depends chiefly on this group. Coccoliths are abundant in the ooze on the bottoms of oceans. They occur as fossils as far back as the Cambrian, being particularly abundant in certain Cretaceous deposits.

Family 6. **Phaeothamnionacea** [Phaeothamnionaceae] Pascher in Süsswasserfl. Deutschland 2: 113 (1913). Family *Chrysotrichaceae* Pascher (1914). Family *Nematochrysidaceae* Pascher (1925). Brown organisms, minute, marine, epiphytic, filamentous, reproducing by zoospores bearing paired unequal flagella. *Nematochrysis,* the filaments unbranched; *Phaeothamnion,* the filaments branched. These organisms are believed to represent the transition between the Chrysocapsacea and the typical brown algae.

There is a family Amphimonadidae or Amphimonadaceae of unwalled colorless flagellates with paired supposedly equal flagella. They appear to belong to the kingdom of plants, in the neighborhood of *Chlamydomonas* and *Polytoma.* If, however, future study shows their flagella actually to be respectively pantoneme and acroneme, they are to be placed in the present order.

Order 2. **Silicoflagellata** Borgert in Zeit. wiss. Zool. 51: 661 (1891).
Chromomonadina Klebs in Zeit. wiss. Zool. 55: 394 (1893).
Order *Chromomonadina* Blochmann Mikr. Tierwelt ed. 2. Abt. I: 57 (1895).
Subclass *Chrysomonadineae* Engler in Engler and Prantl Nat. Pflanzenfam. I Teil, Abt. 1a: iv (1900).
Order *Chrysomonadales* Engler Syllab. ed. 3: 7 (1903).
Chrysomonadinae; Euchrysomonadinae, with order *Chromulinales; Chrysocapsinae;* and *Rhizochrysidinae* Pascher in Süsswasserfl. Deutschland Heft 2 (1913).
Chrysomonadales, Chrysocapsales, Chrysosphaerales, and *Chrysotrichales* Pascher in Ber. deutschen bot. Gess. 32: 158 (1914).
Order *Chrysomonadina* Doflein Lehrb. Prot. ed. 4: 401 (1916).
Order *Chrysomonadida* Calkins Biol. Prot. 258 (1926).
Classes *Chrysomonadineae, Rhizochrysidineae, Chrysocapsineae, Chrysosphaerineae,* and *Chrysotrichineae* Pascher in Beih. bot. Centralbl. 48, Abt. 2: 323 (1931).
Suborders *Euchrysomonadina, Silicoflagellina, Rhizochrysidina,* and *Chrysocapsina* Hall Protozoology 125, 128, 130, 132 (1953).
Organisms of much the character of Ochromonadalea, but producing flagellate stages with a single flagellum, or not producing flagellate stages. The detailed structure of the flagella has seemingly never been determined. Statospores are known to be produced by *Chromulina, Mallomonas,* and (of somewhat exceptional character) by *Hydrurus.* Sexual reproduction has not been observed. Mitosis, with an intranuclear spindle and numerous chromosomes, was observed by Doflein (1916) in *Rhizochrysis.*

This order is supposed to represent the direct ancestry of orders Choanoflagellata and Hyphochytrialea.

1. Neither amoeboid nor truly filamentous.
   2. Not flagellate in the vegetative condition.
      3. Microscopic colonies............Family 1. CHRYSOSPHAERACEA.
      3. Macroscopic gelatinous colonies
         simulating filaments..............Family 2. HYDRURACEA.
   2. Flagellate in the vegetative condition.
      3. Without prominent siliceous structures............................Family 3. CHRYSOMONADINA.
      3. With siliceous scales usually bearing
         bristles........................Family 4. MALLOMONADINEA.
      3. With siliceous internal skeletons.....Family 5. ACTINISCEA.

1. Amoeboid..............................Family 6. CHRYSAMOEBIDA.
1. Filamentous............................Family 7. THALLOCHRYSIDACEA.

Family 1. **Chrysosphaeracea** [Chrysosphaeraceae] Pascher in Arch. Prot. 52: 562 (1925). Family *Naegelliellaceae* Pascher op. cit. 561. Family *Nagelliellidae* Hall Protozoology 133 (1953). Non-motile brown cells, either capable of repeated division into two, thus forming aggregates of indefinite number, or else undergoing multiple division and producing colonies of definite number of cells; mostly known to produce uniflagellate zoospores. *Chrysosphaera, Epichrysis, Chrysospora, Gloeochrysis, Naegelliella,* and other genera.

Family 2. **Hydruracea** [Hydruraceae] West British Freshw. Algae 45 (1904). *Hydrurina* Klebs in Zeit. wiss. Zool. 55: 420 (1893). Family *Hydruridae* Poche in Arch. Prot. 30: 158 (1913). Like Chrysosphaeracea, but the colonies dendroid, growing at the tips, becoming macroscopic; producing tetrahedral zoospores and spheroidal resting cells bearing a unilateral crest. *Hydrurus foetidus,* in mountain streams.

Family 3. **Chrysomonadina** Stein Org. Inf. 3, I Hälfte: x (1878). Family *Chrysomonadidae* Kent Man. Inf. (1880). Family *Chromulinaceae* Engler in Engler and Prantl Nat. Pflanzenfam. I Teil, Abt. 2: 570 (1897). Family *Chromulinidae* Doflein. Brown flagellates with a single anterior flagellum, sometimes producing siliceous granules but without more extensive siliceous structures. Free-swimming, walled: *Chrysococcus, Microglena.* Naked: *Chromulina,* the type genus of *Chrysomonadina,* the generic name *Chrysomonas* being a synonym. Organisms of this genus are rather freely capable of producing pseudopodia and supplementing photosynthetic nutrition by predatism, or, alternatively, of producing gelatinous aggregates of walled non-motile cells (Hofender, 1913; Gicklhorn, 1922). *Chrysapsis* differs from *Chromulina* in having in each cell a single plastid in the form of a network. Solitary attached cells, producing pseudopodia only occasionally: *Lepochromulina.* Bearing whorls of permanent pseudopodia: *Cyrtophora, Pedinella, Palatinella* (Pascher, 1928).

Family 4. **Mallomonadinea** Diesing in Sitzber. Akad. Wiss. Wien Math.-Nat. Cl. 52, Abt. 1: 304 (1866). Family *Mallomonadidae* Kent (1880). Brown uniflagellate free-swimming cells with an armor of siliceous scales usually bearing bristles. *Mallomonas,* solitary cells, the bristle-bearing scales circular. *Conradiella,* the scales of the form of rings about the body. *Chrysosphaerella,* spherical colonies, each cell with two long bristles.

Family 5. **Actiniscea** [Actinisceae] Kützing Phyc. Germ 117 (1845). Family *Dictyochidae* Wallich. Class *Silicoflagellata* (Borgert), orders *Siphonotestales* and *Stereotestales,* and families *Dictyochaceae* and *Ebriaceae* Lemmermann in Ber. deutschen bot. Gess. 19: 254-268 (1901). Division (?) *Silicoflagellatae* Engler. Family *Silicoflagellidae* Calkins Biol. Prot. 263 (1926). Families *Ebriopsidae, Ditripodiidae, Ammodochidae,* and *Ebriidae* Deflandre in Grassé Traité Zool. 1, fasc. 1: 421, 423, 424 (1952). Solitary brown uniflagellate cells with a continuous internal skeleton of silica. Marine, commonest in colder oceans.

The skeletons are not subject to decay and are found as microfossils in chalk and diatomaceous earth. They have been reported from the Silurian and are commonest in certain Cretaceous deposits. Ehrenberg described several fossil species, classifying them as diatoms. The living forms, subsequently discovered, include apparently the same species.

Gemeinhardt (in Rabenhorst, 1930) accounted for the structure of the cells.

They are approximately of radial symmetry, the axis being shorter than the diameter. The skeleton is completely imbedded in protoplasm. It may be a mere ring; or the ring may bear radially projecting spines; or it may be the margin of a more or less complicated basket-shaped network coaxial with the cell. Numerous brown plastids lie near the surface of the protoplast. There is no cell wall. The double cells, like two cells lying face to face, which have occasionally been seen, are not stages of conjugation, but of cell division, in which one daughter cell retains the original skeleton while the other develops a new skeleton in the position of a mirror image of the original one.

Lemmermann and Gemeinhardt accounted for only six genera and twenty-four species, but Gemeinhardt recognized numerous varieties, and it is probable that the number of species has been underestimated. *Mesocaena,* the skeleton a mere ring, smooth or spiny; *Dictyocha, Distephanus, Cannopilus,* the skeleton more or less netted.

Family 6. **Chrysamoebida** [Chrysamoebidae] Poche in Arch. Prot. 30: 157 (1913). Families *Rhizochrysidaceae, Chrysarachniaceae,* and *Myxochrysidaceae* Pascher in Beih. Bot. Centralbl. 48, Abt. 2: 323 (1931). Family *Rhizochrysididae* Hollande in Grassé Traité Zool. 1, fasc. 1: 547 (1952). Families *Rhizochrysidae* and *Myxochrysidae* Hall Protozoology 130, 132 (1953). Amoeboid organisms with brown plastids of the form of one or two parietal films in each cell. *Rhizaster,* an attached organism resembling *Cyrtophora* and *Pedinella* but lacking the flagellum. *Chrysocrinus,* attached to algae, the protoplast covered by a dome-shaped shell punctured by many pores through which project the slender psudopodia. *Chrysamoeba,* a freely moving cell usually with one flagellum; *Rhizochrysis,* similar, without the flagellum. *Myxochrysis,* a large multinucleate form. *Chrysarachnion,* the cells clustered and linked together by strands of protoplasm. *Lagynion,* having an attached vase-shaped lorica from which projects usually a single slender pseudopodium. *Chrysothylakion,* with a retort-shaped lorica from which project many slender pseudopodia, branching and anastomosing. Only the plastids distinguish these organisms from various genera classified as Rhizopoda, Heliozoa, or Sarkodina.

Family 7. **Thallochrysidacea** [Thallochrysidaceae] Pascher (1925). Brown organisms producing definite filaments of walled cells and reproducing by anteriorly uniflagellate zoospores. *Thallochrysis. Phaeodermatium.*

Order 3. **Vaucheriacea** [Vaucheriaceae] Nägeli Gatt. einzell. Alg. 40 (1849). Class HETEROKONTAE and orders *Chloromonadales* and *Confervales* Luther in Bihang Svensk. Vetensk.-Akad. Handl. 24, part 3, no. 13: 19 (1899). Not *Chloromonadina* Klebs (1893); not order *Confervoidea* C. Agardh (1824). *Vaucheriales* Bohlin Gröna Algernas 25 (1901). Order *Vaucheriales* Clements Gen. Fung. 14 (1909). Orders *Heterochloridales, Heterocapsales, Heterococcales, Heterotrichales,* and *Heterosiphonales* Pascher in Hedwigia 53: 10-21 (1912). Division *Heterokontae,* Classes *Heterochloridineae, Rhizochloridineae, Heterocapsineae, Heterococcineae, Heterotrichineae,* and *Heterosiphoneae,* and orders *Rhizochloridales* and *Botrydiales* Pascher in Beih. bot. Centralbl. 48, Abt. 2: 324 (1931). Class *Xanthomonadina* with orders *Heterochloridea* and *Rhizochloridea* Deflandre in Grassé Traité Zool. 1, fasc. 1: 212, 217, 220 (1952). Order *Heterochlorida* Hall Protozoology 133 (1953).

The Classification of Lower Organisms

Organisms producing motile cells with paired unequal flagella which Vlk (1931) found to be respectively pantoneme and acroneme, differing from Ochromonadalea in being of a green or yellow-green color, and in being mostly of algal body type, i. e., walled and non-motile. The cell wall consists usually of two parts which become separate when the cell divides; the two parts are believed to be distantly homologous with the wall and plug of the statospores of Ochromonadalea and Silicoflagellata (Pascher, 1932). The storage products are oil and sometimes leucosin.

As this is the group to which the class name Heterokontae was first applied, it is

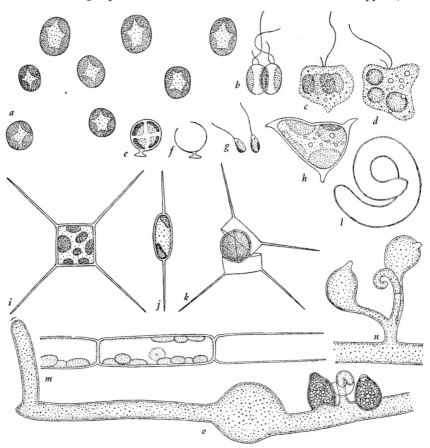

Fig. 10.—Vaucheriacea: **a, b,** *Chlorosaccus fluidus,* cells of the colony and zoo-spores, after Luther (1899). **c, d,** *Chloramoeba heteromorpha* x 1,000 after Bohlin (1897). **e, f, g,** Cell, empty cell, and zoospores of *Characiopsis gibba* x 1,000 after Pascher (1912). **h,** *Dioxys Incus* after Pascher (1932). **i, j, k,** Cell, edge of cell, and statospore of *Pseudotetraedron neglectum* x 1,000 after Pascher (1912). **l,** *Spirodiscus fulvus* x 1,000. **m,** End of a filament of *Tribonema bombycina* x 1,000. **n,** Antheridium and oogonia of *Vaucheria Gardneri* x 100. **o,** Filament of *Vaucheria sessilis* x 100.

the type group of the class. As established by Luther, the class consisted of the new genus *Chlorosaccus* together with a few genera of flagellates (*Vacuolaria* was included in error) and a few transferred from the group of typical green algae. From time to time, other green algae have been transferred, and it has become evident that the group is a fairly extensive one. Green organisms can be recognized as belonging here by a negative reaction to the iodine test for starch, and by the fact that they give a bluish color when heated with hydrochloric acid, instead of a yellow one, as typical green algae do: the difference depends upon differences in the complement of photosynthetic pigments. Bohlin (1901) placed *Vaucheria* here; most authors have not followed him, but Smith (1950) has done so. This genus brings with itself the oldest name for the group as an order.

Mitosis in *Vaucheria* was described by Hanatschek (1932) and Gross (1937). The spindle is intranuclear; Hanatschek saw centrosomes at the poles. The conjugation of equal free-swimming gametes was observed in *Tribonema* and several other genera by Scherffel (1901), and in *Botrydium* by Rosenberg (1930). Vaucheria was one of the organisms by study of which the nature of fertilization was discovered (Pringsheim, 1855). Hanatschek and Gross found that the first two divisions of the nucleus of the zygote are meiotic: the soma is haploid.

This order is believed to represent the direct ancestry of the two following classes, Bacillariacea and Oomycetes.

Pascher (1912, 1925) arranged the green Heterokonta in subordinate groups parallel to those of the typical green algae; and, as the main groups of green algae are treated as orders, he treated these groups also as orders (in 1931 as classes). They are scarcely entitled to such rank: too many of the classes or orders are of single families, and too many of the families are of one or two genera. Here, then, Pascher's classes and orders are suppressed and several of his families are reduced.

1. Not truly filamentous nor producing rhizoids.
   2. The cells walled.
      3. Cells regularly dividing into two, forming gelatinous colonies; occasionally producing small numbers of zoospores.
         4. The colonies globular or irregular, becoming macroscopic......Family 1. CHLOROSACCACEA.
         4. The colonies dendroid, microscopic........................Family 2. MISCHOCOCCACEA.
      3. Cells normally undergoing division into several.
         4. Producing zoospores...........Family 3. CHLOROTHECIACEA.
         4. Producing no motile cells........Family 4. BOTRYOCOCCACEA.
   2. The cells loricate.....................Family 5. STIPITOCOCCACEA.
   2. The cells amoeboid...................Family 6. CHLORAMOEBACEA.
1. Filaments of uninucleate cells................Family 7. TRIBONEMATACEA.
1. Cells becoming highly multinucleate, forming filaments or at least producing rhizoids..... Family 8. PHYLLOSIPHONACEA.

Family 1. **Chlorosaccacea** [Chlorosaccaceae] Smith Freshw. Algae 145 (1933). Family *Heterocapsaceae* Pascher in Hedwigia 53: 13 (1912); there is no corresponding generic name. Gelatinous aggregates of cells which may divide, causing the aggregate to grow to macroscopic dimensions; or may produce one, two, or four zoospores. *Chlorosaccus* Luther, the standard genus of class Heterokonta.

Family 2. **Mischococcacea** [Mischococcaceae] Pascher in Hedwigia 53: 14 (1912). Microscopic colonies of globular cells joined by dichotomously branching gelatinous strands. *Mischococcus*.

Family 3. **Chlorotheciacea** [Chlorotheciaceae] Luther in Bihang Svensk. Vetensk-Akad. Handl. 24, part 3, no. 13: 19 (1899). Families *Chlorobotrydiaceae* and *Sciadiaceae* Pascher in Hedwigia 53: 17 (1912). Family *Halosphaeraceae* Pascher (1925). Family *Ophiocytiaceae* Auctt. Cells solitary, free or attached, capable of reproduction by division to form multiple zoospores, in some examples capable alternatively of producing multiple minute non-motile cells of the same form as the parent. Large free multinucleate cells, more or less globular: *Botrydiopsis, Leuvenia*. Smaller cells, elongate, curved or coiled: *Characiopsis, Spirodiscus. Spirodiscus fulvus* Ehrenberg in Abh. Akad. Wiss. Berlin 1830: 65 (1832) (*nomen nudum*) and Infusionsthierchen 86 (1838), whose identity has been a standing puzzle to bacteriological systematists, is an older name of *Ophiocytium parvulum* (Perty) A. Braun (Copeland, 1954). It antedates the generic name *Ophiocytium* Nägeli (1849); new combinations are required for the dozen additional species of this genus. The cells attached: some species of *Characiopsis; Perionella; Dioxys*.

Family 4. **Botryococcacea** [Botryococcaceae] Pascher in Hedwigia 53: 13 (1912). Solitary or colonial cells reproducing strictly by production of non-motile cells. *Botryococcus. Pseudotetraedron*.

Family 5. **Stipitococcacea** [Stipitococcaceae] Pascher in Beih. bot. Centralbl. 48, Abt. 2: 324 (1931). Family *Stipitochloridae* Deflandre in Grassé Traté Zool. 1, fasc. 1: 221 (1952). Amoeboid cells with green plastids, partially enclosed in loricae attached to objects in water. *Stipitococcus*.

Family 6. **Chloramoebacea** [Chloramoebaceae] Luther in Bihang Svensk. Vetensk.-Akad. Handl. 24, part 3, no. 13: 19 (1899). Family *Chloramoebidae* Poche in Arch. Prot. 30: 155 (1913). Families *Heterochloridaceae* and *Rhizochloridaceae* Pascher Süsswasserfl. Deutschland 11: 22, 26 (1925). Families *Heterochloridae, Rhizochloridae, Chlorarachnidae* and *Myxochloridae* Deflandre in Grassé Traité Zool. 1, fasc. 1: 217-222 (1952). Amoeboid organisms with green plastids, without loricae, sometimes swimming by means of paired unequal flagella. *Chloramoeba, Chlorochromonas, Rhizochloris*.

Family 7. **Tribonematacea** [Tribonemataceae] Pascher in Hedwigia 53: 19 (1912). Family *Confervaceae* Luther (1899). Family *Monociliaceae* Smith Freshw. Algae 160 (1933). Green Heterokonta producing filaments of uninucleate cells. The Linnaean genus *Conferva* included a great variety of growths in water. Definite groups were separated from it, one after another, until the residue was a natural group; but this residue cannot be assumed to be the type of *Conferva* L.; that name is to be abandoned as a *nomen confusum*. The remnant in question has become two genera, *Tribonema* Derbes and Solier, 1858, and *Bumilleria* Borzi, 1895. They are unbranched filaments, common in freshwater pools. From typical green algae of similar appearance they are distinguished in the first place by the presence in each cell of several disk-shaped plastids without pyrenoids or with obscure ones. The cell walls, when treated with sulfuric acid, can be seen to consist of two parts like a barrel sawed across the middle. A broken filament ends always with a broken half wall. *Monocilia*, an unfamiliar alga isolated from soil, differs in producing branching filaments.

Family 8. **Phyllosiphonacea** [Phyllosiphonaceae] Wille in Engler and Prantl. Nat. Pflanzenfam. I Teil, Abt. 2: 125 (1890). Family *Vaucheriaceae* (Nägeli) Areschoug (1850), preoccupied by order VAUCHERIACEAE Nägeli. Family *Botrydiaceae* Luther (1899). Heterokonta whose bodies are highly multinucleate single cells, filamentous or anchored by filamentous rhizoids. *Botrydium* is found on damp soil as dark green globes, sometimes as much as 2 mm. in diameter, anchored by much-branched colorless rhizoids. *Vaucheria* is a familiar alga on damp earth or in fresh water. It consists of irregularly branching filaments, green where exposed to light, colorless where growing downward and serving as rhizoids. The reproductive cells are cut off by walls. The end of an aerial filament, cut off in this fashion, may as a whole act as a spore. In water, the protoplast of such a cell may escape as an exceptionally large zoospore with as many pairs of flagella as the nuclei within it. Antheridia are brief branches, each releasing many minute sperms each with two unequal flagella. Oogonia are globular cells, multinucleate during development, but containing only one functional nucleus when mature. *Phyllosiphon* is of much the same structure as *Vaucheria,* but is parasitic in seed plants, particularly Araceae. It reproduces, apparently, only by the breaking up of the protoplast to produce minute non-flagellate spores.

Order 4. **Choanoflagellata** [Choano-Flagellata] Kent Man. Inf. 1: 36 (1880).
    Order *Bicoecidea* Grassé and Deflandre in Grassé Traité Zool. 1, fasc. 1: 599 (1952).
Non-pigmented flagellates, usually attached, each cell bearing a single flagellum of the type called pantacroneme, with lateral appendages and a terminal whip-lash; the cell bearing also a protoplasmic collar, usually surrounding the base of the flagellum. The collar is a means of nutrition. Bacteria and other scraps of organic matter, driven against it by the beating of the flagellum, adhere and are carried to the interior of the cell by flow of the cytoplasm of which it consists.
    It is probable that the pantacroneme flagellum is a variant of the pantoneme flagellum, and that this order belongs naturally in class Heterokonta. It may have evolved from Silicoflagellata; or it may be that the collar is a modified flagellum, and that the group evolved from order Ochromonadalea.
    Most authors have recognized more than one family of choanoflagellates, but genera are not very numerous and one family seems sufficient to accommodate them. Family **Bicoekida** Stein Org. Inf. 3, I Hälfte: x (1878). Family *Craspedomonadina* Stein 1. c. Families *Bikoecidae, Codonosigidae, Salpingoecidae,* and *Phalansteriidae* Kent op. cit. Families *Codonoecina* and *Bikoecina* Bütschli in Bronn Kl. u. Ord. Thierreichs 1: 814, 815 (1884). Families *Bicoecaceae, Craspedomonadaceae,* and *Phalanasteriaceae* Senn in Engler and Prantl Nat. Pflanzenfam. I Teil, Abt. la: 121, 123, 129 (1900). Family *Gymnocraspedidae* Grassé Traité Zool. 1, fasc. 1: 590 (1952). Characters of the order. Cells naked, solitary: *Monosiga;* colonial: *Codosiga* James-Clark (*Codonosiga* Stein), *Sphaeroeca.* Cells imbedded in gelatinous matter, the collars contracted: *Phalanseterium.* Loricate: *Salpingoeca, Bicosoeca, Poteriodendron.*
    The choanoflagellates were discovered by James-Clark (1866, 1868), who made at the same time the discovery that certain internal cavities of sponges are lined by minute cells (choanocytes) of the same structure as the choanoflagellates. From these observations he drew the conclusion that sponges are a sort of flagellates distinguished by the production of exceptionally large and elaborate colonies. Kent

The Classification of Lower Organisms

described *Proterospongia Haeckeli* as a colonial organism of amoeboid and choano-flagellate cells in a common matrix; he regarded it as a transitional form, important as evidence of the evolution of sponges from choanoflagellates. According to Duboscq and Tuzet (1937) it is no organism, but a stage in the development of an individual sponge from one which has been damaged. In spite of this, the hypothesis that the choanoflagellates represent the evolutionary origin of the sponges, and accordingly of the entire animal kingdom, continues to appear tenable.

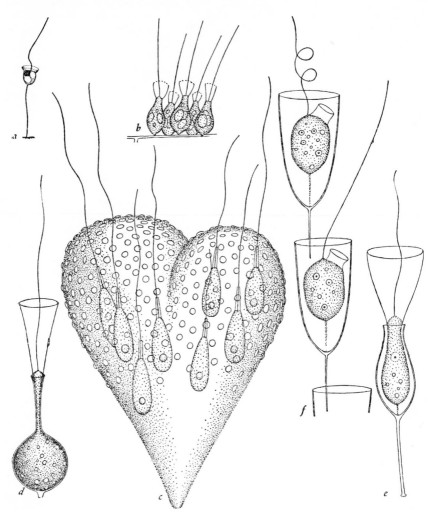

Fig. 11.—CHOANOFLAGELLATA: **a, b,** *Monosiga* spp.; **c,** *Phalanasterium digitatum;* **d,** *Salpingoeca ampullacea;* **e,** *Salpingoeca Clarkii;* **f,** *Poteriodendron petiolatum.* **c** x 500, the remainder x 1,000. **c-f** after Stein (1878).

Order 5. **Hyphochytrialea** [Hyphochytriales] Bessey Morph. and Tax. Fungi 69 (1950).

Order *Anisochytridiales* Karling in American Jour. Bot. 30: 641 (1943), not based on a generic name.

Non-pigmented organisms with walled cells, parasitic or saprophytic, the protoplasm with numerous granules not of a shining appearance, producing zoospores with single anterior pantoneme flagella.

The naked zoospores come to rest upon appropriate hosts or substrata. Ordinarily, in parasitic species, the protoplast of the zoospore makes its way to the interior of a cell of the host. It swells and develops a thin wall. The resulting structure may be called a center. In most members of the group, the center gives rise to a system of slender rhizoids; in some species, these give rise to further centers like the original one. Karling studied the cytology particularly in *Anisolpidium*. There are repeated simultaneous mitoses in the growing centers. Resting nuclei contain conspicuous karyosomes. Dividing ones show about five chromosomes in an intranuclear spindle which ends sharply in centrosomes. Eventually, in the usual course of events, each center produces an exit tube to the exterior. Its contents are released by deliquescence of the tip of the exit tube. Either before this or afterward, the mass of protoplasm undergoes cleavage into uninucleate protoplasts which generate flagella. Sometimes, instead of discharging their contents, the centers are converted into resting spores by the secretion of thick walls (this has been observed in only a few of the species). The resting spores germinate by producing exit tubes and discharging zoospores as ordinary centers do.

The body type which has just been described may be called the chytrid body type; organisms of this body type were formerly assembled as a taxonomic group typified by the genus *Chytridium*. Couch, however, showed that these organisms form three groups distinguished by fundamental differences in type of flagellation. The present group is here given a place implying relationship to order Silicoflagellata.

Karling (1943) accounted for fourteen species. He provided three families; only one is here maintained.

Family **Hyphochytriacea** [Hyphochytriaceae] Fischer in Rabenhorst Kryptog.-Fl. Deutschland 1, Abt. 4: 131 (1892). Families *Anisolpidiaceae* and *Rhizidiomycetaceae* Karling in American Jour. Bot. 30: 641, 643 (1943). Characters of the order. Without rhizoids: *Anisolpidium* on brown algae; *Roesia* on *Lemna; Cystochytrium* on roots of *Veronica*. With rhizoids from a single center: *Rhizidiomyces* and *Latrostium* on green algae, aquatic fungi, and the empty exoskeletons of insects. With multiple centers: *Hyphochytrium* and *Catenariopsis,* on fungi and other hosts.

Class 2. **BACILLARIACEA** Engler and Prantl

Homalogonata Lyngbye Tent. Hydrog. Danicae 177 (1819).
Order DIATOMEAE C. Agardh Syst. Alg. xii (1824).
Division (of order *Algae*) *Diatomaceae* Harvey in Mackay Fl. Hibern. 166 (1836).
Family BACILLARIA Ehrenberg Infusionsthierchen 136 (1838).
Series (of class *Algae*) *Diatomaceae* Harvey Man. British Alg. 15 (1841).
Abtheilung (of class *Isocarpeae*) *Diatomaceae* Kützing Phyc. Germ. 54 (1845).
Stamm Diatomea Haeckel Gen. Morph. 2: xxv (1866).
Division (of class *Algae*) *Diatomaceae* Rabenhorst Kryptog.-Fl. Sachsen 1: 1 (1863).

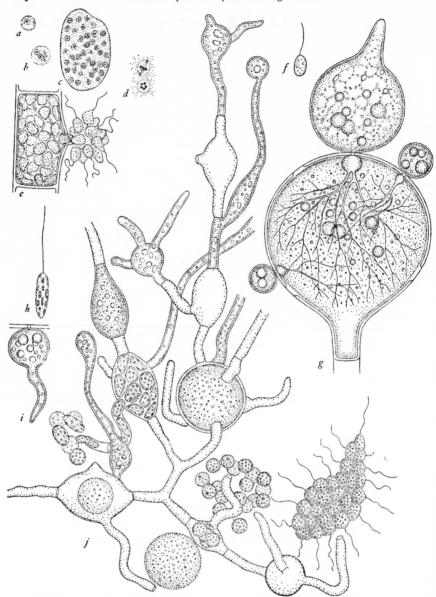

Fig. 12.—Hyphochytrialea: **a-e**, *Anisolpidium Ectocarpii;* **a-c**, individuals developing in cells of *Ectocarpus;* **d**, mitotic figures x 2,000; **e**, cell of *Ectocarpus* filled by a mature individual discharging spores. **f**, **g**, *Rhizidiomyces apophysatus;* **f**, zoospore; **g**, oogonium of *Achlya* parasitized by three individuals. **h**, **i**, **j**, *Hyphochytrium catenoides;* **h**, zoospore; **i**, young individual; **j**, mature individual with filaments, sporangia, and zoospores in various stages of development. All after Karling (1943, 1944, 1939). x 1,000 except as noted.

Class BACILLARIACEAE Engler and Prantl Nat. Pflanzenfam. II Teil: 1 (1889).
Subdivision and class *Bacillariales* Engler Syllab. 6 (1892).
Hauptclasse Diatomeae Haeckel Syst. Phylog. 1: 90 (1894).
Subclass *Bacillariales* Engler in Engler and Prantl Nat. Pflanzenfam. Teil I, Abt.
   la: v (1900).
Class *Bacillarieae* Wettstein Handb. syst. Bot. 1: 74 (1901).
Class *Bacillarioideae* Bessey in Univ. Nebraska Studies 7: 283 (1907).
Class *Diatomeae* Schaffner in Ohio Naturalist 9: 447 (1909).
Abteilung Bacillariophyta Engler.
Abteilung (of *Stamm Chrysophyta*) *Diatomeae* Pascher in Beih. bot. Centralbl.
   48, Abt. 2: 324 (1931).
Class *Bacillariophyceae* Auctt.

Unicellular (occasionally filamentous or colonial) organisms without flagella in
the vegetative condition, each cell with one, two, or more plastids, brown, varying to
yellow or exceptionally to bluish or colorless, and bearing a siliceous shell of two
parts. Globules of oil and granules of something called volutin (the "red granules
of Bütschli," apparently protein) are present. Other granules in some examples are
said to be of leucosin.

These organisms, the diatoms, are very common. There are some 5300 species.
Microscopic examination of the bottoms of fresh water ponds reveals usually more of
diatoms than of any other kind of organisms. Diatoms are frequent prey of many kinds
of predators, from amoebas to whales. In using fish-liver oils as a source of vitamin
D, man adds himself to a long chain of predators of which it is believed that diatoms
are the usual ultimate prey.

The shells of diatoms are not subject to decay. In certain places which were in
the geologic past arms of the sea, there are enormous deposits of diatom shells in
the form of a white earth. The oldest deposits are of the Cretaceous age. Thus it appears that diatoms are a modern offshoot, no more ancient than the flowering plants.
Diatomaceous earth is mined for various uses. It is an effective insulating material,
and was the inert material first used in connection with nitroglycerine in the manu-
facture of dynamite.

The two parts of the shell of a diatom are called valves. They fit one over the
other "like the parts of a pill box" (ZoBell, 1941, objects to this traditional simile,
on the ground that in current language a pillbox is a concrete structure with loop-
holes). The shells consist basically of something of the nature of pectin heavily impregnated with silica and characteristically sculptured. The cells appear markedly
different in different aspects: the aspect which is in effect top or bottom view is
called valve view, and that which is in effect side view is called girdle view. When a
cell divides, each of the daughter cells receives one of the valves and generates an
additional valve fitting within it. Diatoms in culture undergo a gradual diminution
in size; there is an old hypothesis that this is caused by the fact that one of each pair
of sister cells receives a slighly smaller valve than the other.

Lauterborn (1896) described mitosis in *Surirella* and other diatoms. He found a
centrosome, with radiating strands, near the nucleus. At the beginning of mitosis,
the centrosome generates a disk-shaped structure which enters the nucleus and grows
in such fashion as to become a cylinder extending through it. The cylinder is recognizably a spindle, but the chromosomes, instead of appearing within it, form a ring-
shaped mass about its middle and divide into two ring-shaped masses which move
along it to its extremities. The nuclear membrane ceases to be recognizable early in

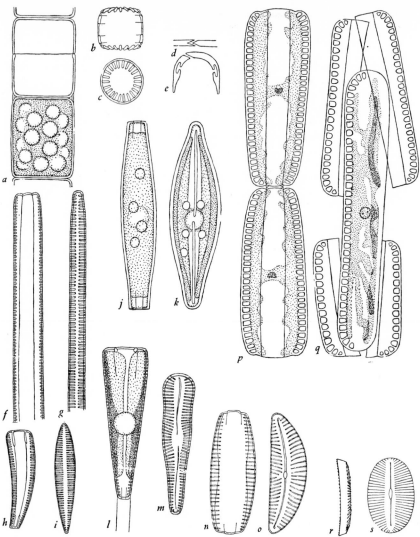

FIG. 13.—BACILLARIACEA: **a**, *Melosira* sp., a living cell and an empty one. **b, c**, Girdle and valve views of cell of *Cyclotella* sp. **d, e**, Sections of a valve of *Pinnularia* sp., highly magnified, after Otto Müller (1896); **d**, about half-way between the middle and the end, **e**, near the end. **f, g**, Girdle and valve views of *Synedra* sp. **h, i**, Girdle and valve views of *Rhoicosphenia curvata*. **j, k**, Girdle and valve views of *Navicula* sp. **l, m**, Girdle and valve views of *Gomphonema* sp. (the former showing the gelatinous stalk by which the cell is attached). **n, o**, Girdle and valve views of *Cymbella* sp. **p, q**, *Surirella saxonica* after Karsten (1900); **p**, two cells joined before conjugation; **q**, zygote; x 250. **r, s**, Girdle and valve views of *Cocconeis* sp. x 1,000 except as noted.

the process, but the nuclear cavity remains distinct until the chromosomes have reached the ends of the spindle. The nuclear sap and the spindle are then absorbed by the cytoplasm, but not until the spindle has budded off a new centrosome from each end.

Subsequent authors, as Karsten (1900), Geitler (1927), Iyengar and Subrahmanyan (1942, 1944), and Subrahmanyan (1947), have not seen as full a series of stages as Lauterborn did. They have found centrosomes in at least some diatoms, and have confirmed the point that the spindle is a cylinder which is surrounded by the chromosomes instead of including them.

The same authors have described sexual processes in *Surirella, Cymbella, Cocconeis, Cyclotella,* and *Navicula.* In *Surirella saxonica* as described by Karsten, pairs of the wedge-shaped cells become attached by little bodies of slime at the narrow ends. Each nucleus divides twice, producing four, of which three are digested by the cytoplasm. The two protoplasts then move in amoeboid fashion out of their shells and they and their nuclei unite. The zygote protoplast grows to a size much greater than that of the parent cells and secretes a membrane which becomes silicified. The resulting cell is called an auxospore.

In most kinds of diatoms, each cell produces two gametes. In some, the cells pair and proceed to produce auxospores individually, without conjugation. Karsten supposed the latter examples to represent a stage in the evolution of sexual reproduction under some *zwingender Nothwendigkeit*: much more probably, they are products of degeneration. In *Cyclotella,* Iyengar and Subrahmanyan found the production of auxospores to involve autogamous karyogamy: the nucleus of a solitary cell undergoes meiosis; two of the haploid nuclei are digested, and the two which remain fuse with each other. It is evident that all diatoms are diploid in the vegetative condition.

The filamentous green Heterokonta *Tribonema* and *Bumilleria* are closely similar to the diatom *Melosira,* and it may reasonably be supposed that they represent the evolutionary origin of the group.

Diatoms are preserved for study by violent methods which destroy the protoplasts, and the classification is based strictly on characters of the shells. So uniform is the group that Schütt (in Engler and Prantl, 1896) treated it as a single family. He provided an elaborate subsidiary classification involving two main groups. Subsequent scholars have found his system essentially sound as a representation of nature, but have raised the main groups to the rank of orders and the minor ones in corresponding degree.

Order 1. **Disciformia** [Disciformes] Kützing Phyc. Germ. 112 (1845).
Order *Appendiculatae* Kützing l. c.
Centricae Schütt in Engler and Prantl Nat. Pflanzenfam. I Teil, Abt. 1b: 57 (1896).
Order *Centricae* Campbell Univ. Textb. Bot. 90 (1902).
Order *Eupodiscales* Bessey in Univ. Nebraska Studies 7: 284 (1907).
Diatoms basically of radial symmetry, which, however, is often distorted; not motile in the vegetative condition; plastids numerous in the cells.

These are the more primitive diatoms. The majority are marine. Three types of reproductive cells are known to be produced by them.

Occasionally, in mass catches of material from the ocean, diatoms are found whose protoplasts have undergone repeated division within the shell and produced numerous little naked protoplasts. These protoplasts are said to bear flagella; whether one or two, equal or unequal, is not certainly known. They are supposed to escape and function as zoospores, but Karsten (1904), on rather scant evidence, supposed them to be gametes.

A protoplast may contract and form a shell within its former shell. The new shell consists like the old one of two parts, one fitting within the other. The outer shell is usually more or less elaborately sculptured, while the inner is smooth. It is supposed that the outer shell is deposited between outer and inner masses of protoplasm, and that the entire protoplast then withdraws to the interior and deposits the inner shell in the opening. It is in this manner that the statospores of chrysomonads are formed. The resting cells of diatoms as just described are believed to be homologous with them, and are called by the same term.

As a third manner of producing a reproductive cell, a protoplast may expand, force apart the valves of its shell, and deposit an enlarged shell about itself. The resulting spore is called an auxospore. As noted, Iyengar and Subrahmanyan found the production of auxospores in *Cyclotella* to involve sexual processes.

Schütt divided the Centricae into three groups with names in *-oideae* (presumably subfamilies) and these into nine groups with names in *-eae* (presumably tribes). Subsequent authorities have made of Schütt's groups a varying number of families. The minimum tenable number of families is three, corresponding to Schütt's subfamilies.

Family 1. **Coscinodiscea** [Coscinodisceae] Kützing Phyc. Germ. 112 (1845). Family *Melosireae* Kützing op. cit. 66. Families *Melosiraceae* and *Coscinodiscaceae* West British Freshw. Alg. 274, 276 (1904). *Melosira*, in fresh water, the shells feebly silicified, the cells joined end to end in filaments. *Cyclotella*, separate drum-shaped cells in fresh water. *Coscinodiscus*, the cells disk-shaped. *Triceratium*, cells of the form of 3-, 4-, or 5-sided prisms with abbreviated axes.

Family 2. **Rhizosoleniacea** [Rhizosoleniaceae] West British Freshw. Alg. 278 (1904). The cells, circular or elliptic in cross section, becoming elongate by intercalation of ring-shaped bands of wall between the valves. *Rhizosolenia. Corethron.*

Family 3. **Biddulphiea** [Biddulphieae] Kützing Phyc. Germ. 115 (1845). Families *Biddulphiaceae* and *Chaetoceraceae* Auctt. Cells laterally compressed, elliptic in valve view, oblong or rhombic in girdle view. Cells of *Biddulphia*, solitary or colonial, are familar as epiphytes on marine algae. *Chaetoceros*, the cells with a long spine at each corner, frequently united valve to valve in filaments, abundant in subpolar oceans.

Order 2. **Diatomea** [Diatomeae] C. Agardh Syst. Alg. xii (1824).
Tribe *Striatae* with orders *Astomaticae* and *Stomaticae,* and tribe *Vittatae* also with orders *Astomaticae* and *Stomaticae,* Kützing Phyc. Germ. (1845).
Pennatae Schütt in Engler and Prantl Nat. Pflanzenfam. I Teil, Abt. 1b: 101 (1896).
Order *Pennatae* Campbell Univ. Textb. Bot. 90 (1902).
Order *Naviculales* Bessey in Univ. Nebraska Studies 7: 284 (1907).
Diatoms basically of isobilateral symmetry, occasionally so skewed as to be dorsiventral or asymmetric; valves usually punctured by a longitudinal cleft called the raphe, or bearing a marking of some sort, called the pseudoraphe, in the same position; exhibiting, when possessed of a true raphe, a gliding motion; cells usually with two plastids.

The motion of the pennate diatoms is a gliding upon surfaces, with frequent reversal, in either direction of the long axis of the cell. It depends upon the flow of a stream of exposed protoplasm. This is the opinion of Max Schultze (1865), Otto Müller (1889, 1896), and Lauterborn (1896); there have been other hypotheses. Müller showed that the true raphe, without which the motion does not occur, is an actual opening. The raphe is not a simple crack; it enters the wall obliquely and bends at a sharp angle to come from another oblique direction to the interior. Its proportions vary along its length, and it is interrupted at the middle of the valve by a knob, the central granule, projecting inward from the valve.

The pennate diatoms do not produce flagellate cells nor statospores, but they produce auxospores, usually by sexual processes. The majority inhabit fresh water.

Eleven families are currently recognized.

a. *Without raphes.*

Family 1. **Fragilariea** [Fragilarieae] (Harvey) Kützing Phyc. Germ. 62 (1845). Family *Fragilariaceae* West British Freshw. Alg. 285 (1904). Cells symmetrical with respect to three planes, without internal partitions. *Fragilaria. Synedra.*

Family 2. **Tabellariea** [Tabellarieae] Kützing op. cit. 110. Family *Tabellariaceae* West op. cit. 281. Cells symmetrical with respect to three planes, with longitudinal internal partitions. *Tabellaria.*

Family 3. **Bacillaria** Ehrenberg Infusionsthierchen 136 (1838). Family *Diatomaceae* West op. cit. 284. Cells symmetrical with regard to three planes, with transverse internal partitions, solitary, or joined valve to valve in ribbons, or corner to corner in zig-zag chains. *Diatoma.*

Family 4. **Meridiea** [Meridieae] Kützing op. cit. 61. Family *Meridionaceae* West op. cit. 283. Cells symmetrical with regard to two planes, wedge-shaped both in valve and in girdle view, with transverse internal partitions, often joined valve to valve in fan-shaped colonies which are sometimes so extended as to produce spiral filaments. *Meridion.*

b. *With raphes, the valves of each cell alike.*

Family 5. **Naviculea** [Naviculeae] Kützing op. cit. 90. Family *Naviculaceae* Rabenhorst Kryptog.-Fl. Sachsen 1: 33 (1863). This is the most numerous family of diatoms. In most of the genera the cells are narrowly rectangular in girdle view, narrowly elliptic in valve view, being of the shape of flat-bottomed boats. *Navicula, Pinnularia,* etc. In other genera, as *Gyrosigma* and *Pleurosigma,* the cells are so skewed as to be sigmoid in valve view.

Family 6. **Gomphonemea** [Gomphonemeae] Kützing op. cit. 87. Family *Gomphonemaceae* West op. cit. 297. Cells wedge-shaped. *Gomphonema.*

Family 7. **Cymbellea** [Cymbelleae] (Harvey) Kützing op. cit. 84. Family *Cocconemaceae* West op. cit. 298. Cells with two planes of symmetry, in valve view crescent-shaped or approximately so. *Cymbella. Rhopalodia.*

Family 8. **Eunotiea** [Eunotieae] Kützing op. cit. 57. Family *Eunotiaceae* West op. cit. 287. Cells curved as in the preceding family, the raphes reduced to brief clefts near the ends of the valves. *Eunotia.*

Family 9. **Nitzschiacea** [Nitzschiaceae] West op. cit. 301. Cells asymmetric in valve view, the raphe along one margin. *Nitzschia. Hantschia.*

Family 10. **Surirellea** [Surirelleae] Kützing op. cit. 70. Family *Surirellaceae* West op. cit. 303. Each cell with two marginal raphes. *Surirella.*

c. *The two valves of each cell unlike, one with a raphe, one with a pseudoraphe.*

Family 11. **Achnanthea** [Achnantheae] Kützing op. cit. 81. Families *Achnanthaceae* and *Cocconeidaceae* West op. cit. 289, 290. *Achnanthes, Rhoicosphenia, Cocconeis.*

Class 3. OOMYCETES Winter

Class OOMYCETES Winter in Rabenhorst Kryptog.-Fl. Deutschland 1, Abt. 1: 32 (1879).

Phycomyceten de Bary Vergl. Morph. Pilze 142 (1884), in part.

Class *Phycomycetes* Engler and Prantl Nat. Pflanzenfam. II Teil: 1 (1889), in part.

Reihe Oomycetes Fischer in Rabenhorst Kryptog.-Fl. Deutschland 1, Abt. 4: 310 (1892).

Stamm *Phykomycophyta* Pascher in Beih. bot. Centralbl. 48, Abt. 2: 330 (1931), in part.

Biflagellatae Sparrow Aquatic Phycomycetes 487 (1943).

Organisms of fungal or chytrid body type, that is, non-pigmented saprophytes or parasites whose bodies are walled filaments or cells with or without rhizoids; the walls consisting partially of cellulose; reproducing asexually by zoospores with paired unlike flagella which are, so far as is known, respectively pantoneme and acroneme, and usually sexually by fertilization, the eggs being distinct cells within the oogonia. The regularly cited example and evident standard genus of the group is *Saprolegnia.*

Conventional botanical classification recognizes within the group of Fungi a subordinate group named Phycomycetes, which is in turn divided into Oomycetes and Zygomycetes, the former including the chytrids. This arrangement suggests an evolutionary series, originating perhaps among non-pigmented flagellates, and leading through chytrids, typical Oomycetes, and Zygomycetes to the typical fungi. It does not now appear tenable. Couch (1939) pointed out differences between Oomycetes and Zygomycetes which make any direct connection between them appear quite improbable; and his observations on flagella showed that only a small minority among organisms of chytrid body type have anything to do with the proper Oomycetes.

There is an old hypothesis (Sachs, 1874) that *Vaucheria* may represent the direct ancestry of *Saprolegnia.* This hypothesis could not be taken seriously while *Saprolegnia* and its allies were known to produce heterokont zoospores, while *Vaucheria* was supposed to be a typical isokont green alga. Now it again appears probable. It implies that in the present group the fungal body type is more primitive than the chytrid.

The Oomycetes may be organized as three orders.

1. Of fungal body type, i.e., consisting of filaments.
    2. Essentially aquatic.........................Order 1. SAPROLEGNINA.
    2. Mostly not aquatic, parasitic on higher plants.....................................Order 2. PERONOSPORINA.
1. Of chytrid body type, i.e., the cells not elongated to filamentous form, though sometimes proliferating or producing rhizoids................Order 3. LAGENIDIALEA.

Order 1. **Saprolegnina** [Saprolengninae] Fischer in Rabenhorst Kryptog.-Fl. Deutschland 1, Abt. 4: 311 (1892).

Order *Eremospermeae* and suborder *Mycophyceae* Kützing Phyc. Gen. 146 (1843), in part.

Order *Oosporeae* Cohn in Hedwigia 11: 18 (1872), in part.

Order *Oomycetes* and suborder *Saprolegniineae* Engler Syllab. 24 (1892).

Order *Saprolegniineae* Campbell Univ. Textb. Bot. 153 (1902).

Order *Siphonomycetae* Bessey in Univ. Nebraska Studies 7: 286 (1907).

Order *Saprolegniales* Auctt.

Order *Leptomitales* Kanouse in American Jour. Bot. 14: 295 (1927).

Aquatic Oomycetes, filamentous, saprophytic or facultatively parasitic, the zoospores diplanetic (exhibiting two periods of swimming) or giving evidence of an ancestral diplanetic condition. The old ordinal names Eremospermeae and Oosporeae designated miscellaneous collections of groups in which this one was listed at or near the beginning. Either one, if taken up, would be applied here, but it seems better to treat them as *nomina confusa*.

1. Filaments not constricted.........................Family 1. SAPROLEGNIEA.
1. Filaments constricted at intervals.
   2. Filaments not differentiated into basal
     and reproductive parts...................... Family 2. LEPTOMITEA.
   2. Filaments differentiated into basal and
     reproductive parts.......................... Family 3. RHIPIDIACEA.

Family 1. **Saprolegniea** [Saprolegnieae] Kützing Phyc. Gen. 157 (1843). Family *Saprolegniaceae* Cohn in Hedwigia 11: 18 (1872). Aquatic Oomycetes consisting of branching filaments of essentially uniform diameter without crosswalls other than those which set apart differentiated reproductive structures.

These well-known organisms are called water molds. According to Coker (1923) there are about eighty definitely recognizable species. They may be parasitic on fishes or saprophytic on organic remains in water or soil. In almost any body of soil or of fresh water they may be found by "baiting," in former practice with dead flies, currently with hemp seeds.

Mitosis has rarely been observed in the vegetative filaments, the nuclei being very minute. Eggs are produced in large globular multinucleate oogonia borne at the ends of filaments. The nuclei in the developing oogonia become enlarged and undergo a single flare of concurrent mitoses (Davis, 1903; Couch, 1932). The sharp-pointed spindles, ending in centrosomes, are formed within the nuclear membrane. The membrane disappears toward the end of the mitotic process, and a nucleolus, which has persisted to this stage, undergoes solution in the cytoplasm. The chromosome numbers (Ziegler, 1953) are 3, 4, 5, 6, or 7.

Within each oogonium there appear one or a few minute bodies called coenocentra. One nucleus becomes associated with each coenocentrum; all others break down and disappear. Each surviving nucleus with the cytoplasm associated with it becomes organized as an egg. When several eggs are produced, they share all of the cytoplasm of the oogonium; when only one egg is produced, some of the cytoplasm is left outside of it.

Sperms are produced in small multinucleate antheridia borne at the tips of filaments in contact with oogonia. Typically, each individual bears both oogonia and antheridia. Some species are capable of self-fertilization; others exist as two kinds of individuals, each capable of fertilizing the other; some occur as distinct male and

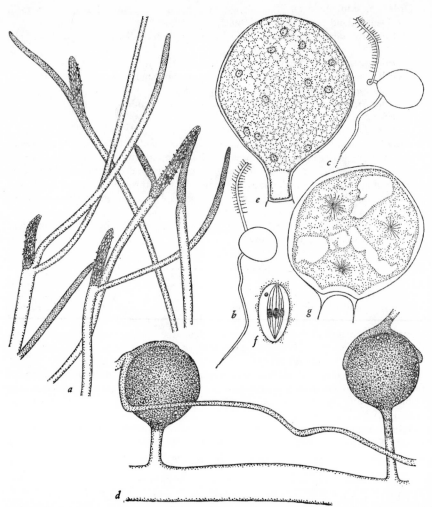

Fig. 14.—Oomycetes: **a**, Filaments and sporangia of *Dictyuchus* sp. x 50.
b, **c**, Zoospores of the second stage of swimming, of *Achlya caroliniana* and *Saprolegnia ferax*, after Couch (1941) x 1,000. **d**, Oogonia and antheridia of *Dictyuchus*
x 400. **e**, **f**, **g**, *Saprolegnia mixta* after Davis (1903): **e**, developing oogonium with
numerous nuclei x 500; **f**, metaphase of nuclear division x 2,000; **g**, developing
oogonium in which most of the nuclei have undergone degeneration; a few have
become associated with coenocentra, and the cytoplasm is undergoing cleavage to
produce eggs about these.

female individuals. Parthenogenesis (reproduction by eggs which have not been fertilized) is rather common in this group. There are no swimming sperms: nuclei from the antheridia reach the eggs through fertilization tubes, or by migration through the periplasm.

Ziegler found that the first nuclear divisions of the nucleus of the zygote are meiotic: all cells except the zygotes are haploid.

The organs of asexual reproduction are cylindrical sporangia terminal on the filaments. Within these the multinucleate protoplasts undergo cleavage into minute uninucleate spores. It is chiefly by details of the behavior of the sporangia and spores (the latter diplanetic, monoplanetic, or not swimming at all) that the dozen genera are distinguished. Diplanetism is the character of zoospores which are not directly infective; they undergo encystment, and the cysts release infective zoospores. During the first stage of swimming, the spores are pear-shaped, with the nucleus drawn out into a beak toward the narrow anterior end, where the flagella are attached. Spores released from cysts for a second period of swimming are bean-shaped, with the flagella attached laterally, each connected through a separate rhizoplast to the nucleus, which lies at some distance from the cell membrane (Cotner, 1930). No explanation of this behavior, whether by phylogeny, genetics, physiology, or competitive advantage, is known. The apparent trend of evolution is to eliminate it. Monoplanetic spores in the present group are usually released from the sporangia as naked protoplasts which undergo encystment and emerge subsequently as flagellate spores of the second form.

Saprolegnia releases diplanetic spores through circular pores in the tips of sporangia in which the spores are formed in several rows; new sporangia develop within empty old ones. Organisms which differ from *Saprolegnia* only in producing new sporangia beside, instead of within, the old ones, were formerly assigned to *Achlya,* but are now called *Isoachlya. Leptolegnia* differs from *Saprolegnia* and *Isoachlya* in forming spores in a single row. In *Achlya* proper, the spores are discharged without flagella, to encyst and swim only once. In *Thraustotheca* the monoplanetic spores are released by irregular breakdown of the distal part of the sporangium. In *Dictyuchus* the spores become encysted before discharge; their protoplasts escape in the form of secondary swarmers through individual pores in the wall of the sporangium. Salvin (1942) found that cultures while growing release into the medium substances which affect the type of sporangium produced, so that a given culture may be while young of the character of *Achlya,* and later of the character of *Thraustotheca* or *Dictyuchus.*

Family 2. **Leptomitea** [Leptomiteae] Kützing Phyc. Gen. 150 (1843). Family *Leptomitaceae* Schröter in Engler and Prantl Nat. Pflanzenfam. I Teil, Abt. 1: 101 (1893). Oomycetes consisting of filaments which are constricted at intervals, but are not differentiated into a basal cell and reproductive branches. In sewage or on organic matter decaying in water. *Leptomitus, Apodachlya, Apodachlyella,* with some seven known species. The numbers of species and degree of distinction of this family and the following do not appear to justify the proposed establishment of a separate order for them.

Family 3. **Rhipidiacea** [Rhipidiaceae] Sparrow in Mycologia 34: 116 (1942). Saprophytes resembling the Leptomitea, the body differentiated into a main part, the basal cell, rhizoids of limited growth, and slender branches bearing the reproductive structures. *Sapromyces, Araiospora, Rhipidium, Mindeniella,* with perhaps a dozen known species.

Order 2. **Peronosporina** [Peronosporinae] Fischer in Rabenhorst Kryptog.-Fl. Deutschland 1, Abt. 4: 383 (1892).
Suborder *Peronosporineae* Engler in Engler and Prantl Nat. Pflanzenfam. I Teil, Abt. 1: iv (1897).
Order *Peronosporineae* Campbell Univ. Textb. Bot. 155 (1902).
Order *Peronosporales* Auctt.

Mostly parasites on terrestrial plants, but including also aquatic parasites and a few saprophytes, the bodies filamentous, reproducing sexually by fertilization, the eggs solitary in the oogonia, reproducing asexually chiefly by conidia, that is, by air-born cells cut off from the ends of the filaments. The conidia are homologous with the sporangia of the Saprolegnina: they germinate in most examples by release of zoospores (which show no signs of diplanetism), but in the more highly evolved examples they give rise to filaments. Ferris (1954) found the zoospores of *Phytophthora* to bear the paired flagella, respectively pantoneme and acroneme, which are typical of Phaeophyta.

In the multinucleate oogonia of most members of the group, single flares of mitoses occur. The sharp-pointed spindles, described in some accounts as ending in centrosomes, are formed within the persistent nuclear membrane, which undergoes constriction during the final stages of mitosis. A coenocentrum appears (this structure was first described as occurring in *Albugo,* by Stevens, 1899); in general, one nucleus becomes associated with it, and is thus selected as the egg nucleus, the remaining nuclei being cast out to undergo disolution in a body of periplasm. The antheridium develops in contact with the oogonium, and fertilization is accomplished by the growth of a fertilization tube through the periplasm to the egg (Davis, 1900; Stevens, 1899, 1901, 1902).

In *Albugo Bliti* and *A. Tragopogonis,* Stevens observed two flares of simultaneous mitoses in the oogonium and antheridium. If this phenomenon were general in the group one would confidently identify it as meiosis. The single coenocentrum attracts many nuclei; the fertilization tube delivers a large number of sperm nuclei; thus multiple karyogamy occurs within a single cell. The further history of the resulting peculiar zygote, containing many nuclei which are not by any evident necessity genetically uniform, is unknown.

This order is evidently a specialized offshoot of the preceding. The family Pythiacea is a good example of a transition group; many authorities have assigned it to the preceding order.

1. Producing solitary globular sporangia or conidia at the ends of scarcely specialized filaments; mostly aquatic......................Family 1. PYTHIACEA.
1. Producing conidia usually in clusters at the ends of specialized filaments (conidiophores); parasites on land plants.
    2. Conidiophores brief, unbranched, the conidia in chains........................ Family 2. ALBUGINACEA.
    2. Conidiophores elongate, usually branched, the conidia solitary or clustered, not in chains............................. Family 3. PERONOSPORACEA.

Family 1. **Pythiacea** [Pythiaceae] Schröter in Engler and Prantl Nat. Pflanzenfam. I Teil, Abt. 1: 104 (1893). Aquatic parasites and saprophytes releasing zoospores from globular reproductive structures terminal on the filaments, together with parasites attacking land plants under moist conditions. The reproductive structures act as sporangia if formed in water, as conidia if formed in air. *Pythium,* saprophytic on plant remains in water or parasitic on algae or higher plants, includes some forty species (Matthews, 1931). The few other genera include perhaps a dozen species. *Zoophagus* produces specialized branches which serve as traps for rotifers which are parasitized and killed.

Family 2. **Albuginacea** [Albuginaceae] Schröter op. cit. 110. Parasites of higher plants, called white rusts, the masses of conidia which push up and burst through the epidermis being of a white color. *Albugo.*

Family 3. **Peronosporacea** [Peronosporaceae] Cohn in Hedwigia 11: 18 (1872). Parasites of higher plants, called downy mildews. The ovoid conidia are produced solitary or in clusters, not in chains, on elongate conidiophores, usually branched, projecting through the stomata of the hosts. This numerous group includes the agents of some of the most important diseases of cultivated plants. *Plasmopara viticola,* causing downy mildew of grapes. *Phytophthora infestans,* the cause of the blight of potatoes which produced the Irish famine of 1846. *Peronospora,* the many species attacking many kinds of plants.

Order 3. **Lagenidialea** [Lagenidiales] Karling in American Jour. Bot. 26: 518 (1939).

Suborder *Ancylistineae* Engler in Engler and Prantl Nat. Pflanzenfam. I Teil, Abt. 1: iv (1897), for the most part, not as to the type genus *Ancylistes.*

Order *Ancylistales* Auctt., in part.

Oomycetes of chytrid body type, parasites consisting of walled cells which are more or less isodiametric, sometimes proliferating or producing rhizoids, but not forming extensive branched filaments. The cells become multinucleate. Mitotic figures of *Olpidiopsis* as described by Barrett (1912) and McLarty (1941) are quite as in the preceding orders, with sharp-pointed intranuclear spindles apparently with centrosomes at the poles. In the usual course of events, each cell develops an exit tube to the exterior of the host, and the protoplast becomes divided into uninucleate cells which escape as unequally biflagellate zoospores. Fertilization, by the migration of the protoplast of one cell into another, has been observed; the zygote becomes a thick-walled resting spore.

1. Internal parasites without rhizoids.
   2. The cells not proliferating.
      3. The zoospores diplanetic.......... Family 1. ECTROGELLACEA.
      3. The zoospores not diplanetic....... Family 2. OLPIDIOPSIDACEA.
   2. The cells proliferating.
      3. Marine....................... Family 3. SIROLPIDIACEA.
      3. Fresh-water................... Family 4. LAGENIDIACEA.
1. External parasites with rhizoids............Family 5. THRAUSTOCHYTRIACEA.

Family 1. **Ectrogellacea** [Ectrogellaceae] Scherffel in Arch. Prot. 52: 6 (1925). *Ectrogella, Eurychasma, Eurychasmidium, Aphanomycopsis,* with about a dozen known species, attacking diatoms and red and brown algae.

Family 2. **Olpidiopsidacea** [Olpidiopsidaceae] Sparrow in Mycologia 34: 116 (1942). *Olpidiopsis* and a few other genera, with some thirty known species, attacking water molds, green algae, red algae, and other aquatic organisms.

Family 3. **Sirolpidiacea** [Sirolpidiaceae] Sparrow l. c. *Sirolpidium* and *Pontisma,* each with one species, attacking marine algae, respectively green and red.

Family 4. **Lagenidiacea** [Lagenidiaceae] Schröter in Engler and Prantl Nat. Pflanzenfam. I Teil, Abt. 1: 89 (1893). *Lagenidium, Myzocytium,* and *Lagenocystis*[1], with some twenty known species, attacking green algae, rotifers, pollen which has fallen into water, and the roots of grasses.

Family 5. **Thraustochytriacea** [Thraustochytriaceae] Sparrow op. cit. 115. The single species *Thraustochytrium proliferum* Sparrow was found as solitary cells external on certain marine green algae and red algae which are penetrated by means of branching rhizoids. Reproduction is by release of naked protoplasts which become laterally biflagellate after a period of rest.

Class 4. MELANOPHYCEA (Ruprecht) Rabenhorst

Order *Fucacées* Lamouroux in Ann. Mus. Hist. Nat. Paris 20: 28 (1813).

Fucoideae C. Agardh Synops. Alg. Scand. ix (1817).

Order Fucoideae C. Agardh Syst. Alg. xxxv (1824).

Division (of order *Algae*) *Melanospermeae* Harvey in Mackay Fl. Hibern. 157 (1836).

Series (of order *Algae*) *Melanospermeae* Harvey Man. British Alg. 1 (1841).

Order *Pycnospermeae* and tribe *Angiospermeae* Kützing Phyc. Gen. 333, 349 (1843).

Class *Fucoideae* J. Agardh Sp. Alg. 1: 1 (1848).

Melanophyceae Ruprecht in Middendorff Sibir. Reise 1, part 2: 200 (1851).

Class Melanophyceae Rabenhorst Kryptog.-Fl. Sachsen 1: 275 (1863).

Stamm Fucoideae Haeckel Gen. Morph. 2: xxxv (1866).

Series (*Reihe*) *Phaeophyceae* Hauck in Rabenhorst Kryptog.-Fl. Deutschland 2: 282 (1885).

Class *Phaeophyceae* Engler and Prantl Nat. Pflanzenfam. II Teil: 1 (1889).

Class *Dictyotales* Engler in Engler and Prantl Nat. Pflanzenfam. I Teil, Abt. 2: ix (1897).

Classes *Phaeosporeae, Tetrasporeae,* and *Cyclosporeae* Bessey in Univ. Nebraska Studies 7: 288, 290 (1907).

Class *Dictyoteae* Schaffner in Ohio Naturalist 9: 448 (1909).

Subclass *Melanophyceae* Setchell and Gardner in Univ. California Publ. Bot. 8: 387 (1925).

Classes *Isogeneratae, Heterogeneratae* (with subclasses *Haplostichinae* and *Polystichinae*) and *Cyclosporeae* Kylin in Kungl. Fysiog. Sällsk. Handl. n. f. 44, no. 7: 91 (1933).

Filamentous or thallose Phaeophyta, yellow to brown in color and living by photosynthesis, producing reproductive cells with paired unequal flagella.

These are the typical brown algae. They are almost exclusively marine, being abundant along with red and green algae on most coasts, and particularly abundant farther toward the poles than the red and green groups. The lower brown algae are branched filaments of microscopic dimensions, commonly epiphytic on other algae. More highly developed examples are thallose and anchored to rocks. Some of these, particularly the ones whose English name is kelp, reach great sizes and considerable elaboration of structure. Papenfuss (in Smith, 1951) gives the number of genera as about 240, and that of known species as about fifteen hundred.

---

[1]**Lagenocystis** nom. nov. *Lagena* Vanterpool and Ledingham in Canadian Jour Res. 2: 192 (1930), non Parker and Jones 1859. **L. radicicola** (Vanterpool and Ledingham) comb. nov.

The cells are walled chiefly with readily hydrolyzable modified polysaccharides. Algin, the soda extract of kelps, consists of chains of oxidized mannose units. A polysaccharide of the sugar fucose, with a sulfate radicle to each sugar unit, is also present. A small percentage of cellulose is present, apparently as the immediate investment of each protoplast. A glycogen- or dextrin-like dextrosan, laminarin, is stored (Miwa, 1940; Tseng, 1945). The plastids contain chlorophylls *a* and *c* (Strain, in Franck and Loomis, 1949) and carotin; xanthophyll is also present in the more primitive examples. In all examples, there is an additional carotinoid called fucoxanthin, which produces the brown color. The analytic process of separating the pigments yields also a sterol, fucosterol, not found in green plants; but this substance, and fucoxanthin, are found in chrysomonads, green Heterokonta, and diatoms (Carter, Heilbron, and Lythgoe, 1939).

Cytological study of a considerable variety of brown algae (Swingle, 1897; Farmer and Williams, 1896; Mottier, 1898, 1900; Simons, 1906; Yamanouchi, 1909, 1912; McKay, 1933) has shown that the spindle and chromosomes appear within an intact nuclear membrane which disappears during the later stages of division. A centrosome, usually with radiating rays, is present outside of the membrane at each pole of the spindle. In *Stypocaulon,* a comparatively primitive brown alga, Swingle found the centrosome to be a permanent structure, dividing as a preliminary to each division of the nucleus. In the generality of brown algae, the centrosomes appear *de novo* as division begins.

Swimming cells are produced by primitive brown algae as spores and as morphologically undifferentiated gametes; in the most advanced brown algae, such cells are produced only as sperms. The flagella are attached laterally. The anterior flagellum is the longer except in order Fucoidea (Kylin, 1916). Longest (1946) found in *Ectocarpus* that the anterior flagellum is pantoneme, and the posterior one acroneme. The swimming cells are without walls, and contain, beside the nucleus, usually one plastid and a light-sentitive speck, the stigma or eyespot. They are quite small. No system of structures linking the nuclei, centrosomes, and flagella has been discovered.

Thuret (1850) discovered that most brown algae produce swimming cells from structures of two different sorts, which he named (1855) respectively plurilocular sporangia and unilocular sporangia. The difference between them is this. In the developing plurilocular structure, each division of the nucleus is followed by division of the protoplast and deposition of a wall, with the result that the swimming cells emerge from separate walled spaces. In the unilocular structure, the nucleus divides repeatedly before the protoplast divides; the protoplast then undergoes cleavage to produce swimming cells which emerge from a single walled space. A number of studies (Clint. 1927; Higgins, 1931; Knight, 1923, 1929) have shown that the first two nuclear divisions in the unilocular structure are normally meiotic. Unilocular structures occur normally only on diploid individuals and release haploid swimming cells. A few exceptional species, however, are known to bear unilocular structures which produce swimming cells without the intervention of meiosis.

In *Ectocarpus siliculosus* as studied by Berthold (1881) at Naples, the swimming cells from unilocular structures are spores which give rise to haploid individuals. In the same species as studied in the Irish Sea by Knight (1929), they were found to act as gametes, conjugating and giving rise to diploid individuals. Diploid and haploid individuals of *Ectocarpus* are alike, and *E. siliculosus* may be said to have a facultatively complete homologous life cycle. The haploid individuals produce plurilocular reproductive structures; the swarmers from these act either as spores, re-

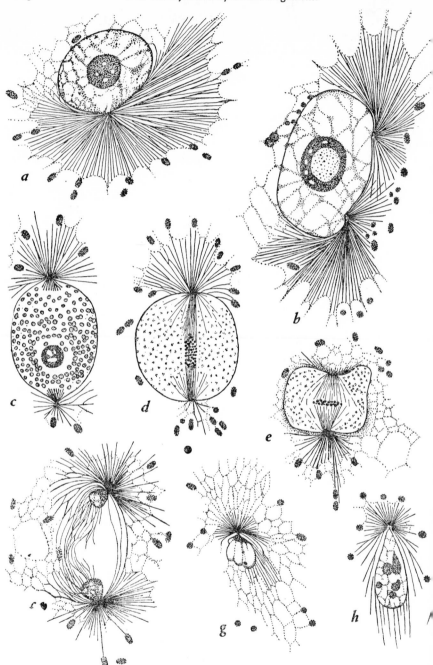

F<small>IG</small>. 15.—Stages of nuclear division in *Stypocaulon* x 1,000 after Swingle (1897).

producing the haploid stage, or as gametes, initiating the diploid stage. The diploid individuals produce both plurilocular and unilocular reproductive structures. The swarmers from the former are spores, reproducing the diploid body. The swarmers from the latter act either as spores, giving rise to haploid individuals, or as gametes, reproducing the diploid body.

It is believed that the brown algae arose by evolution from order Ochromonadalea. Filamentous organisms with a facultatively complete homolgous life cycle, as just described, are believed to be primitive among them: such organisms appear to be the starting point of evolution in many features. The filaments have become differentiated and woven into thalli, and thalli of tridimensionally placed cells have been produced. The haploid and diploid stages have become differentiated. The plurilocular and unilocular structures have undergone specialization. Even in the most primitive brown algae, there is a physiological differentiation of gametes; this has evolved into extreme morphological differentiation. Every one of these evolutionary changes appears to have occurred in more than one line of descent; research is constantly revealing intermediate examples and rather free parallel evolution.

Conservative classification, such as that of Fritsch (1945), recognizes as orders a comparatively primitive miscellany followed by a series of small derived groups marked by distinctive specializations. Features of the life cycle, as applied to classification by Taylor (1922), Oltmanns (1922), Svedelius (1929) and Kylin (1933), are not reliable as marks of natural groups. Kylin provided three classes (one of them divided into two subclasses) and twelve orders. His system appears to provide an excessive number of subdivisions of high category within a moderately small group exhibiting no very profound evolutionary gaps. Tentatively, the seven orders distinguished as follows may be recognized.

1. Producing spores, that is, cells which germinate without syngamy.
  2. All spores bearing flagella.
    3. Having an alternation of haploid
and diploid stages which are alike,
both being filamentous; or else completely lacking one of these stages.
      4. The filaments uniseriate...........Order 1. PHAEOZOOSPOREA.
      4. The filaments becoming pluri-
seriate........................Order 2. SPHACELARIALEA.
    3. Not as above.
      4. Haploid stage thallose, not dis-
tinctly less highly developed
than the diploid stage.............Order 5. CUTLERIALEA.
      4. Haploid stage filamentous, dis-
tinctly less highly developed
than the diploid stage.
        5. Diploid stage filamentous;
or, if partially or com-
pletely thallose, the thal-
lose part with apical growth.....Order 4. SPOROCHNOIDEA.
        5. Diploid stage thallose, its
growth intercalary............Order 6. LAMINARIEA.
  2. Producing large non-motile spores...........Order 3. DICTYOTEA.

1. Producing no spores; all individuals diploid
and reproducing exclusively sexually............Order 7. FUCOIDEA.

Order 1. **Phaeozoosporea** [Phaeozoosporeae] Hauck in Rabenhorst Kryptog.-Fl. Deutschland 2: 312 (1885).
Order *Syntamiidae* Areschoug in Act. Reg. Soc. Upsala 14: 387 (1850), in part; a *nomen confusum.*
Order *Ectocarpeae* J. Agardh Sp. Alg. 1: 6 (1848), preoccupied by family ECTOCARPEAE Kützing (1843).
Section (of *Algae Zoosporeae*) *Phaeosporeae* Thuret in Ann. Sci. Nat. Bot. sér. 3, 14: 233 (1850).
Order *Phaeosporeae* Wettstein Handb. syst. Bot. 1: 173 (1901).
Order *Ectocarpales* Bessey in Univ. Nebraska Studies 7: 288 (1907).
Order *Phaeosporales* and suborder *Ectocarpineae* Taylor in Bot Gaz. 74: 435, 436 (1922).
Microscopic brown algae of the form of undifferentiated uniseriate branching filaments, mostly with distinct haploid and diploid stages (exceptionally lacking the former), the stages distinguishable only by the limitation of unilocular reproductive structures to the diploid stage, the gametes morphologically uniform.
The order is typified by *Ectocarpus,* which is by coincidence also the theoretical ancestral type of the brown algae, the living organism which supposedly represents the evolutionary origin of the group. Recent systems of classification limit this order, formerly construed as extensive, to this genus and a few others, as *Pylaiella* and *Streblonema,* which make up the family **Ectocarpea** [Ectocarpeae] Kützing (family *Ectocarpaceae* Cohn).

Order 2. **Sphacelarialea** [Sphacelariales] (Oltmanns) Engler and Gilg Syllab. ed. 9 u. 10: 27 (1924).
Order *Sphacelarieae* J. Agardh Sp. Alg. 1: 27 (1848), preoccupied by family SPHACELARIEAE Kützing (1843).
SPHACELARIALES Oltmanns Morph. u. Biol. Alg. ed. 2, 2: 2 (1922).
Brown algae distinguished from the Ectocarpea only by features of the vegetative structure, namely that the filaments have large apical cells, and that the cells cut off from them divide lengthwise without increasing considerably in thickness, with the result that the filaments consist of tiers of cells. The life cycle is the same as in Ectocarpea. Family **Sphacelariea** [Sphacelarieae] Kützing (family *Sphacelariaceae* Cohn) includes *Sphacelaria* and *Stypocaulon.* A few other families have been segregated.

Order 3. **Dictyotea** [Dictyoteae] Greville Alg. Brit. 46 (1830).
Tribe *Dictyoteae* Harvey in Mackay Fl. Hibern. 159 (1836).
Family *Dictyoteae* Kützing Phyc. Gen. 337 (1843).
Order *Dictyotaceae* Hauck in Rabenhorst Kryptog.-Fl. Deutschland 2: 302 (1885).
Class *Dictyotales* Engler in Engler and Prantl Nat. Pflanzenfam. I Teil, Abt. 2: ix (1897).
Akinetosporeae Oltmanns Morph u. Biol. Alg. 1: 473 (1904).
Order *Tilopteridales* and Class *Tetrasporeae* with order *Dictyotales* Bessey in Univ. Nebraska Studies 7: 290 (1907).
Series *Aplanosporeae* Setchell and Gardner in Univ. California Publ. Bot. 8: 649 (1925).

Filamentous or thallose brown algae with haploid and diploid stages equally developed, producing large spores without flagella, solitary or few in the sporangia. Here are placed two families, Tilopteridea and Dictyotacea.

Family **Tilopteridea** [Tilopterideae] Cohn is a small group, apparently known only from European coasts. They are evidently closely related to the Ectocarpea. They consist of branching filaments which may become pluriseriate. In *Haplospora* (poorly known; but *Tilopteris* and other genera are even more so), the haploid stage bears both plurilocular structures, releasing minute swimming cells of the structure usual in brown algae, and unilocular structures which release their contents as single uninucleate protoplasts without flagella. The diploid stage bears only unilocular structures which release their contents as single quadrinucleate non-motile spores. It is inferred that the swimming cells from the plurilocular structures are sperms, and that the protoplasts released from the unilocular structures on haploid bodies are eggs, capable, however, of reproducing the haploid stage if not fertilized; further, that the nuclei of the quadrinucleate spores released by diploid individuals are haploid, and become on germination the nuclei of as many cells of the haploid body. The Tilopteridea are believed to represent the evolutionary transition between Ectocarpea and the following family.

Family **Dictyotacea** [Dictyotaceae] (Hauck) Kjellmann includes about twenty genera, *Dictyota, Zonaria, Padina,* etc., with about one hundred species which are commonest on the coasts of warmer oceans. They are thalli of moderate size, erect and dichotomously branched or appressed and fan-shaped. They grow by the division of a single apical cell or a row of apical cells in each branch. The cells multiplying behind the apical cells become differentiated into two tissues, superficial small cells rich in plastids and internal larger ones with fewer plastids, forming in different species single or multiple layers of cells.

The life cycle has been studied by Mottier (1898, 1900), Williams (1898), and Haupt (1932). There are distinct male haploid individuals, female haploid individuals, and diploid individuals, all of the same vegetative structure. The males produce sperms from clusters of densely packed plurilocular antheridia. The females produce eggs solitary in large oogonia solitary or clustered on the thalli. The eggs are without flagella. The diploid individuals produce unilocular sporangia of much the same structure as the oogonia. In *Zonaria,* each sporangium produces eight non-motile spores; in *Dictyota,* each one produces four.

Order 4. **Sporochnoidea** [Sporochnoideae] Greville Alg. Brit. 36 (1830).
Order *Chordarieae* Greville op. cit. 44.
Order *Chordariaceae* Haeckel Gen. Morph. 2: xxxv (1866).
Orders *Desmarestiales* and *Chordariales* Setchell and Gardner in Univ. California Publ. Bot. 8: 554, 570 (1925).
Order *Sporochnales* Sauvageau in Compt. Rend. 182: 364 (1926).

Brown algae producing motile spores, the haploid stage reduced to scant undifferentiated filaments, the diploid stage filamentous or thallose, when thallose with apical growth. *Ralfsia* is an exception to the formal characters of the order: it has a haploid stage of the same structure as the diploid. This is a rather miscellaneous assemblage, rather arbitrarily separated from Phaeozoosporea on the one hand and from Laminariea on the other.

The haploid body of the form of a short-lived body of a few undifferentiated filaments, like a reduced *Ectocarpus,* bearing gametangia reduced to single cells, has been demonstrated by Kylin (1933, 1934, 1937) in a wide variety of genera, as *Ascocyclus, Desmotrichum, Mesogloia, Eudesme, Leathesia,* and *Stilophora*. In the more primitive examples, the gametes are not visibly differentiated; in more advanced ones, as *Carpomitra* and *Desmarestia,* different haploid bodies produce respectively smaller sperms and larger eggs, the latter non-motile.

There is a series of families, Ralfsiacea, Myrionematacea, Myriogloiacea, Mesogloiacea, and others, in which the diploid body consists of filaments differentiated into different types. In the simplest of these, the germinating zygote produces in the first place a minute thallus-like plate, generally epiphytic on other algae, one cell thick, and consisting obviously of branched filaments of limited growth. From this plate grow erect filaments. Some of these are simply cylindrical and appear nutritive in function; others are attenuate, and may function in protection or in absorbing materials from the water; yet others bear the reproductive structures, unilocular or plurilocular or both.

In the more advanced families, the diploid body, after passing through a *Ralfsia-* or *Myrionema*-like stage, may produce a compacted column of filaments with a terminal plate of apical cells. Besides adding cells to the column, the apical plate gives rise to a fascicle of attenuate hairs projecting forward. Members of the families Chordariacea, Sporochnea, and Desmarestiacea produce cylindrical or flattened thallose bodies of tridimensionally placed cells differentiated into an outer layer of small actively photosynthetic cells and an inner mass of nearly colorless cells. Superficial hairs, growing in intercalary fashion, may become few, and growth may become restricted to a single apical cell.

By differences in the detailed manner of growth, Setchell and Gardner distinguished two orders among the thalloid forms just mentioned. It is evident, however, that the thallose structure (and, likewise, differentiation of gametes) has developed repeatedly and independently in the present group. Knowledge which would make it possible to divide it into several recognizably natural orders is not yet available.

Order 5. **Cutlerialea** [Cutleriales] Bessey in Univ. Nebraska Studies 7: 289 (1907).
Brown algae producing motile spores, the haploid and diploid bodies being macroscopically visible thalli, alike or different.

This is a small group, of one family, **Cutleriacea**, with two genera, *Zanardinia* and *Cutleria*, known chiefly from the Mediterranean. In *Zanardinia*, both haploid and diploid bodies are erect and rather freely branched. In *Cutleria*, the haploid bodies are of this description, while the diploid bodies are appressed and fan-shaped. The distinct diploid bodies of *Cutleria* were originally named as a different genus, *Aglaozonia*. Falkenberg (1879) first showed that *Cutleria* and *Aglaozonia* are stages of the same thing; Yamanouchi showed that they are respectively a haploid stage with 24 chromosomes and a diploid stage with 48.

The growing margins of the thalli consist of laterally compacted filaments growing by the divisions of a band of meristematic cells which produce free hairs in the distal direction and a continuous body of cells in the proximal direction. The latter cells are capable of further division, and produce a body several cells thick, with small cells rich in plastids on the surface and larger ones with fewer plastids in the interior.

Haploid individuals bear clusters of stalked plurilocular structures of two types, almost always on different individuals, the larger ones consisting of fewer cells which release eggs, the smaller of more numerous cells which release sperms. Both kinds of gametes are flagellum-bearing cells of the type usual in brown algae. The eggs are capable of germination without fertilization, reproducing the haploid stage. Diploid individuals bear clusters of unilocular sporangia.

It is only in the life cycle that the Cutlerialea are decidedly different from higher Sporochnoidea such as *Desmarestia*. Their evolutionary origin is explicable by the hypothesis of a single mutation which enabled the haploid stage to exhibit the comparatively complicated morphology of the diploid stage, instead of being rudimentary as in all Sporochnoidea except *Ralfsia* (and the exceptional life cycle of *Ralfsia* would be explained by a similar mutation in some primitive example of Sporochnoidea, such as *Myrionema*).

Order 6. **Laminariea** [Laminarieae] Greville Alg. Brit. 24 (1830).
Order *Pycnospermeae* Kützing Phyc. Gen. 333 (1843).
Order *Laminariaceae* Haeckel Gen. Morph. 2: xxxv (1866).
Laminariales Oltmanns Morph. u. Biol. Alg. ed. 2, 2: 2 (1922).
Order *Laminariales* Engler and Gilg Syllab. ed. 9 u. 10: 27 (1924).
Order *Dictyosiphonales* Setchell and Gardner in Univ. California Publ. Bot. 8: 586 (1925).
Order *Punctariales* Kylin in Kungl. Fysiog. Sällsk. Handl. n. f. 44, no. 7: 93 (1933).

Brown algae with motile spores, the haploid stages reduced to microscopic dimensions, the diploid stages thallose, growing in intercalary fashion.

This numerous group, like the preceding small one, is evidently a specialized offshoot from order Sporochnoidea. The familiar examples are the kelps, whose large diploid bodies are differentiated into definite members. Kylin considered his order Punctariales to represent the transition to the kelps. They are thallose, without differentiation of members, but their microscopic and reproductive characters, as observed in *Soranthera* by Angst (1926, 1927), tend to confirm Kylin's opinion, and they are accordingly included in the same order with the kelps. Papenfuss (1947) pointed it out that the Punctariales of Kylin are essentially the same group as the Dictyosiphonales of Setchell and Gardner.

Sauvageau (1915) first showed that the reproduction of kelps is sexual. The grossly visible individuals produce zoospores; these, on germination, produce microscopic filamentous haploid individuals, generally of distinct sexes, releasing gametes from unicellular gametangia. The eggs are without flagella, and it is characteristic of them that in emerging from the oogonia they become attached at the opening (Kylin, 1916, 1933; Myers, 1928; McKay, 1933; Kanda, 1936; Hollenberg, 1939). The same things are true in *Soranthera,* except that the eggs, although much larger than the sperms, are also flagellate.

The visible bodies of kelps consist of three kinds of members, holdfasts (hapteres), being stout root-like growths by which the individuals are anchored to rocks, and stalks and blades comparable to stems and leaves. Growth is most active at the summits of the stalks. The histology is the same in all members (A. I. Smith, 1939). There is a superficial photosynthetic tissue of small cells rich in plastids; on the holdfasts and stalks, this tissue is meristematic, adding cells to the tissue within and increasing the thickness. Internally there is a cortex of larger cells with fewer plastids. In the center there is a medulla containing trumpet fibers, filaments whose cells are expanded where they meet and marked by pit-pairs. In the trumpet fibers of *Nereocystis* there are actual perforations from cell to cell. The trumpet fibers are not quite

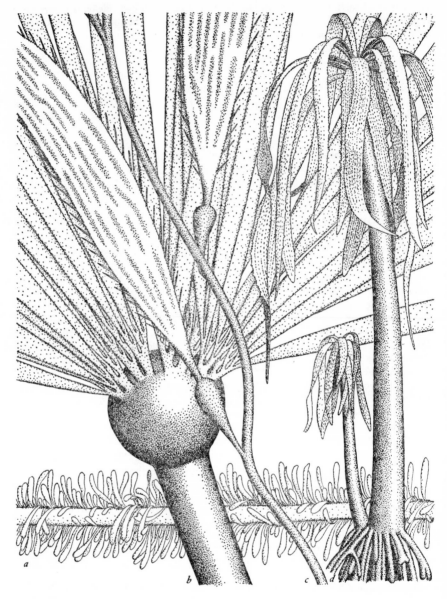

FIG. 16.—Familiar kelps of Pacific North America: **a**, *Egregia Menziesii;* **b**, *Nereocystis Luetkeana;* **c**, *Macrocystis pyrifera;* **d**, *Postelsia palmaeformis.* All approximately x ½.

perfectly analogous to the sieve tubes of higher plants; the nuclei remain alive. The minute zoospores are produced in unilocular sporangia. These occur on the surface of the body in dense masses, intermingled with, and protected while young by, specialized sterile hairs.

Individuals of *Laminaria* consist simply of hapteres, a stalk, and one or more terminal blades. In various other genera, growth occurs in such fashion as to cause the blades to split at the base. With further growth, the splits extend to the margins of the blades and increase their number, while intercalary growth at the transitions between the stalks and the blades produces elongation and branching of the stalks. Early explorers described the stalks of *Macrocystis pyrifera* as reaching prodigious lengths, matters of hundreds of meters, and these accounts have been repeated in textbooks down to recent times. Frye, Rigg, and Crandall (1915) found a maximum length of somewhat less than fifty meters. The stalks are dichotomously branched to a moderate extent and bear series of blades, each with a pear-shaped pneumatocyst or float at the base. The stalks of *Nereocystis Luetkeana* also were said to be extremely long, but the recent observers did not find them to attain fifty meters. They are unbranched and bear a single large float from which spring several blades which may exceed four meters in length. This great organism is an annual, growing and dying within a year. *Postelsia palmaeformis*, called the sea palm, grows on rocks exposed to surf. It has erect stalks some 30 cm. tall bearing many pendant linear blades. *Egregia Menziesii* has flattened stalks many meters long with fringes of floats and blades along the margins. *Laminaria* is widely distributed. *Macrocystis* occurs on the northwest coast of North America and in southern oceans. The other kelps which have been mentioned are confined to the northwest coast of North America.

On coasts where they occur, kelps are used as fertilizer. They have been used commercially as sources of potash, as much as 1-3% of the fresh weight being K as K_2O (Cameron, 1915); they have been used also as sources of iodine. These uses are not economic at most times.

Setchell and Gardner divided the proper kelps, of which there are about one hundred species, into four families. The groups of less elaborate structure which appear properly to be placed in the same order are treated by Papenfuss (under Dictyosiphonales) as six families.

Order 7. **Fucoidea** [Fucoideae] C. Agardh Syst. Alg. xxxv (1824).
  FUCOIDEAE C. Agardh Synops. Alg. Scand. ix (1817).
  Tribe *Angiospermeae* Kützing Phyc. Gen. 349 (1843).
  Order *Cyclosporeae* Areschoug in Act. Roy. Soc. Upsala 13: 248 (1847).
  Order *Fucaceae* J. Agardh Sp. Alg. 1: 180 (1848).
  Order *Sargassaceae* Haeckel Gen. Morph. 2: xxxv (1866).
  Order *Fucales* Bessey in Univ. Nebraska Studies 7: 290 (1907).
  Order *Cyclosporales* and suborder *Fucineae* Taylor in Bot. Gaz. 74: 439 (1922).
  Class *Cyclosporeae* Kylin in Kungl. Fysiog. Sällsk. Handl. n. f. 44, no. 7: 91
    (1933).

Thallose brown algae, producing no spores, diploid in all stages except the gametes; the latter being sperms, whose posterior flagellum is longer than the anterior one, and non-motile eggs. The genus *Fucus* L. is to be construed as the type genus of order Fucoidea, class Melanophycea, and phylum Phaeophyta.

Two families are usually recognized (others have been segregated). In family **Fucea** [Fuceae] Kützing (family *Fucaceae* Cohn), called the rockweeds, the bodies are flat dichotomously branching thalli. In family **Sargassea** [Sargasseae] Kützing there is a differentiation of holdfasts, stalks, blades, and floats. Growth is by division of a single apical cell in each branch or member. There are the usual two tissues, a superficial photosynthetic tissue of small cells and an inner tissue of larger cells which pull apart to produce a spongy or fibrous mass.

The gametangia are borne, mixed with sterile hairs, in pits called conceptacles. These are clustered, in the Fucea near the tips of branches which have ceased to grow (these tips are swollen, and are called receptacles), in the Sargassea on special branches. Rarely, oogonia and antheridia occur in the same conceptacles; not infrequently, they occur in different conceptacles on the same individuals; commonly, they occur on different individuals. Male and female conceptacles may be distinguished by color, the male being orange-yellow, the female of the same dark color as the thalli.

Male conceptacles are full of branching hairs bearing minute antheridia. In each antheridium, the original single nucleus undergoes six successive simultaneous divisions, producing sixty-four nuclei. These become the nuclei of sperms. Female conceptacles contain fewer, larger, oogonia, in which the nuclei divide three times, pro-

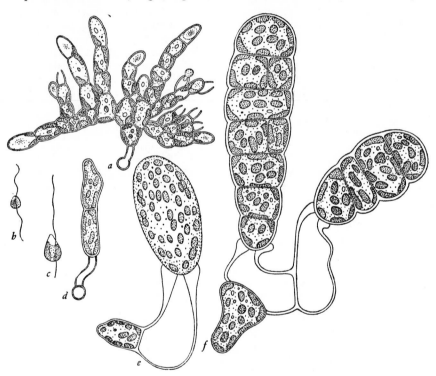

Fig. 17.—Microscopic reproductive structures of *Laminaria yezoensis* after Kanda (1938) : **a**, male haploid individual releasing sperms; **b**, sperm; **c**, zoospore; **d**, female haploid individual of three cells; **e**, female individual with an egg extruded from the oogonium and attached in the mouth **f**, female individual with two young diploid individuals attached at the mouths of oogonia.   All x 1,000.

ducing eight. In *Fucus,* these become the nuclei of as many eggs. In other genera, the number of functional eggs is reduced by degeneration of some of them, or of some of the nuclei before cell division. In *Sargassum,* Kunieda (1928) found each oogonium to produce a single egg in which seven nuclei undergo dissolution while one remains to function.

The first two nuclear divisions in each gametangium are meiotic. Farmer and Williams (1896) and Strasburger (1897) showed that the bodies are diploid; Yamanouchi (1909) first gave a full account of the meiotic process. The haploid chromosome number of *Fucus vesiculosus* is 32. In *Sargassum Horneri* Kunieda found it to be 16.

By a swelling of colloidal material in the conceptacles, the gametangia are forced out into the water, where they burst and release the gametes. *Fucus* was one of the first organisms in which syngamy was observed. Thuret (1855) saw multitudes of sperms swarm about the eggs, and showed that without sperms the eggs would not develop. This much had already been observed in frogs and certain fishes; the discovery that the essential process is the union of just one sperm with the egg was not made until later. The growing zygotes give rise directly to diploid thalli.

The gametangia of the Fucoidea appear to be homologous with the unilocular sporangia of other brown algae. In the gametangia, as in unilocular sporangia, the meiotic divisions are followed by a few divisions of the haploid nuclei: the Fucoidea are not quite perfect examples of the reduction of the haplod stage to the gametes only. As to which other brown algae may have provided their evolutionary origin, there is no very satisfactory hypothesis; *Sporochnus* shows certain resemblances.

Chapter VII
PHYLUM PYRRHOPHYTA

Phylum 3. PYRRHOPHYTA Pascher

Order ASTOMA Siebold in Siebold and Stannius Lehrb. vergl. Anat. 1: 10 (1848).
Order *Phytozoidea* Perty Kennt. kleinst. Lebensf. 161 (1852).
Flagellata Cohn in Zeit. wiss. Zool. 4: 275 (1853).
Orders *Flagellata* and CILIO-FLAGELLATA Claparède and Lachmann Etudes Infus. 1: 73 (1858).
Suborder MASTIGOPHORA Diesing in Sitzber. Akad. Wiss. Wein Math.-Nat. Cl. 52, Abt. 1: 294 (1866).
Stämme Flagellata and *Noctilucae* Haeckel Gen. Morph. 2: xxv, xxvi (1866).
Class *Flagellata* Kent Man. Inf. 1: 27, 211 (1880).
Class MASTIGOPHORA and orders *Flagellata, Dinoflagellata,* and CYSTOFLAGELLATA Bütschli in Bronn Kl. u. Ord. Thierreichs 1, Abt. 2, *Inhalt* (1887).
Class *Peridineae* Wettstein Handb. syst. Bot. 1: 71 (1901).
Divisions *Flagellatae* and *Dinoflagellatae* Engler Syllab. ed. 3: 6, 8 (1903).
Pyrrhophyta, Eugleninae, and *Chloromonadinae* Pascher in Ber. deutschen Bot. Gess. 32: 158 (1914).
Stämme PYRRHOPHYTA and *Euglenophyta,* and *Abteilungen Cryptophyceae, Desmokontae,* and *Dinophyceae,* Pascher in Beih. bot. Centralbl. 48, Abt. 2: 325, 326 (1931).
Division PYRRHOPHYTA G. M. Smith Freshw. Algae 10 (1933).
Protistes trichocystifères ou progastréades Chadefaud in Ann. Protistol. 5: 323 (1936).
Phyla *Pyrrhophycophyta* and *Euglenophycophyta* Papenfuss in Bull. Torrey Bot. Club 73: 218 (1946).

Unicellular or colonial organisms, typically with brown or green plastids, flagellate, the flagella solitary or more than one and unequal, the cells marked by grooves or pits and sometimes containing trichocysts, i. e., minute structures which lie close to the cell membrane and eject thread-like bodies when stimulated.

The organisms included here are the ones conventionally treated as four orders of pigmented flagellates, cryptomonads, dinoflagellates, euglenids, and chloromonads. These groups include organisms of the same varied body types, algal, amoeboid, and chytrid, that occur in other groups in which the flagellate body type is construed as typical. *Peridinium* may be considered to be the type of the phylum.

Deflandre (1934) designated as stichoneme (*stichonématé*) the type of flagellum which bears a single file of appendages, and which had been discovered by Fischer (1894) in *Euglena.* Petersen (1929) reobserved the stichoneme flagellum of *Euglena,* and found it also in other euglenids, *Phacus* and *Trachelomonas.* Deflandre found that one flagellum is stichoneme in various further euglenids (but not in all), and also in the dinoflagellate *Glenodinium.* This is the only report of a stichoneme flagellum outside of the euglenid group. The fine structure of the flagella of cryptomonads and chloromonads has not been determined.

In some cryptomonads, as *Chilomonas,* the cells contain granules which stain blue with iodine; if these are not starch, one knows not what to call them. Dinoflagellates produce a so-called starch which gives a reddish color with iodine, and many of them have walls of a so-called cellulose which gives a reddish color with zinc chlor-iodide. The euglenids store granules of a white solid believed not to be starch and called paramylum.

The plastids of cryptomonads and dinoflagellates are of various colors, off-color green, yellow, brown, bluish, or red. Those of dinoflagellates contain chlorophylls *a* and *e*; the latter is an exceptional chlorophyll which occurs also in *Tribonema*. Euglenids and chloromonads are typically of the same bright green color as typical plants, and the euglenids are known to have the same chlorophylls, *a* and *b,* as typical plants (Strain, in Franck and Loomis, 1949).

The groups here brought together exhibit family resemblances in details of the mitotic process, so far as these are known. The nuclear membrane usually persists through the process. In many examples the chromosomes appear to be present at all times, and are quite numerous, elongate, and of the appearance of strings of beads. In mitosis, quite as one would assume, they divide lengthwise; the point had been disputed, and was established by Hall (1923, 1925, 1937) and Hall and Powell (1928). There is a neuromotor apparatus consisting of a centrosome at or near the nuclear membrane together with one or more rhizoplasts connecting it to as many blepharoplasts at the bases of the flagella. No spindle has been seen, unless the peculiar structure, seen in *Noctiluca* outside of and next to the dividing nucleus, is such. The centrosomes may lie at the sides of the dividing nucleus instead of at its ends. In the euglenids and some dinoflagellates the nucleus contains a nucleolus-like body which does not disappear during mitosis, but divides as the chromosomes do.

There are few reports of sexual processes in this group.

Pascher (1914) united the crytomonads and dinoflagellates in a group which he named Pyrrhophyta. He and those who follow him leave the euglenids as an isolated group. Tilden (1933) placed the four groups of flagellates with which we are here concerned in division Chrysophyceae, while leaving the Phaeophyceae as a distinct division. Her arrangement does not appear to be contrary to nature: the cryptomonads are apparently not very far removed from the chrysomonads. The different arrangement here maintained, by which the brown algae instead of the cryptomonads and so forth are placed in the same phylum with the chrysomonads, is believed to have the advantage that that phylum at least is well marked by character.

Chadefaud (1936) proposed a group consisting of the four groups of flagellates here under consideration together with the Infusoria: this on the ground that the Infusoria also have deeply indented cells containing trichocysts. He did not give to his proposed group a place in the taxonomic system by assigning it to a category and giving it a Latin name: he called it by the French common names *protistes trichocystifères* and *progastréades*. He suggested two ideas: that if a cell marked by a considerable indentation should become divided into many cells forming two layers, respectively superficial and against the indentation, the resulting structure would be a gastrula; and that the gastrula, and, in fact, the kingdom of animals, might have come into existence in this fashion. Perhaps because of novelty, these ideas seem far-fetched. So far as it concerns flagellates, Chadefaud's grouping appears sound and has been followed in giving limits to the present phylum.

The phylum is treated as a single class.

Class **MASTIGOPHORA** (Diesing) Bütschli

Classes *Cryptomonadineae, Rhizocryptineae, Cryptocapsineae, Cryptococcineae, Desmomonadineae, Desmocapsineae, Dinoflagellatae, Rhizodininae, Dinocap-*

sineae, Dinococcineae, Dinotrichineae, Euglenineae, and *Euglenocapsineae*
Pascher in Beih. bot. Centralbl. 48, Abt. 2: 325, 326 (1931).
Classes *Chloromonadina, Euglenoidina,* and *Cryptomonadina* Hollande in Grassé
Traité Zool. 1, fasc. 1: 227, 238, 285 (1952).
Further synonymy as of the name of the phylum.
Characters of the phylum.
There are about one thousand known species. Clearly, thirteen classes for their
accommodation, as proposed by Pascher, are excessive; perhaps one goes too far
in the other direction in making the entire group a single class. The type of the
class is the euglenid *Astasia.* This is true because the family *Astasiaea* was listed
first in the earliest appearance of the traditional group Flagellata or Mastigophora
in due taxonomic form, as order Astoma Siebold. If the euglenids are set apart,
taking with them the class name Mastigophora, the remaining larger class will be
called Peridinea [Peridineae] Wettstein.

The traditional four orders are tenably natural; but that of dinoflagellates includes
about four-fifths of the species, while the chloromonad group is very inconsiderable.
The system will be more convenient if the former order is divided into three, and if
the latter is included in the euglenid order. The resulting five orders are distinguished
as follows:

1. Pigmentation if present brown, olive, or the
   like; flagella normally two.
    2. Flagella at the anterior end of the cell,
       not moving in longitudinal and trans-
       verse grooves.
        3. Not walled in the flagellate con-
           dition, flagella not markedly dif-
           ferentiated, or not differentiated
           as anterior and circumferential......... Order 1. Cryptomonadalea.
        3. Usually walled in the flagellate
           condition; flagella respectively an-
           terior and circumferential............ Order 2. Adiniferidea.
    2. Flagella attached laterally, respectively
       longitudinal and circumferential, moving
       in grooves impressed upon the cells.
        3. Not walled in the flagellate con-
           dition............................Order 3. Cystoflagellata.
        3. Flagellate cells with a wall usually
           of articulated plates................ Order 4. Cilioflagellata.
1. Pigmentation if present typically bright
   green, flagella normally solitary, sometimes
   two or more.............................. Order 5. Astoma.

Order 1. **Cryptomonadalea** [Cryptomonadales] Engler Syllab. ed. 3: 7 (1903).
Subclass *Cryptomonadineae* Engler in Engler and Prantl Nat. Pflanzenfam.
I Teil, Abt. 1a: iv (1900).
Cryptophyceae, including *Phaeocapsales* and *Cryptococcales,* Pascher in Ber.
deutschen bot. Gess. 32: 158 (1914).
Order *Cryptomonadinae* Pascher Süsswasserfl. Deutschland 1: 28 (1914).
Order *Cryptomonadina* Doflein Lehrb. Prot. ed. 4: 417 (1916).

Order *Cryptomonadida* Calkins Biol. Prot. 265 (1926).
Orders *Cryptocapsales* and *Cryptococcales* Pascher in Beih. bot. Centralbl 48,
Abt. 2: 325 (1931).

Solitary (exceptionally colonial) cells, usually with one or two plastids of various colors, usually observed in the motile condition, then naked, of dorsiventral (exceptionally isobilateral) symmetry, with two anterior flagella which are not markedly differentiated or not respectively anterior and circumferential.

The resting nucleus contains a karyosome, i. e., a globule which occupies most of

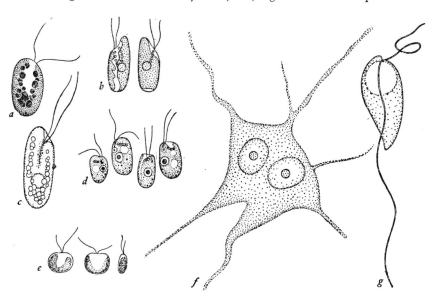

Fig. 18.—a, *Cryptomonas* sp. b, *Rhodomonas baltica* after Kylin (1935). c, *Chilomonas Parmecium.* d, *Cyathomonas* sp. e, *Sennia* sp. f, Vegetative cell, and g, zoospore of *Paradinium Pouchetii* after Chatton (1920). All x 1,000.

its volume and contains most of the chromatin. Dangeard (1910) and Bĕlăr (1916) have observed details of mitosis. The numerous chromosomes appear within an intact nuclear membrane and form a disk- or drum-shaped figure with its axis at right angles to the axis of the cell. No granule more massive than the chromosomes persists and divides with them.

About thirty species are known. They may be treated as five families.
1. Flagellate cells elongate, with one plane
   of symmetry.
   2. Not parasitic, flagella not markedly dif-
      ferentiated.
      3. Non-motile in the vegetative con-
         dition.......................Family 1. CRYPTOCOCCACEA.
      3. Flagellate in the vegetative con-
         dition.......................Family 2. CRYPTOMONADINA.

      3. Amoeboid in the vegetative con-
dition. . . . . Family 3. PARAMOEBIDA.
      2. Parasitic amoeboid organisms, the flag-
ella of swimming stages respectively
anterior and trailing. . . . . .Family 4. PARADINIDA.
      1. Flagellate cells with two planes of symmetry. . . . .Family 5. NEPHROSELMIDACEA.

Family 1. **Cryptococcacea** [Cryptococcaceae] Pascher in Beih. Bot. Centralbl. 48, Abt. 2: 325 (1931). Family *Phaeocapsaceae* West British Freshw. Alg. 48 (1904), in part; *Phaeocapsa* is a chrysomonad. Family *Phaeoplakaceae* Pascher l. c. Solitary or clustered cells, non-motile in the vegetative condition, reproducing by flagellate cells of cryptomonad type. *Phaeococcus, Cryptococcus, Phaeoplax. Chrysidella* includes yellowish cells called zooxanthellae, internally symbiotic in Radiolaria, Rhizopoda, sponges, coelenterates, and rotifers. It is believed that the supposed zoospores of various amoeboid organisms are actually flagellate reproductive cells of *Chrysidella* escaping at certain stages of the life cycles of their hosts.

Family 2. **Cryptomonadina** Ehrenberg Infusionsthierchen 38 (1838). Family *Chilomonadidae* Kent Man. Inf. 1: 423 (1880). Family *Cryptomonadaceae* Engler Syllab. ed. 3: 7 (1903). Family *Chilomonadaceae* Lemmermann 1909. Family *Cryptomonadidae* Poche in Arch. Prot. 30: 159 (1913). Flagellate in the vegetative condition, the two flagella not markedly differentiated, springing from the anterior end of the cells, usually from the mouth of a pit lined by granules of some sort. *Cryptomonas* and *Cryptochrysis* have brown or yellow plastids; *Chromomonas* and *Cyanomonas* have blue ones; *Rhodomonas* has red ones. *Chilomonas* is a colorless saprophyte familiar in infusions. The colorless *Cyathomonas,* also from infusions, was shown by Ülehla (1911) to be related to *Chilomonas.*

Family 3. **Paramoebida** [Paramoebidae] Poche in Arch. Prot. 30: 173 (1913). Schaudinn (1896) discovered the sole known species, *Paramoeba Eilhardi,* in an aquarium of sea water. It is an amoeboid organism with the peculiarity that each cell contains beside the nucleus an additional body which divides when the nucleus does. The cell may form about itself a shell of debris, and within this may undergo division into many cells which escape as pigmented swarmers resembling cells of *Cryptomonas.*

Family 4. **Paradinida** [Paradinidae] Chatton in Arch Zool. Exp. Gen. 59: 444 (1920). The sole known species, *Paradinium Poucheti,* is a parasite in the body cavity of copepods. The amoeboid cells are linked together by slender pseudopodia so as to form a network. The reproductive cells have a shorter anterior flagellum and a longer trailing flagellum.

Family 5. **Nephroselmidacea** [Nephroselmidaceae] Pascher Süsswasserfl. Deutschland 2: 110 (1913). Family *Nephroselmidae* Calkins Biol. Prot. 267 (1926). Cells isobilateral. Cells disk-shaped, the flagella on the margin: *Sennia.* Cells laterally extended, bean- or kidney-shaped, the indentation anterior and bearing the flagella: *Protochrysis, Nephroselmis.*

    Order 2. **Adiniferidea** Kofoid and Swezy in Mem. Univ. California 5: 108 (1921).
    Suborder *Adinida* Bütschli in Bronn Kl. u. Ord. Thierreichs 1: 1001 (1885).
    Suborder *Prorocentrinea* Poche in Arch. Prot. 30: 160 (1913).
    Desmokontae, including *Desmomonadales* and *Desmocapsales,* Pascher in Ber. deutschen bot. Gess. 32: 158 (1914).

Division *Desmokontae;* classes *Desmomonadineae* and *Desmocapsineae;* and orders *Desmomonadales, Prorocentrales,* and *Desmocapsales* Pascher in Beih. bot. Centralbl. 48, Abt. 2: 325 (1931).

Suborder *Prorocentrina* Hall Protozoology 142 (1953).

Solitary cells, mostly flagellate in the vegetative condition, the flagellate stages either naked or bearing a close wall of two valves, with two flagella at the anterior end, one extending forward while the other is bent circumferentially and causes the cell to whirl while swimming.

The few known organisms of this group may be treated as a single family.

Family **Adinida** Bergh in Morph. Jahrb. 7: 273 (1882). Family *Prorocentrinen* Stein Org. Inf. 3, II Hälfte: 8 (1883). Family *Prorocentrina* Bütschli in Bronn Kl. u. Ord. Thierreichs 1: 1002 (1885). Family *Prorocentraceae* Schütt in Engler and Prantl Nat. Pflanzenfam. I Teil, Abt. 1b: 6 (1896). *Prorocentridae* Kofoid in Bull. Mus. Comp. Zool. Harvard 50: 164 (1907). Family *Prorocentridae* Poche in Arch. Prot. 30: 160 (1913). *Desmocapsa, Haplodinium, Desmomastix, Pleuromonas, Exuviaella, Prorocentrum;* minute brown organisms, mostly marine.

Order 3. **Cystoflagellata** (Haeckel) Bütschli in Bronn Kl. u. Ord. Thierreichs 1, Abt. 2, *Inhalt* (1887).

Tribe [group of families] *Gymnodinioidae* Poche in Arch. Prot. 30: 161 (1913).

Classes *Rhizodininae, Dinocapsineae, Dinococcineae,* and *Dinotrichineae;* orders *Gymnodiniales, Rhizodiniales, Dinocapsales, Dinococcales,* and *Dinotrichales* Pascher in Beih. bot. Centralbl. 48, Abt. 2: 326 (1931).

Suborders *Gymnodinina, Dinocapsina,* and *Dinococcina* Hall Protozoology 143, 147, 149 (1953).

Haeckel (1866) made of *Noctiluca* alone a phylum under the name of Noctilucae. He had the carelessness, as it appears, to publish in the same work the synonymous phylar name Myxocystoda as a label in a phylogenetic diagram. In 1873 he used a third name, Cystoflagellata, and Bütschli took this up; in the text of the *Klassen und Ordnungen* ambiguously as an *Unterabtheilung* or *Ordnung,* in the table of contents definitely as an order. Allman (1872) had shown that *Noctiluca* belongs to the present group. Bütschli did not agree with this opinion, but it is evidently correct, and Haeckel's name becomes the valid one for the order to which *Noctiluca* belongs.

Typical members of the present order are naked motile cells with brown plastids. The two flagella are attached near the equator of the cell. One of them extends in a posterior direction, in a groove called the sulcus. The other extends horizontally about the cell (generally to the right, in the reversed image seen in the microscope), lying in a groove called the girdle. The part of the cell anterior to the girdle is called the epicone, the part posterior to it, the hypocone. From the typical structure as thus described, there are, as will be seen, many deviations.

The species, of which more than three hundred are known, may be treated as nine families.

1. Relatively unspecialized; having stages freely propelled by two flagella, with a single girdle, no tentacles, and unspecialized eyespots or none; not parasitic; commonly pigmented.
    2. Walled and non-motile in the vegetative condition..............................Family 1. PHYTODINIACEA.
    2. Flagellate in the vegetative condition........Family 2. GYMNODINIACEA.

1. Not as above, always without photosynthetic
   pigments.
     2. Amoeboid.............................Family 3. DINAMOEBIDINA.
     2. Flagellate or free-floating.
       3. With multiplied girdles, without
       tentacles or specialized light-sensi-
       tive organelles.......................Family 4. POLYKRIKIDA.
       3. With one girdle or none.
         4. Cells more or less isodiamet-
         ric.
           5. With prominent light-sen-
           sitive organelles, some-
           times with tentacles...........Family 5. POUCHETIIDA.
           5. Without light-sensitive or-
           ganelles, with tentacles.
             6. Not exceptionally
             large................... Family 6. PROTODINIFERIDA.
             6. Reaching exceptional
             sizes, to 1 mm. in di-
             ameter.................. Family 7. NOCTILUCIDA.
         4. Cells dome-shaped................Family 8. LEPODISCIDA.
     2. Parasitic................................Family 9. BLASTODINIDA.

Family 1 . **Phytodiniacea** [Phytodiniaceae] Schilling in Pascher Süsswasserfl. Deutschland 3: 61 (1913). Family *Phytodinidae* Calkins Biol. Prot. 277 (1926). *Dinocapsales, Dinocapsaceae, Dinococcales, Dinotrichales,* and *Dinotrichaceae* Pascher in Ber. deutschen bot. Gess. 32: 158 (1914). Orders *Dinocapsales, Dinococcales,* and *Dinotrichales,* and families *Gloeodiniaceae, Hypnodiniaceae, Dinotrichaceae,* and *Dinocloniaceae* Pascher in Beih. bot. Centralbl. 48, Abt. 2: 326 (1931). Organisms with numerous yellow to brown plastids, walled and non-motile in the vegetative condition, reproducing by gymnodinioid zoospores. Some fifty species are known; it is only recently that Thompson (1949) has found several of these in America. Cells multiplying in a gelatinous matrix: *Gloeodinium.* Cells solitary, dividing into several which escape usually in the flagellate condition; with smooth ellipsoid walls: *Phytodinium, Stylodinium;* anvil-shaped, stalked and with two horns: *Raciborskya;* tetrahedral, with horns at each corner: *Tetradinium;* with a ring of about six horns: *Dinastridium.* Tending to produce filaments; marine: *Dinothrix, Dinoclonium.*

Family 2. **Gymnodiniacea** [Gymnodiniaceae] Schütt in Engler and Prantl Nat. Pflanzenfam. I Teil, Abt. 1b: 2 (1896). Subfamily *Gymnodinida* Bergh in Morph. Jahrb. 7: 274 (1882). *Gymnodinidae* Kofoid in Bull. Mus. Comp. Zool. Harvard 50: 164 (1907). Family *Gymnodiniidae* Poche in Arch. Prot. 30: 162 (1913). The typical unarmored dinoflagellates, free-swimming, with sulcus and girdle, without tentacles or a conspicuous light-sensitive organelle, commonly with photosynthetic pigments.

The genus which is most numerous in species is *Gymnodinium* Stein. It includes both pigmented and non-pigmented species, mostly marine, occasional in fresh water, the girdles nearly equatorial and forming nearly complete circles. The cells readily become encysted, and the cysts may grow to large sizes, reaching diameters of 0.5 mm. These cysts have been taken for a distinct genus *Pyrocystis.* Observed in darkness, the protoplasm in the cysts is seen to become luminous in response to disturbance of the medium; they are among the agents of phosphorescence at sea. In *Gymnodinium Lunula* the protoplast of each large globular cyst undergoes division into several protoplasts which do not immediately become flagellate; each of them becomes crescent-shaped, deposits a cell wall, and is released by dissolution of the wall of the parent cyst. In the crescent-shaped cysts, the protoplasts divide into several which develop flagella and escape as typical gymnodinioid cells.

In *Hemidinium* the girdle forms less than a complete circle; in *Amphidinium,* the girdle is close to the anterior end of the cell; in *Gyrodinium,* it forms a steep left spiral; in *Cochliodinium* it forms a left spiral of more than one and one half turns.

Family 3. **Dinamoebidina** nom. nov. Order *Rhizodiniales* and family *Amoebodiniaceae* Pascher (1931), not based on generic names. Non-pigmented amoeboid organisms producing crescent-shaped cysts which germinate by releasing gymnodinioid zoospores. *Dinamoebidium varians* Pascher (1916; originally *Dinamoeba,* but there is an earlier genus of this name, and the author changed it).

Family 4. **Polykrikida** [Polykrikidae] Kofoid and Swezy in Mem. Univ. California 5: 395 (1921). Family *Polydinida* Bütschli (1885), not based on a generic name. There is a single genus *Polykrikos,* of only three known species. They are colorless predatory organisms of such a structure as might be produced if a cell of *Gymnodinium* were repeatedly to enter upon division and fail to complete it. Each elongate cell bears a single extended sulcus and a series of girdles; with each girdle are associated the usual two differentiated flagella. Of nuclei there are usually half as many as of girdles. The cells contain structures called nematocysts, whose development and structure was studied by Chatton (1914). Each nematocyst consists of a conical wall, with a peculiar operculum at the broad end, surrounding a minute cavity containing fluid and a coiled thread. Nematocysts are supposed to be homologous with trichocysts, and to contribute to protection, or to the capture of prey; the points seem not fully established. They occur only in this family and the following.

Family 5. **Pouchetiida** [Pouchetiidae] Kofoid and Swezy in Mem. Univ. of California 5: 414 (1921). Each of the gymnodinioid cells contains a light-sensitive apparatus, the ocellus, consisting of a pigmented area and of one or more transparent globes, of unknown composition, serving as lenses. Most species have nematocysts. *Protopsis, Pouchetia,* etc.; *Erythropsis,* in warm seas, with a prominent tentacle.

Family 6. **Protodiniferida** [Protodiniferidae] Kofoid and Swezy in Mem. Univ. California 5: 111 (1921). Family *Pronoctilucidae* Lebour Dinofl. Northern Seas 10 (1925). Predatory organisms, the cells subglobular, without ocellus or nematocysts, but with a tentacle. *Pronoctiluca* Fabre-Domergue 1889 (*Protodinifer* Kofoid and Swezy 1921); *Oxyrrhis* Dujardin.

Description of the neuromotor apparatus and process of division in *Oxyrrhis marina* by Hall (1925) provides part of the authority for, and is in good conformity to, the remarks on mitosis included above in the description of the phylum. The nucleus contains a prominent internal body (endosome) which does not contain the material of the chromosomes and does not disappear during mitosis. A centrosome, close outside the nuclear membrane, is connected by two rhizoplasts to blepharoplasts at the bases of the flagella. When a cell is to divide, the centrosome divides; the daughter centrosomes do not necessarily lie at the poles of the nucleus where the chromosomes assemble. Each daughter centrosome appears to generate one rhizoplast, blepharoplast, and flagellum to complete the neuromotor apparatus of a cell. In due course, the endosome, nucleus, and cell undergo constriction.

Family 7. **Noctilucida** [Noctilucidae] Kent Man. Inf. 1: 396 (1880). The single species *Noctiluca scintillans* (Mackartney) Kofoid and Swezy (1921; usually known as *N. miliaris* Suriray) is a predatory marine organism, the subglobular cells reaching dimensions exceeding 1 mm., luminescent when stimulated and accordingly contributing to phosphorescence at sea. Each cell is marked by an extensive depression representing the sulcus; the girdle is obsolete. A part of the area of the sulcus functions as a cytostome. A tooth in the sulcus represents the transverse flagellum. Present are a longitudinal flagellum, minute in proportion to the cell, and a prominent tentacle.

Mitosis in *Noctiluca* has been studied by Calkins (1899), van Goor (1918), and Pratje (1921). Adjacent to the nucleus there is a body of differentiated cytoplasm, as large as the nucleus, called by Calkins the attraction sphere. Before mitosis, the tentacle and flagellum are absorbed. The attraction sphere becomes elongate and its central part becomes converted into fibers. The nucleus becomes appressed to, and curved about, the bundle of fibers, and the numerous elongate chromosomes assemble against this. The two curved margins of the nucleus draw apart along the bundle of fibers, appearing to draw the daughter chromosomes with them. Division is completed by constriction of the nucleus and disappearance of the fibers, leaving a daughter attraction sphere in association with each daughter nucleus. This peculiar mitotic process is probably of no phylogenetic significance, being, like the organism in which it occurs, an aberrant by-product of evolution.

Nuclear division may be followed by division of the cell into two, the entire process requiring from twelve to twenty-four hours. Alternatively, the nucleus may divide repeatedly, each division requiring from three to four hours; the numerous nuclei produced are budded off from the cell in small uniflagellate spores. Ischikawa (1891) saw conjugation of pairs of cells, and van Goor stated that this is a preliminary to the production of spores; Pratje, on the other hand, could find no evidence of conjugation. The spores are believed to give rise by direct growth to cells like the original one.

Family 8. **Leptodiscida** [Leptodiscidae] Kofoid 1905. Large dome-shaped predatory marine organisms with small flagella or none. *Leptodiscus* R. Hertwig (1877) was placed by Bütschli in order Cystoflagellata as the sole genus in addition to *Noctiluca; Craspedotella* is a comparatively recent discovery of Kofoid.

Family 9. **Blastodinida** [Blastodinidae] Chatton in Arch. Zool. Exp. Gen. 59: 442 (1920). *Ordre Blastodinides* Chatton in Compt. Rend. 143: 981 (1906). Families *Apodinidae, Haplozoonidae, Oodinidae,* and *Syndinidae* Chatton op. cit (1920). Dinoflagellates which are parasitic chiefly in copepods and tunicates, also in other animals and in diatoms. As a general rule, after the parasite has grown to a certain size, and a multiplication of nuclei has taken place, a part of the protoplast undergoes division to form gymnodinioid zoospores, while the remainder resumes growth in the host. *Schizodinium, Blastodinium, Apodinium, Chytriodinium,* etc.

Order 4. **Cilioflagellata** Claparède and Lachman Etudes Inf. 1: 394 (1858).
  Family PERIDINAEA Ehrenberg Infusionsthierchen 249 (1838).
  Family *Dinifera* Bergh in Morph. Jahrb. 7: 273 (1882).
  Order *Dinoflagellata* Bütschli in Bronn Kl. u. Ord. Thierreichs 1, Abt. 2: *Inhalt* (1887).
  Subclass *Peridiniales* Engler in Engler and Prantl Nat. Pflanzenfam. I Teil, Abt. 1b: v (1896).

Class *Peridineae* Wettstein Handb. syst. Bot. 1: 71 (1901).
Division *Dinoflagellata* Engler Syllab. ed. 3: 8 (1903).
Dinophyceae and *Dinoflagellatae* Pascher in Ber. deutschen bot. Gess. 32: 158 (1914).
Order *Diniferidea* and tribe [group of families] *Peridinioidae* Kofoid and Swezy in Mem. Univ. California 5: 106, 107 (1921).
Order *Dinoflagellida* Calkins Biol. Prot. 267 (1926).
Division *Dinophyceae*, Class *Dinoflagellatae*, and order *Peridiniales* Pascher in Beih. bot. Centralbl. 48. Abt. 2: 326(1931).
Suborder *Peridinina* Hall Protozoology 144 (1953).
This order is very close to the preceding; its members are distinguished only by the presence, while the cells are in the flagellate condition, of cell walls, consisting in most examples of separable plates. The name Cilioflagellata is evidence of an early error of observation: the circumferential flagellum was mistaken for a whorl of cilia. This name and most of its synonyms were published as applying both to the preceding order and this. For almost all of these names the type or obvious standard example is *Peridinium*, with the effect that the names belong to the present order.
There are about five hundred species, prevalently marine. Five families may be recognized.
Family 1. **Peridinaea** Ehrenberg Infusionsthierchen 249 (1838). Family *Peridinidae* Kent Man. Inf. 1: 441 (1880). Family *Peridiniaceae* Schütt in Engler and Prantl Nat. Pflanzenfam. I Teil, Abt. 1b: 9 (1896). *Ceratiidae* Kofoid in Bull. Mus. Comp. Zool. Harvard 50: 164 (1907). The typical dinoflagellates, of numerous genera and species. The distinctions among them are largely matters of the detailed arrangement of the plates making up the walls. *Glenodinium*, the plates scarcely distinguishable. *Peridinium, Goniodoma, Goniaulax, Ceratium, Oxytocum*, etc. The cells of certain species in various genera are ornamented with prominent horns; in *Ceratium* especially the epitheca is drawn out into one long horn, and the hypotheca into one, two, or three. *Goniaulax* becomes abundant at certain seasons, is eaten by shellfish, and renders them poisonous.
The neuromotor apparatus (much as in *Menoidium*) and the process of nuclear and cell division in *Ceratium Hirundinella* were described by Entz (1921) and Hall (1925). Many nuclei lack the endosome; if present, it disappears during mitosis, as does also the nuclear membrane. The daughter centrosomes lie at the sides of the blunt-ended mitotic figure. When nuclear division is complete, the protoplast expands and then becomes constricted in such fashion that each daughter cell receives certain plates of the wall; each daughter cell then secretes the plates which it lacks.
Zederbauer (1904) reported conjugation in *Ceratium*. He saw an elongate protoplast with each of its ends covered by a complete cell wall. Dividing cells are of quite different appearance.
Families **Ptychodiscida, Cladopyxida**, and **Amphilothida** of Kofoid (1907, the names in the feminine; explicitly made families by Poche, 1913) are minor segregates from Peridinaea.
Family 5. **Dinophysida** (Bergh) Bütschli in Bronn Kl. u. Ord. Thierreichs 1: 1009 (1885). Subfamily *Dinophysida* Bergh in Morph. Jahrb. 7: 273 (1882). The limits of the plates obscure; girdle near the anterior end; sulcus and girdle bordered by prominent flanges. Strictly marine, mostly in warmer oceans. *Dinophysis, Oxyphysis, Amphisolenia, Triposolenia*, etc.

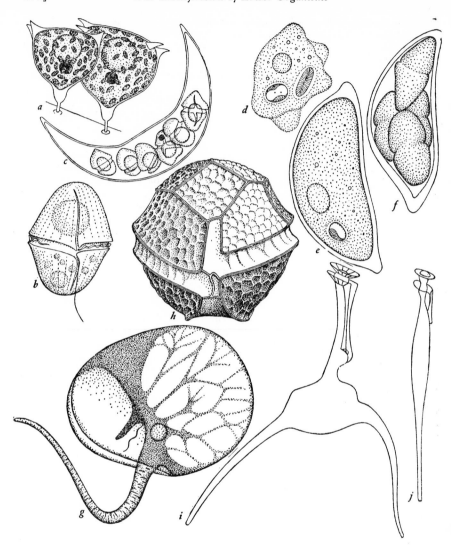

Fig. 19.—a, *Tetradinium javanicum* x 1,000 after Thompson (1949). b, *Gymnodinium striatum* x 500 after Kofoid & Swezy (1921). c, *Gymnodinium Lunula*, flagellate cells forming in a cyst x 500, after Kofoid & Swezy *op. cit.* d, e, f, *Dinamoebidium varians;* amoeboid vegetative cell, cyst, and production of gymnodinioid zoospores x 1,000 after Pascher (1916). g, *Noctiluca scintillans* x 100 after Allman (1872). h, *Peridinium cinctum* x 1,000. i, *Triposolenia Ambulatrix* x 500 after Kofoid (1907). j, *Amphisolenia laticincta* after Kofoid, *op. cit.*

Order 5. **Astoma** Siebold in Siebold and Stannius Lehrb. vergl. Anat. 1: 10 (1848).
Order *Phytozoidea* Perty Kennt. kleinst. Lebensf. 161 (1852), in part.
Order *Flagellata* Claparède and Lachmann Etudes Inf. 1: 73 (1858), in part.
Order *Flagellato-Eustomata* Kent Man. Inf. 1: 36 (1880).
Suborder *Euglenoidina* Bütschli in Bronn Kl. u. Ord. Thierreichs 1: 818 (1884).
Abtheilung (suborder) *Chloromonadina* Klebs in Zeit. wiss. Zool. 55: 391 (1893).
Order *Euglenoidina* Blochmann Mikr. Tierwelt 1, ed. 2: 50 (1895).
Subclasses *Chloromonadineae* and *Euglenineae* Engler in Engler and Prantl Nat. Pflanzenfam. I Teil, Abt. 1a: v, vi (1900).
Orders *Euglenales* and *Chloromonadales* Engler Syllab. ed. 3: 7 (1903).
Orders *Eugleninae* and *Chloromonadinae* Pascher Süsswasserfl. Deutschland 1: 29 (1914).
Orders *Euglenida* and *Chloromonadida* Calkins Biol. Prot. 283, 285 (1926).
Mostly solitary flagellate cells of fresh water, unwalled and capable of contraction and writhing movement; the anterior end of each cell (in the flagellate condition) penetrated by a pit, the reservoir or cytopharynx, into which contractile vacuoles open; having one flagellum, or two, usually unequal, or more, one flagellum of each cell usually being stichoneme; mostly producing a solid storage product, not staining blue with iodine, called paramylum.

Jahn (1946) reviewed this group. He recognized four families, to which one more, to include the chloromonads, is to be added.

1. Producing paramylum.
  2. Flagellum with a swelling near the base,
    usually single but formed of two parts
    which join below the swelling; cells
    mostly pigmented.
      3. Non-motile and walled in the vege-
        tative condition...................Family 1. Colaciacea.
      3. Flagellate in the vegetative con-
        dition.............................Family 2. Euglenida.
  2. Flagellum not swollen and usually not
    forked near the base; cells not pig-
    mented.
      3. Cells without internal rod-shaped
        structures; flagella stichoneme..........Family 3. Astasiaea.
      3. Cells with internal rod-shaped struc-
        tures; flagella acroneme or simple.......Family 4. Anisonemida.
1. Not producing paramylum, storing oil............Family 5. Coelomonadina.

Family 1. **Colaciacea** [Colaciaceae] Smith Freshw. Alg. 617 (1933). Family *Colaciidae* Jahn in Quart. Rev. Biol. 21: 264 (1946). Euglenoid organisms which are walled and non-motile in the vegetative condition. There is a single genus *Colacium,* producing dendroid colonies.

Family 2. **Euglenida** Stein Org. Inf. 3, I Hälfte: x (1878). Family *Euglenina* Bütschli in Bronn Kl. u. Ord. Thierreichs 1: 820 (1884). Family *Euglenaceae* Engler in Engler and Prantl Nat. Pflanzenfam. I Teil, Abt. 2: 570 (1897). Solitary motile cells, mostly with abundant green plastids, the flagella with swellings near the base, mostly solitary and forked below the swelling. Jahn recognized twelve genera. *Eutreptia* has two flagella; *Euglenamorpha* has three. Members of the latter genus

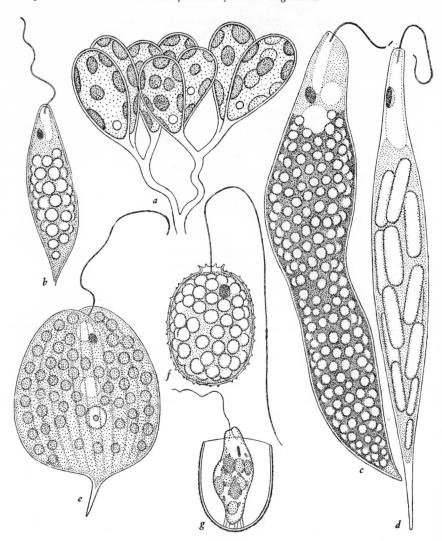

Fig. 20.—**a**, *Colacium Arbuscula* after Stein (1878). **b**, *Euglena viridis.* **c**, *Euglena Spirogyra.* **d**, *Euglena acus.* **e**, *Phacus* sp. **f**, *Trachelomonas* sp. **g**, *Klebsiella alligata* after Pascher (1931). All x 1,000.

are entozoic in frog tadpoles; some of them are non-pigmented. Three genera having the typical single flagella are among the most familiar of flagellates. *Euglena* has fusiform to cylindrical cells freely capable of writhing changes in shape. *Phacus* has flattened cells with a rigid membrane. In *Trachelomonas,* the protoplast lies loose in a rigid lorica which is often ornamented with spines; variations in the form and ornamentation of the lorica have made it possible to distinguish a large number of species.

There are accounts of mitosis in *Euglena* by Keuten (1895), Baker (1926), Ratcliffe (1927) and Hall and Jahn (1929). All observers have seen within the nucleus a large globule which divides as the nucleus does and appears to guide the separating chromosomes. Keuten applied to it the term nucleolo-centrosome; the implications of this term are not confidently to be accepted, and the body will better be called by the neutral term endosome. Ratcliffe's account of mitosis in *Euglena Spirogyra* is the most detailed. It appears that division is initiated when the endosome buds off a small granule which migrates to a position just within the nuclear membrane and divides. The resulting granules may be regarded as centrosomes. The nucleus moves forward within the cell and comes into contact with the cell membrane at the bottom of the reservoir. Each centrosome appears to generate, just within the cell membrane, a granule recognizable as a blepharoplast; the nucleus then withdraws from the cell membrane, but the centrosomes remain connected to the blepharoplasts by rhizoplasts. The flagellum, already split at the base, divides throughout its length into two; a new flagellum-base grows out from each blepharoplast and becomes fused to one of the halves of the old one not far from the base of the latter. Meanwhile, within the intact nuclear membrane, the chromosomes and endosome are dividing. The centrosomes are at the sides of the dividing nucleus. No spindle has been recognized. Nuclear division is completed by constriction of the membrane. The cell divides by constriction which proceeds longitudinally from the anterior end. The centrosomes and rhizoplasts disappear, to be replaced during the next division by new ones.

Hall and Hall and Schoenborn (in several papers, 1938, 1939) have reported experiments on nutrition in *Euglena.* All species are capable of photosynthesis. Some of them, surprisingly, have lost the capacity to synthesize amino acids which usually accompanies photosynthesis; and there are transitional species in which some individuals possess the capacity to make amino acids and others do not, evidently as heritable characters.

Family 3. **Astasiaea** Ehrenberg Infusionsthierchen 100 (1838). Family *Astasiidae* Kent Man. Inf. 1: 375 (1880). Family *Astasiina* Bütschli in Bronn. Kl. u. Ord. Thierreichs 1: 826 (1884). Family *Astasiaceae* Senn in Engler and Prantl Nat. Pflanzenfam. I Teil, Abt. 1a: 177 (1900). Colorless organisms. Deflandre found the flagella stichoneme, as to the single flagella of *Astasia* and *Menoidium,* and as to one of the two flagella of *Distigma.* Hall and Jahn (1929) found the flagella not swollen near the base. The internal rod-shaped structures which characterize the following family are absent.

Bělǎr (1915) described mitosis in *Astasia,* and Hall (1923) described it in *Menoidium.* There is a blepharoplast at the base of the flagellum, and some preparations show a rhizoplast connecting this to a centrosome immediately outside the nuclear membrane. The blepharoplast divides during the early stages of mitosis, and the flagellum appears to divide lengthwise. The daughter centrosomes mark the loci toward which the dividing chromosomes move. The chromosome number appears to be 12. A dividing endosome like that of *Euglena* is present.

The Classification of Lower Organisms

Scytomonas pusilla Stein (*Copromonas subtilis* Dobell) occurs in the intestines of frogs and toads. When cast out with the feces, it exhibits conjugation as a preliminary to encystment (Dobell, 1908).

Family 4. **Anisonemida** [Anisonemidae] Kent Man. Inf. 1: 429 (1880). Families *Pernamina* and *Anisonemina* Bütschli in Bronn Kl. u. Ord. Thierreichs 1: 824, 828 (1884). Family *Peranemaceae* Senn in Engler and Prantl Nat. Pflanzenfam. I Teil, Abt. 1a: 178 (1900). Family *Heteronemidae* Calkins Biol. Prot. 285 (1926). Each cell of these colorless organisms bears one conspicuous anterior flagellum; most of them bear also a less conspicuous trailing flagellum. The trailing flagellum of *Peranema* is grown fast to the cell membrane, and is detected only with difficulty (Hall,

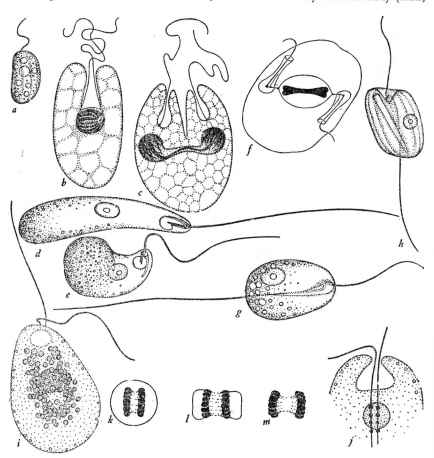

Fig. 21.—**a**, *Menoidium incurvum.*   **b, c**, Stages of mitosis in *Menoidium incurvum* x 2,000 after Hall (1923).   **d, e**, *Peranema trichophorum.*   **f**, Stage of division in *Peranema trichophorum* after Hall (1934).   **g**, *Anisonema truncatum.*   **h**, *Entosipon sulcatum.*   **i-m**, *Vacuolaria viridis*: **i**, cell; **j**, neuromotor apparatus after Fott (1935); **k-m**, stages of mitosis x 2,000 after Fott, *op cit.*   x 1,000 except as noted.

1934). Deflandre was unable to find appendages on the flagella of members of this family. As in other members of the order, the flagella spring from a deep anterior pit in the cell; in this family, the pit is a functional cytopharynx (Hall, 1933). The cytoplasm of *Peranema* contains three brief rods, the pharyngeal rods or *Staborgane,* lying near the cytopharynx; their function is unknown. Each cell of *Urceolus,* of *Anisonema,* and of *Heteronema* contains a single conspicuous rod extending the length of the body. Hall and Powell (1928) and Hall (1934) described the mitotic process in *Peranema,* which is much as in *Menoidium.*

Family **Coelomonadina** Bütschli in Bronn Kl. u. Ord. Thierreachs 1: 819 (1884). Family *Vacuolariaceae* Luther in Bihang Svensk. Vetensk-Akad. Handl. 24, part 3, no. 13: 19 (1889). Family *Chloromonadaceae* Engler Syllab. ed 3: 7 (1903). Family *Thaumatonemidae* Poche in Arch. Prot. 30: 155 (1913). Family *Chloromonadidae* Hollande in Grassé Traitè Zool. 1, fasc. 1: 235 (1952); family *Thaumatomonadidae* Hollande op. cit. 686. Unicellular organisms, mostly green, with two differentiated flagella springing from a large reservoir, producing globules of oil but no solid storage product. Klebs apologized for erecting the *grössere Abtheilung* Chloromonadina for the single genus *Vacuolaria,* and in fact, this genus differs from other members of the present order only in one conspicuous character, the failure to produce paramylum. Fott (1935) studied the cytology of *Vacuolaria.* From the base of each flagellum, a rhizoplast extends into the cytoplasm, but fails to come into contact with the nucleus. Several granules or swellings, not definitely identifiable as blepharoplasts or centrosomes, are distributed along the length of each rhizoplast. In mitosis, which takes place within an intact nuclear membrane, the numerous subglobular chromosomes form a blunt-ended figure much as in *Chilomonas.* Genera believed to be allied to *Vacuolaria* include the green flagellate *Goniostomum; Chysophaeum* Lewis and Bryan (1941), a marine organism forming non-motile yellow dendroid colonies of macroscopic dimensions; and the colorless flagellate *Thaumatomastix* Lauterborn (originally named *Thaumatonema,* but there is among plants an older genus of this name).

Chapter VIII
PHYLUM OPISTHOKONTA

Phylum 4. OPISTHOKONTA, phylum novum

Chytridieae de Bary in Bot. Zeit. 16, Beil. 96 (1858).

Family *Chytridieen* de Bary and Woronin (1864).

Family CHYTRIDIACEAE Cohn in Hedwigia 11: 18 (1872).

CHYTRIDINEAE Schröter in Engler and Prantl Nat. Pflanzenfam. I Teil, Abt. 1: 62 (1892).

Series (*Reihe*) ARCHIMYCETES (CHYTRIDINAE) A. Fischer in Rabenhorst Kryptog.-Fl. Deutschland 1, Abt. 4: 11 (1892).

Suborders *Chrytidiineae* and *Monoblepharidineae* Engler in Engler and Prantl Nat. Pflanzenfam. I Teil, Abt. 1: iii, iv (1897).

Order CHYTRIDINEAE Campbell Univ. Textb. Bot. 152 (1902).

Classes ARCHIMYCETAE and *Monoblepharideae* Schaffner in Ohio Naturalist 9: 447, 449 (1909).

Class ARCHIMYCETES Gäumann Vergl. Morph. Pilze 15 (1926).

Uniflagellatae Sparrow Aq. Phyc. 21 (1943).

Parasites and saprophytes of simple structure (filamentous, of uniform diameter or tapering; or unicellular, with or without rhizoids, i. e. tapering filamentous outgrowths), with cell walls of chitin, containing no cellulose; producing motile cells with solitary posterior acroneme flagella. Type, *Chytridium Olla* Braun. From ὀπίσθιος, rearward, and κοντός, oar.

Chytrid is the English form of the generic name *Chytridium*, from Greek χυτρίς, a jug. Braun (1856) applied this name to a colorless unicellular organism found attached to green algae whose cells are penetrated by rhizoids which draw food from them and kill them. By chytrids we mean organisms of body types of the general nature of that of *Chytridium*. All such organisms were formerly treated as a single taxonomic group. Couch (1938, 1941) showed that the organisms of chytrid body type form three markedly distinct groups distinguished by types of flagellation. The proper chytrids, those which legitimately constitute a taxonomic group, are marked by swimming cells with solitary posterior acroneme flagella, and further by lack of cellulose in the cell walls. The group thus marked includes, beside organisms of chytrid body type, a few organisms of the filamentous body type of the typical fungi.

The cytoplasm of members of this group is described as peculiarly lustrous and as containing shining globules. In mitosis (seen repeatedly, as by Dangeard, 1900, Stevens and Stevens, 1903, Wager, 1913, and Karling, 1937), the sharp-pointed spindle forms within the intact nuclear membrane. Some observers have seen centrosomes at the poles. The nuclear membrane disappears toward the end of the process.

The formation of motile cells (zoospores and sometimes gametes) occurs in enlarged cells. In these cells there are repeated simultaneous nuclear divisions. After the last of these, uninucleate protoplasts, each one containing, ordinarily, one of the above-mentioned shining globules, are separated by cleavage. On each of these protoplasts a flagellum grows from the cell membrane at the point nearest that part of the nucleus which represents a pole of the previous mitotic spindle. Among the Blastocladiacea, the nucleus lies against the cell membrane and the flagellum appears to spring from a granule within it (Cotner, 1930; Hatch, 1935). Similarly, in *Clado-*

chytrium, it appeared to Karling (1937) that the nucleolus generates the flagellum. Within the developing swimming cell a body of granules assembles and produces a "cap," prominent in stained material, on the anterior side of the nucleus, that is, on the side away from the flagellum.

Nowakowski (1876) observed sexual processes in *Polyphagus,* and Scherffel (1925) observed them in many other chytrids. Sexual processes were known in *Monoblepharis* from the discovery of this genus, and have been studied in detail in *Allomyces* by Emerson (1939, 1941) and Emerson and Wilson (1949).

The group thus characterized is of fewer than three hundred known species. One takes no satisfaction in making it a phylum, but feels constrained to do so by its isolation. Note has been taken that other groups including organisms of chytrid body type, as Hyphochytrialea, Lagenidialea, and Phytomyxida, have nothing to do with the proper chytrids. Furthermore, it will not do to thrust the proper chytrids in with the groups of colorless flagellate and amoeboid organisms treated below as phylum Protoplasta. One does not trust that group as natural, but it has a morphological continuity which would be defaced by the addition of this one.

Vischer, 1945, coined the name *Opistokonten* for organisms whose motile cells have posterior flagella. Gams (1947) listed as such the green organisms *Pedilomonas* and *Chlorochytridion;* the choanoflagellates; the proper chytrids; the Sporozoa (the whole group by virtue of such examples as have flagellate stages); and the proper animals. He inferred that these groups make up a major natural group derived from the lowest green algae. This interesting hypothesis must as yet be treated as far-fetched. *Pedilomonas* is scarcely known; it was described by Korschikoff, 1923, as a green flagellate of somewhat the appearance of a *Chlamydomonas* lacking one of its flagella. The flagella of the choanoflagellates are pantacroneme instead of acroneme. There remains a striking resemblance between the motile cells of the proper chytrids and the sperms of animals. The nuclear cap of the former is quite similar, in development and structure, to the beak of the latter.

The Opisthokonta are reasonably treated as a single class.

Class **ARCHIMYCETES** (A. Fischer) Schaffner

Synonymy of the phylum.
Characters of the phylum.
Previous authors have arranged these organisms in a sequence from strictly unicellular forms to typically filamentous forms. In the following treatment, this sequence is reversed. The course of the evolution of the group is unknown, and it seems reasonable to place the body types in the same sequence as among the Oomycetes. The class is treated as two orders, Monoblepharidalea, essentially filamentous, and Chytridinea, unicellular or producing filaments which taper or are swollen at intervals.

Order 1. **Monoblepharidalea** [Monoblepharidales] Sparrow in Mycologia 34: 115 (1942).
Suborder *Monoblepharidineae* Engler and Prantl Nat. Pflanzenfam. I Teil, Abt. 1: iv (1897).
Blastocladiineae Petersen in Bot. Tidsskr. 29: 357 (1909).
Order *Blastocladiales* Sparrow l. c.
Opisthokonta whose bodies consist of filaments of uniform diameter, or are of types apparently immediately derived from this. Saprophytes in fresh water or soil, chiefly on vegetable remains. There are two families.

Family 1. **Monoblepharidacea** [Monoblepharidaceae] A. Fischer in Rabenhorst Kryptog.-Fl. Deutschland 1, Abt. 4: 378 (1892). *Gonapodiineae* and *Gonapodiaceae* Petersen in Bot. Tidsskr. 29: 357 (1909). Producing extensive coenocytic filaments, non-septate but with false septa of cytoplasm, anchored by rhizoids, reproducing asexually by zoospores produced in sporangia which are usually terminal on the filaments, the gametes produced in smaller antheridia and larger oogonia which are in the more familiar forms terminal and subterminal on the filaments, the branches commonly proliferating below them, the eggs without flagella.

The species, about a dozen, form three genera. In *Monoblepharis,* the zygote, being the entire protoplast of the oogonium, moves out of the oogonium through a terminal pore, becomes attached in the opening, and develops a thick wall. In *Monoblepharella* the zygote, retaining the flagellum of the sperm, swims for a time before becoming encysted. *Gonapodya* resembles *Monoblepharella* (Johns and Benjamin, 1954). *Myrioblepharis* Thaxter is believed not to be an organism; it is described as something which might be produced if sporangia of *Monoblepharis* were parasitized by an infusorian.

Family 2. **Blastocladiacea** [Blastocladiaceae] Petersen in Bot. Tidsskr. 29: 357 (1909). Coenocytic filaments, in some examples of a false appearance of septation, of the body type of the Rhipidiacea, i. e., differentiated into a basal cell anchored by rhizoids and distal branches bearing reproductive structures, sometimes so reduced that the basal cell bears, or is itself, the reproductive structure; the reproductive structures including thin-walled zoosporangia, thick-walled resting spores which germinate by releasing zoospores, and gametangia; the gametes morphologically uniform or larger and smaller, all bearing flagella.

These organisms are not familiar, although they are readily isolated by baiting pond water, or tap water to which soil has been added, with hemp seeds or pieces of fruit. There are four genera, *Allomyces, Blastocladia, Blastocladiella,* and *Sphaerocladia,* with about twenty-five known species. *Allomyces* is of interest for varied life cycles, and *Blastocladia* for a peculiar type of metabolism.

The first known species of *Allomyces, A. Arbuscula,* was discovered by Butler (1911) on dead flies in water in India. The individuals are of the appearance of minuscule shrubs, the branches divided by pseudosepta punctured in the middle and ending in series of varicolored reproductive structures. Ordinary sporangia are colorless, resting spores are brown, mature antheridia are pink, and mature oogonia dull gray. Kniep (1929), in discovering the second species, *A. javanicus,* found that the individuals are of two types, one bearing sporangia and resting spores, the other oogonia and antheridia. Thus this organism has a complete life cycle of morphologically homologous haploid and diploid individuals. Kniep supposed that meiosis occurs in the resting spores, and Emerson and Wilson (1949) established the point. The chromosome number (n) of *A. Arbuscula* is 7; that of *A. javanicus* var. *macrogynus* and of *A. cystogenes* is 14.

The life cycle of *A. Arbuscula* is the same as that of *A. javanicus.* In *A. cystogenes,* the haploid stage consists merely of the zoospores from the resting spores; these become encysted and germinate by releasing isogametes. Thus this species has a life cycle essentially of the advanced type characteristic of animals. There are further species of *Allomyces* in which a sexual cycle is believed not to occur.

In *Blastocladia* the basal cell bears directly multiple reproductive structures. Organisms of this genus are less easily cultured than *Allomyces;* they require several vitamins of the B group (Cantino, 1948). They tolerate oxygen, but do not require it.

They convert sugars to lactic and succinic acids, producing no CO_2; the acids, if not neutralized, check the growth of cultures (Emerson and Cantino, 1948; Cantino, 1949). *Blastocladia* appears to have lost the capacity to carry on the aerobic stages of energesis, thus reverting to the type of metabolism characteristic of the supposedly most primitive bacteria.

In *Blastocladiella,* the basal cell bears a single reproductive structure. Different species have the same three types of life cycle which occur in *Allomyces* (Couch and Whiffen, 1942). In *Sphaerocladia* the vegetative body is reduced to the unicellular condition which is characteristic of the following order rather than of this. The life cycle is of the complete homologous type.

Order 2. **Chytridinea** [Chytridineae] (Schröter) Campbell Univ. Textb. Bot. 152 (1902).

Orders *Myxochytridinae* and *Mycochytridinae* A. Fischer in Rabenhorst Kryptog. Fl. Deutschland 1, Abt. 4: 20, 72 (1892), not based on generic names.

Order *Chytridiales* Auctt.

Further synonymy as of the name of the phylum.

Opisthokonta which consist entirely or largely of more or less isodiametric bodies called centers: the centers may send out filaments more slender than themselves, generating at their ends further centers; or may be capable only of producing rhizoids, i. e., tapering absorptive filaments; or may be by themselves complete individuals.

The chytrids are commonly thought of as prevalently parasitic on algae and higher plants. They attack also rotifers, insects, nematodes, and other minute animals; some parasitize other chytrids (Karling, 1942, 1948). It is probable, however, that the majority of the group are saprophytic on organic remains. Some have been cultured with no other organic food than cellulose (Haskins, 1939); new forms have been discovered by baiting with, and culturing on, chitin (Karling, 1945; Hanson, 1946) or keratin (Karling, 1946, 1947).

The following varieties of vegetative structure may be noted. (a) A zoospore, settling upon the surface of an appropriate host or substratum, may penetrate this by means of a walled filament which develops a terminal center; the center then sends out rhizoids, and also filaments which generate further centers. (b) Development may be as above except that only one center is formed. The body thus described is of the *Entophlyctis* type of Sparrow (1943). (c) The zoospore may itself become the single center, penetrating its host or substratum only by rhizoids. The resulting body is of the *Chytridium* type if the center is in contact with the host or substratum, of the *Rhizidium* type if it is not. (d) The protoplast of the zoospore may migrate into the protoplast of the host and there become a center without rhizoids; the resulting body is of the *Olpidium* type. To the varied bodies thus described, the following terminology is applicable:

Pluricentric, with more centers than one; *monocentric,* with a single center.

Intramatrical, the center developing within the substratum or host; alternatively, in a host, *endobiotic.*

Extramatrical or *epibiotic,* contrary to the foregoing.

Eucarpic, the center not constituting the entire body; *holocarpic,* the center constituting the entire body.

The center regularly remains uninucleate during the vegetative phases and then becomes the seat of successive simultaneous nuclear division, of cleavage, and of the maturation of zoospores. Thus it is converted into a sporangium. In many forms, the

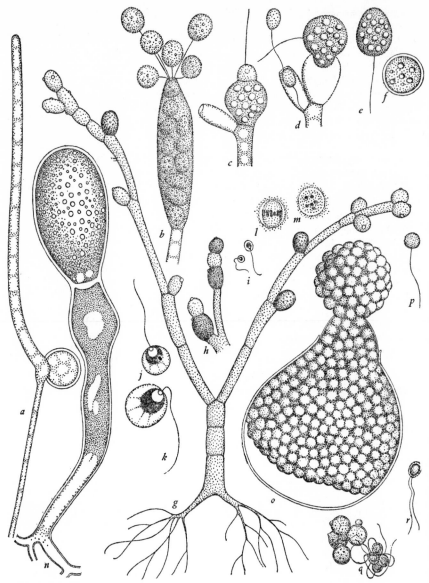

FIG. 22.—MONOBLEPHARIDALEA: **a-f**, *Monoblepharella Taylori* x 1,000 after Springer (1945); **a**, germinating spore producing a filament and a rhizoid; **b**, sporangium releasing spores; **c**, empty antheridium and sperm uniting with egg; **d**, sperms escaping from antheridium and zygote escaping from oogonium; **e**, swimming zygote; **f**, encysted zygote. **g-i**, *Allomyces javanicus* x 100 after Kniep (1929); **g**, asexual

(*Continued bottom p. 115*)

proximal part of the system of rhizoids develops a large swelling called the apophysis. In other forms, the center generates the sporangium as an outgrowth. In these circumstances, the center is sometimes called an apophysis, but were better called a presporangium. The sporangium discharges its spores, usually, through one or more tubes which grow forth from it. The tube may open through a differentiated cap, the operculum; the production of opercula appears to mark a natural subordinate group.

Syngamy occurs in different chytrids in most of the possible fashions, by union of like or unlike swimming cells, by the union of a swimming cell with a stationary one, or by the establishment of contact by growth. The zygote regularly becomes a thick-walled resting spore (asexual resting spores are also of frequent occurrence). Resting spores germinate by producing zoospores. Meiosis has not been observed, but is believed to occur during the first nuclear divisions in the germinating zygote; the life cycle is apparently of the primitive type, in which all cells except the zygote are haploid (*Physoderma*, or at least some of its species, is believed to be exceptional).

Sparrow (1943) recognized nine families. One of these does not appear tenable; the remainder are distinguished as follows:

1. Sporangia not opening through opercula.
  2. Eucarpic, i. e., producing rhizoids and sometimes other filaments, the centers not constituting the entire body.
    3. Pluricentric.....................Family 1. Cladochytriacea.
    3. Monocentric.
      4. Germinating spores generating the center as a distinct body......................Family 2. Phlyctidiacea.
      4. Zoospores themselves becoming centers, and subsequently sporangia or presporangia......Family 3. Rhizidiacea.
  2. Holocarpic, i. e., without rhizoids, the individual consisting entirely of one or more centers.
    3. Centers becoming presporangia, each one generating a cluster of sporangia......................Family 4. Synchytriacea.
    3. Centers proliferating, giving rise to linear series of sporangia..........Family 5. Achlyogetonacea.
    3. Each center becoming one sporangium........................Family 6. Olpidiacea.

individual with light sporangia and dark resting cells with pitted walls; **h**, branch of sexual individual, the oogonia larger and darker than the antheridia; **i**, gametes. **j-m**, *Allomyces Arbuscula* after Hatch (1935); **j**, **k**, gametes, x 1,000; **l**, **m**, mitotic figures in the gametangia, x 2,000. **n-r**, *Blastocladiella cystogena*, x 500, after Couch and Whiffen (1942); **n**, individual producing a resting spore; **o**, resting spore germinating by release of numerous naked protoplasts; these become flagellate zoospores, **p**, which subsequently encyst; **q**, the protoplast of each cyst divides to produce four gametes; **r**, young zygote with the flagella of both gametes.

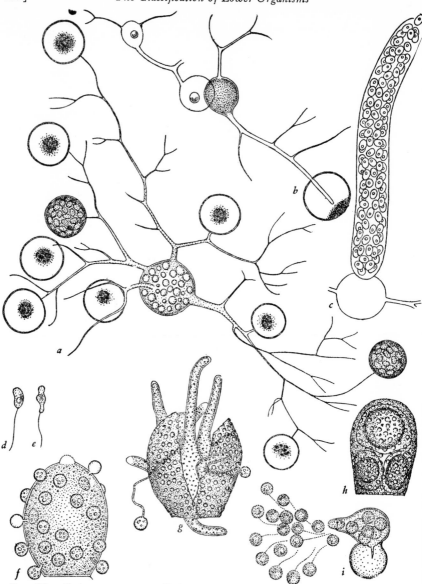

FIG. 23.—CHYTRIDINEA: **a-c**, *Polyphagus Euglenae* attacking cells of *Euglena*, x 400, after Nowakowski (1876); in figure **b**, two individuals have made contact and a zygote is developing at the point of junction; **c**, sporangium. **d-i**, *Olpidium Allomycetos* attacking *Alomyces anomalus*, x 1,000, after Karling (1948); **d**, **e**, zoospores; **f**, sporangium of the host beset with many parasites; **g**, **h**, resting cells of the host containing respectively sporangia and resting cells of the parasite; **i**, germination of resting cell.

1. Sporangia opening through opercula.
   2. Pluricentric.........................Family 7. Nowakowskiellacea.
   2. Monocentric.........................Family 8. Chytridiacea.
Family 1. **Cladochytriacea** [Cladochytriaceae] Schröter in Engler and Prantl Nat. Pflanzenfam. I Teil, Abt. 1: 80 (1892). Family *Hyphochytriaceae* (*Cladochytriaceae*) A. Fischer in Rabenhorst Kryptog.-Fl. Deutschland 1, Abt. 4: 131 (1892), in part. Family *Physodermataceae* Sparrow Aq. Phyc. 304 (1943). Pluricentric chytrids, the sporangia not operculate. The members of this family are of the same body type (designated by Karling, 1931, the *rhizomycelium*) as the anisochytrid *Hyphochytrium* and the Nowakowskiellacea of the present order. In most Cladochytriacea the rhizomycelium includes pairs of swollen cells ("turbinate organs") which give a false appearance of conjugation. There are some forty known species, mostly of two genera, *Cladochytrium*, saprophytic in vegetable remains, and *Physoderma* (including *Urophlyctis*), parasitic in higher plants. Sparrow (1946, 1947) discovered in certain species of *Physoderma* an alternation of morphologically distinguishable generations, both on the same hosts; the generations are presumably haploid and diploid, but this has not been established by observation of syngamy and meiosis. *Polychytrium* grows well only on chitin (Ajello, 1948).

Family 2. **Phlyctidiacea** [Phlyctidiaceae] Sparrow in Mycologia 34: 114 (1942). Family *Sporochytriaceae* (*Rhizidiaceae, Polyphagaceae*) subfamily *Metasporeae* A. Fischer in Rabenhorst Kryptog.-Fl. Deutschland 1, Abt. 4: 85 (1892). Monocentric eucarpic chytrids, the centers developed at the ends of filaments which grow from the zoospores, sporangia without opercula. These are the most familiar chytrids. There are more than one hundred species. Many are parasitic, on blue-green and green algae, diatoms, pollen grains, nematodes, and other minute fresh-water life; others are saprophytic, on cellulose, chitin, or keratin. *Rhizophidium*, the most numerous genus; *Phlyctidium, Phlyctorhiza, Entophlyctis, Diplophlyctis, Loborhiza*, etc.

Family 3. **Rhizidiacea** [Rhizidiaceae] Schröter in Engler and Prantl Nat. Pflanzenfam. I Teil, Abt. 1: 75 (1892). Family *Sporochytriaceae* (*Rhizidiaceae, Polyphagaceae*) A. Fischer in Rabenhorst Kryptog.-Fl. Deutschland 1, Abt. 4: 85 (1892) and subfamily *Orthosporeae* op. cit. 124. Monocentric eucarpic chytrids, the zoospores enlarging and becoming centers, which in turn become sporangia or presporangia; the sporangia without opercula. A moderate number of species, parasitic on blue-green or green algae, flagellates, or diatoms; or chitinophilous, saprophytic in the shed exoskeletons of insects. *Rhizidium, Siphonaria, Asterophlyctis, Polyphagus*, etc. *Polyphargus Euglenae* Nowakowski (1876) is a classic example. The centers lie free in the water, parasitizing cysts of *Euglena* through freely branching and widely spreading rhizoids. Most centers act as presporangia. Syngamy occurs when a rhizoid from one center makes contact with another center. The protoplasm of the latter migrates into the tip of the rhizoid, which swells and becomes a resting spore.

Family 4. **Synchytriacea** [Synchytriaceae] Schröter op. cit. 71. Family *Merolpidiaceae* (*Synchytriaceae*) A. Fischer op. cit. 45. Holocarpic chytrids, the intramatrical cell unwalled in the vegetative condition, becoming a presporangium or a resting spore, either of which gives rise to a cluster of sporangia. *Synchytrium*, parasitic on higher plants; *Micromycopsis* on Conjugatae.

Family 5. **Achlyogetonacea** [Achlyogetonaceae] Sparrow in Mycologia 34: 114 (1942). Chytrids without rhizoids, the intramatrical center proliferating and producing a linear series of centers, each of which becomes a sporangium without an

118] *The Classification of Lower Organisms*

operculum. *Achlyogeton,* in green algae, diatoms, and nematodes; of very much the appearance of certain Lagenidialea.

Family 6. **Olpidiacea** [Olpidiaceae] Schröter op. cit. 67. Family *Monolpidiaceae* (*Olpidiaceae*) A. Fischer op. cit. 20. Holocarpic chytrids, each individual a single intramatrical parasitic center, naked until the reproductive phase, when it becomes a sporangium without an operculum. *Olpidium,* attacking blue-green and green algae, diatoms, flagellates, *Allomyces, Vampyrella,* rotifers, and nematodes. *Rozella,* attacking Oomycetes and producing spiny resting spores, has been confused with certain Lagenidialea. The genera *Sphaerita* and *Nucleophaga* of Dangeard, including intracellular parasites of amoebas and Infusoria, have been placed in this family; it seems more probable that they should be placed among bacteria of family Rickettsiacea.

Family 7. **Nowakowskiellacea** [Nowakowskiellaceae] Sparrow in Mycologia 34: 115 (1942). Family *Megachytriaceae* Sparrow Aq. Phyc. 378 (1943). Pluricentric chytrids, the sporangia with opercula. A moderate number of saprophytes on material of green algae and higher plants. *Nowakowskiella, Megachytrium,* etc. *Zygochytrium* was described by Sorokin, 1874, as living on decaying insects, producing multiple operculate sporangia, and exhibiting a conjugation of filaments to produce zygotes much like those of Zygomycetes. It has apparently not been reobserved.

Family 8. **Chytridiacea** [Chytridiaceae] Cohn in Hedwigia 11: 18 (1872). Family *Chytridieen* de Bary and Woronin in Berichte Verhandl. Naturf. Gess. Freiburg 3 (Heft 2): 46 (1864). Monocentric eucarpic chytrids, the sporangia operculate. Some fifty species, the majority parasitic on fresh water algae. *Chytridium,* etc. *Catenochytridium,* saprophytic in cast-off exoskeletons of insects.

Chapter IX
PHYLUM INOPHYTA
Phylum 5. INOPHYTA Haeckel

Order FUNGI L. Sp. Pl. 1171 (1753).

Hysterophyta Link, 1808.

Classes *Fungi* and *Lichenes* Bartling Ord. Nat. 4 (1830).

Regnum Mycetoideum Fries Syst. Myc. 1: lvi (1832).

Class *Lichenes* and section *Hysterophyta* with class *Fungi* Endlicher Gen. Pl. 11, 16 (1836).

Stamm INOPHYTA Haeckel Gen. Morph. 2: xxxvi (1866).

Subdivision *Fungi* Engler and Prantl Nat. Pflanzenfam. II Teil: 1 (1889).

Division *Eumycetes* Engler Syllab. ed. 3: 25 (1903).

Phylum *Carpomyceteae* Bessey in Univ. Nebraska Studies 7: 249 (1907).

Stamm *Mycophyta* Pascher in Beih. bot. Centralbl. 48, Abt. 2: 330 (1931).

Kingdom *Mycetalia* Conard Plants of Iowa iv (1939).

Phylum *Eumycophyta* Tippo in Chron. Bot. 7: 205 (1942).

Parasites and saprophytes without flagellate stages, the bodies filamentous, the walls containing no cellulose.

This group represents the conventional division or subdivision Fungi of the kingdom of plants, excluding, of course, the bacteria, Oomycetes, chytrids, and Mycetozoa. The name Fungi, used as a scientific name, is properly to be applied, by authority of Linnaeus, to an order. *Agaricus campestris* L. will be recognized as the standard species of the phylum and of the order.

Those who study Inophyta are accustomed to use, for soma and filament respectively, the terms mycelium and hypha. The walls of the hyphae are believed to consist of pectic material. A small percentage of chitin is usually present (Schmidt, 1936); cellulose is totally absent (Thomas, 1928; Nabel, 1939; Castle, 1945). The organism *Basidiobolus,* having hyphae walled with cellulose, is tentatively retained among Inophyta as an exception.

The multiplication and dissemination of those organisms is by spores, of various types, scattered in the air. Most Inophyta produce two or more kinds of spores, some of them asexually, others as features of a sexual cycle. Spores produced within cases are called endospores, and the cases sporangia. Other spores are produced externally, commonly by constriction of the ends of hyphae. Spores thus produced are called conidia, and the hyphae or other structures which bear them, conidiophores. Spores are commonly produced not directly on the mycelium but on macroscopic structures of various types, all of which may be called by the familiar term fruit. The common mushroom as we see it is a fruit; it is the temporary spore-producing structure of an organism whose soma consists of filaments living saprophytically in the soil below.

It is expedient to mention at this point the growths called lichens, which are traditionally treated as a taxonomic group, either subordinate to Fungi or of the same rank. Lichens are gelatinous or thallose growths, usually of an impure green color, common everywhere, terrestrial or epiphytic, as on stones, trees, or fence posts. The microscope, in the hands of de Bary and others, showed that they consist of cells of two types, colorless filaments like those of Inophyta, and pigmented cells of quite the character of those of certain algae. De Bary (in Hofmeister, 1866) concluded that some lichens are not organisms but combinations of totally diverse organisms. Presently (1868) he was convinced by the work of Schwendener, soon (1868) published under his own name, ". . . dass die Flechten sammt und sonders keine selbstständigen Pflanzen seien, sondern Pilze aus der Abtheilung der Ascomyceten, denen die fraglichen Algen—deren Selbstständigkeit ich also nicht bezweifle—als Nährpflanzen dienen." In 1879 de Bary coined the term symbiosis to designate the association of different kinds of organisms. In de Bary's usage the term included parasitism; in general usage, it means association to mutual advantage. The lichens are a classic example of symbiosis.

Clearly, the group of lichens is not to be maintained; the algal components are known to have natural places among algae, and the inophyte components are to be assigned to their natural places among Inophyta, almost all in various orders of class Ascomycetes. This has already been done by Clements (1909) and Clements and Shear (1931). The numerous names which students of lichens have given to them are to be applied to the inophyte components.

Another common example of symbiosis involving inophytes is furnished by at least some of those which live on or in the tissues of higher plants without killing them (Kelley, 1950). They occur mostly on roots. Frank (1885) coined the term mycorhiza to designate the combination of roots and inophytes; it will be more convenient to hold that this term designates the inophyte component of the combination. Such mycorhizae as cover the growing tips of roots are helpful to their hosts by serving as agents of absorption.

Jones (1951) estimated the number of species of Inophyta as 40,000. This is surely an extreme underestimate. Martin (1951) gives reason for believing the number to be about as great as that of flowering plants, of the order of 300,000.

The early classifications of "fungi," as by Persoon (1801) and Fries (1821-1832), were based on gross characters. They presented, along with recognizable groups whose names are to be applied in order of priority, others which were mere random assemblages, and whose names are to be abandoned as *nomina confusa.* De Bary (in Hofmeister, 1866; 1884), having applied comparatively modern methods, established a dozen groups (under German names). These, so far as they are retained in the present phylum, have been assembled as three classes distinguished by details of the sexual cycle. A fourth class, acknowledgedly artificial, is maintained for the accomodation of the numerous and important fungi whose sexual cycles are unknown. The termination *-mycetes,* of the names of the classes and also of various subordinate groups, is the Greek μύκητες, the plural of μύκης, a mold or mildew. The termination *-mycetae* which some authors have used is a solecism.

1. Reproducing sexually, or by apomictic processes clearly of sexual origin.
    2. The zygote becoming a thick-walled resting cell; fruits none or inconsiderable . . . . Class 1. ZYGOMYCETES.
    2. The zygote not becoming a thick-walled resting cell; mostly producing fruits.
        3. The zygotes giving rise, usually indirectly, to sporangia called asci, each typically containing eight spores called ascospores . . . . Class 2. ASCOMYCETES.

3. The zygotes giving rise indirectly to
conidiophores called basidia, each
bearing typically four conidia
called basidiospores....................Class 4. BASIDIOMYCETES.
1. Not known to reproduce sexually.................Class 3. HYPHOMYCETES.

Class 1. ZYGOMYCETES (Sachs ex Bennett and Thistleton-Dyer) Winter

Zygomyceten Sachs Lehrb. Bot. ed. 4: 248 (1874).
ZYGOMYCETES Bennett and Thistleton-Dyer in Sachs Textb. Bot. English ed. 847 (1875).
Class ZYGOMYCETES Winter in Rabenhorst Kryptog.-Fl. Deutschland 1, Abt. 1: 32 (1879).
Order *Zygomycetes* Engler Syllab. 23 (1892).
Class *Zygomyceteae* Schaffner in Ohio Naturalist 9: 449 (1909).
Inophyta whose zygotes are thick-walled resting cells, in germination giving rise to spores indistinguishable from those produced asexually; hyphae usually without cross-walls; mostly not producing fruits. The standard species is *Mucor Mucedo* L.
Among the Inophyta as here limited, the Zygomycetes appear to be primitive (an alternative hypothesis, that certain Ascomycetes are primitive, will be discussed below). Traditionally, the Zygomycetes are associated with the Oomycetes. The association is probably mistaken, being based merely on similarity of body form: the Zygomycetes are terrestrial instead of aquatic, produce no flagellate cells, have no cellulose in their cell walls (except in *Basidiobolus*), and do not produce female gametes by the cutting out of cells within a cell. In later editions of Engler's Syllabus (1924), one finds most of the chytrids included among the Zygomycetes, instead of in their conventional place among the Oomycetes. The hypothesis thus suggested, that the Opisthokonta may represent the ancestry of the Inophyta, is attractive, but not to present knowledge supported by convincing evidence. Class Zygomycetes and phylum Inophyta must as yet be regarded as of unknown origin and treated as isolated.
There are some 500 known species of Zygomycetes. They form two orders. The bulk of the group, and the typical examples, are order Mucorina. A minority, distinguished by parasitism and by explosively discharged conidia, are order Entomophthorinea.

Order 1. **Mucorina** [Mucorini] Fries Syst. Myc. 3: 296 (1832).
Suborder *Mucorineae* Engler in Engler and Prantl Nat. Pflanzenfam. I Teil, Abt. 1: iv (1897).
Order *Mucorineae* Campbell Univ. Textb. Bot. 158 (1902).
Order *Spirogyrales* (presumably in part only) Clements Gen. Fung. 12 (1909).
Order *Mucorales* Smith Crypt. Bot. 1: 405 (1938).
Order *Zoopagales* Bessey Morph. and Tax. Fungi 117 (1950).
The typical Zygomycetes, mostly saprophytic, not producing explosively discharged conidia (*Pilobolus* produces explosively discharged sporangia).
The asexual reproductive structures of the supposedly primitive Mucorina, as *Mucor* and *Rhizopus,* are solitary globular sporangia terminal on erect hyphae. In the developing sporangium, a dome-shaped basal sterile area, the columella, is set apart by cleavage followed by deposition of a wall. The protoplasm above the

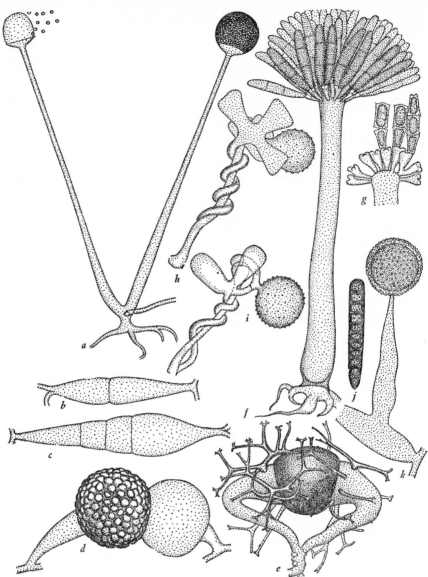

Fig. 24.—Zygomycetes: **a-d**, *Rhizopus nigricans;* **a**, sporangia x 50; **b-d**, pregametes, suspensors and gametes, and zygote x 200. **e**, Zygote of *Phycomyces nitens* after Blakeslee (1904). **f, g**, Conidiophore with young conidia, and mature conidia, of *Syncephalis pycnosperma* after Thaxter (1897). **h, i**, Conjugation of *Syncephalis nodosa* after Thaxter, *op. cit.* **j**, Sporangium of *Syncephalastrum racemosum* after Thaxter, *op. cit.* **k**, Sporangium of *Haplosporangium lignicola* after Martin (1937), x 1,000.

columella undergoes cleavage to form spores, which may remain plurinucleate (Swingle, 1903). Other members of the order exhibit transitions (apparently two distinct series of transitions) from sporangia as just described to typical conidia.

Syngamy occurs when the tips of pairs of hyphae meet and are cut off by crosswalls to act as multinucleate gametes. The process is regarded as conjugation, although the gametes of a pair are usually not of the same size. Conjugation does not occur at random, but, in most Zygomycetes, between branches from hyphae of two mating types, designated plus and minus (the distinction of mating types is not identical with the differentiation of sexes). Zygomycetes were the first group reproducing by conjugation in which a distinction of mating types was discovered; the discovery was by Blakeslee (1904).

Syngamy is preceded by a flare of mitoses in the gametes. The mitotic figures are sharp-pointed, as though centrosomes were present; the haploid chromosome number appears to be 2. The process is not meiotic (Moreau, 1913). After these divisions, the walls between the gametes break down and the nuclei unite in pairs. Unpaired nuclei, presumably contributed in excess by one gamete or the other, undergo dissolution (Keene, 1914, 1919). Ordinarily, the zygote enlarges and becomes a thick-walled resting spore; in some examples, the resting spore forms as an outgrowth on what was one of the gametes. In *Phycomyces, Absidia,* and *Syncephalis,* the hyphae which have produced the gametes, and to which the zygote remains attached, send out branches which form a layer about the zygote. These branches might be interpreted as making up fruits. *Endogone* produces definite fruits of considerable size.

A zygote germinates by production of a hypha bearing a sporangium (Blakeslee, 1906). Meiosis is believed to occur in the course of germination.

While Mucorina in general are saprophytic, some of them are parasitic on others, *Piptocephalis* and *Chaetocladium* on *Mucor,* and *Parasitella* on *Absidia.* Drechsler (1935, 1937) discovered a number of organisms apparently of this group parasitizing amoebas and nematodes in the soil.

The Mucorina may be treated as five families.
1. Not producing macroscopic fruits.
   2. Not parasitic on amoebas or nematodes.
      3. All spores produced in sporangia
         with columellae...................Family 1. MUCORACEA.
      3. Not as above.
         4. Producing sporangia or else
            conidia as outgrowths from a
            knob, homologous with a
            sporangium, solitary on an un-
            branched stalk................Family 2. PIPTOCEPHALIDACEA.
         4. Sporangia or conidia solitary
            and terminal on branches of a
            branched sporangiophore or
            conidiophore; sporangia, if
            produced, without columellae.... Family 3. MORTIERELLACEA.
   2. Parasitic on amoebas or nematodes........Family 4. ZOOPAGACEA.
1. Producing macroscopic fruits................Family 5. ENDOGONACEA.

Family 1. **Mucoracea** [Mucoraceae] Cohn in Hedwigia 11: 17 (1872). Mucorina whose spores are produced exclusively in sporangia with columellae solitary on unbranched sporangiophores. *Mucor* L., typified by *M. Mucedo,* is now limited to a small group mostly saprophytic on manure. *Pilobolus,* another coprophilous genus, is distinguished by sporangiophores which become swollen at the summit, bend toward the light, and discharge the sporangia violently to a distance of several meters. *Rhizopus nigricans,* the common black bread mold; *Phycomyces, Absidia, Sporodinia, Zygorhynchus.*

Family 2. **Piptocephalidacea** [Piptocephalidaceae] Schröter in Engler and Prantl Nat. Pflanzenfam. I Teil, Abt. 1: 132 (1893). Family *Choanephoraceae* Schröter op. cit. 131. Mucorina producing sporangia without columellae, or conidia, in compact clusters terminal on unbranched stalks. *Blakesleea,* transitional between the preceding family and this, may produce solitary sporangia with columellae, or else, as outgrowths from the primordia of sporangia, clusters of minuscule sporangia without columellae. *Cunninghamella,* producing heads of globular conidia; *Syncephalastrum,* with clustered cylindrical sporangia; *Syncephalis* and *Piptocephalis,* producing clustered chains of conidia.

Family 3. **Mortierellacea** [Mortierellaceae] Schröter op. cit. 130. Family *Chaetocladiaceae* Schröter op. cit. 131. Mucorina whose sporangiophores or conidiophores are branched, the sporangia (without columellae) or conidia solitary and terminal on the branches. *Thamnidium, Chaetocladium, Mortierella, Haplosporangium.*

Family 4. **Zoopagacea** [Zoopagaceae] Drechsler in Mycologia 27: 37 (1935). Mucorina parasitic in amoebas or nematodes, producing conidia. The hosts of Zoopagacea inhabit the soil and are infected by contact with hyphae or conidia. From the point of contact, a hypha grows into the host and gives rise to a mycelium; this is in some examples reduced to a single coiled cell. The host being killed, the parasite sends out hyphae which may produce conidia, usually in chains, or else may conjugate and produce zygotes. *Endocochlus, Cochlonema, Bdellospora, Zoopage, Acaulopage, Stylopage.*

Family 5. **Endogonacea** [Endogonaceae] Paoletti in Saccardo Sylloge Fungorum 8: 905 (1889). *Endogonei* Fries. Mucorina saprophytic in soil or wood, producing macroscopic subterranean fruits. The fruits may reach a diameter of 2 cm. Within them, the tips of hyphae are cut off by crosswalls, and develop either into sporangia without columellae or into gametes.

Order 2. **Entomophthorinea** [Entomophthorineae] (Engler) Campbell Univ. Textb. Bot 161 (1902).

Suborder ENTOMOPHTHORINEAE Engler in Engler and Prantl Nat. Pflanzenfam. I Teil, Abt. 1: iv (1897).

Order *Entomophthorales* Smith Crypt. Bot. 1: 408 (1938).

Zygomycetes, mostly parasitic, producing explosively discharged conidia (*Massospora,* while clearly belonging to the group, is an exception to the stated character).

These organisms, although of the general nature of ordinary Inophyta, exhibit cytological characters markedly distinguishing the two families from the generality of Inophyta and from each other. The position here given to them is the customary one; it is doubtful that it is natural.

Family 1. **Entomophthoracea** [Entomophthoraceae] Berlese and de Toni in Saccardo Sylloge 7: 280 (1888). Most species are parasitic in the bodies of insects, whose tissues they replace. The hyphae become divided by crosswalls, and the multinucleate cells thus produced tend to round up and become separate. A well-nourished cell may send forth a hypha which reaches the outer air and whose tip is cut off and discharged in the direction of the light. Martin (1925) and Couch (1939) described the mechanism of discharge. The conidiophore ends in a columella projecting into the base of the conidium. The columella develops a double wall. Increasing pressure within the conidium causes a sudden eversion of the wall on the side of the conidium, and this movement throws the conidium forth to a distance of perhaps 1 mm. Conidia which come down on unfavorable substrata may form and discharge secondary conidia.

Adjacent cells may conjugate, the thick-walled zygote forming either in one of them or as an outgrowth from one of them. Many examples produce thick-walled resting spores without conjugation.

Olive (1906) described the nuclei and the process of mitosis in *Empusa*. The resting nuclei are fairly large, 7-9μ in diameter. In the course of division, two stain-resistant granules are seen, with strands of chromatin radiating from them. These move apart, while the nucleus becomes dumb-bell shaped. The nuclear membrane remains intact and division is completed by its constriction. As Olive remarked, the process is much as in *Euglena*.

Entomophthora, Empusa, and *Massospora* attack insects; the first produces zygotes, while the other two produce asexual resting spores; *Massospora* does not discharge the conidia violently. *Conidiobolus* and *Delacroixia* are saprophytic. *Completoria* attacks the prothallia of ferns. *Ancylistes,* a parasite in the green alga *Closterium,* was formerly included among chytrids or Oomycetes. Berdan (1938) showed that it belongs here; it produces conidia and zygotes quite of the character of the present group, and does not produce zoospores.

Family 2. **Basidiobolacea** [*Basidiobolaceae*] Engler and Gilg Syllab. ed. 9 u. 10: 45 (1924). *Basidiobolus ranarum* Eidam (1886) occurs in the intestinal contents of frogs and toads as uninucleate cells, solitary or in brief filaments, walled with cellulose. In manure the filaments develop into a scant branching mycelium. The protoplasm gathers in the ends of erect hyphae which are cut off as conidia and discharged. Conjugation occurs between adjacent cells of a filament. It is preceded by a single nuclear division in each gamete (Fairchild, 1897). In this process, the nuclear membrane disappears and the numerous minute chromosomes are found in a blunt-ended spindle without centrosomes. Each gamete form a papilla; one of the two nuclei enters the papilla, whose contents, after being cut off by a wall, die and disappear. The gametes and their nuclei unite and the zygote secretes a thick wall.

Class 2. **ASCOMYCETES** (Sachs ex Bennett and Thistleton-Dyer) Winter

Order *Ascosporeae* Cohn in Hedwigia 11: 17 (1872).

Ascomyceten Sachs Lehrb. Bot. ed. 4: 249 (1874).

AscomyCETES Bennett and Thistleton-Dyer in Sachs Textb. Bot. English ed. 847 (1875).

Class AscomyCETES Winter in Rabenhorst Kryptog.-Fl. Deutschland 1, Abt. 1: 32 (1879).

Class *Ascosporeae* Bessey in Univ. Nebraska Studies 7: 295 (1907).

Class *Ascomycetae* Schaffner in Ohio Naturalist 9: 449 (1909).

Inophyta which produce, as a feature of the sexual cycle, sporangia called asci, in which the spores, called ascospores, typically eight in number, are delimited by the manner of cell division called free cell formation, i.e., in such fashion as to exclude a part of the cytoplasm.

The hyphae of Ascomycetes are septate and the cells most often uninucleate.

Most Ascomycetes produce, beside the ascospores, conidia of one type or another. A mycelium may produce a mass of densely woven hyphae with conidia on the surface; such a mass is called an acervulus or sporodochium. Either a mycelium or an acervulus or sporodochium may send up spore-bearing columns called coremia.

Many Ascomycetes produce, either directly from the mycelium or from special structures consisting of interwoven hyphae, globular or flask-shaped structures which produce conidia internally and release them through a pore. These structures are called pycnidia, and the spores pycniospores. In many examples, the pycniospores are capable of functioning as sperms; so far as this is true, the pycnidia may alternatively be called spermagonia, and the pycniospores spermatia.

Hyphae woven into a mass may go into a resting condition, becoming thick-walled, hard, and usually dark in color. The resulting structure is a sclerotium. If a structure of the general nature of an acervulus, sporodochium, or sclerotium gives rise either to pycnidia or to fruits bearing asci, it is called a stroma.

As to asci and ascospores, Dangeard (1893, 1894, 1907) reached definitely the conclusion that they are essentially sexual products. There had been earlier observations, beginning with de Bary, 1863, that there are meetings, coilings together, and fusions of hyphae as a preliminary to the production of asci. Many ascomycetes are of two mating types; this was first discovered of *Glomerella*, by Edgerton (1914). As Dodge (1939) remarks, the mating types are not sexes; in forms producing recognizable male and female reproductive structures, each mating type may produce both.

In Ascomycetes which may be regarded as primitive, differentiated male and female cells are produced. The male cell or antheridium is ordinarily terminal on a hypha. The female cell (constituting, together with other differentiated cells of the same hypha, if any are present, the ascogonium) may be terminal; more often it bears an elongate cell, or a chain of cells, called the trichogyne, and having the function of reaching the antheridium. In some Ascomycetes, antheridia are produced, but syngamy does not take place; the egg is binucleate or multinucleate, and the nuclei within it take the part of gamete nuclei in further development. There are others in which no antheridia are produced. Hansen and Snyder (1943) found, in *Hypomyces Solani* var. *Cucurbitae,* that "any part of the living thallus, ascospores, conidia or bits of the mycelium could act as the male fertilizing agent." There are forms in which fusions take place between undifferentiated hyphal cells; and yet others in which it appears that the paired nuclei involved in sexual processes arise by divisions of a single nucleus originally present in a spore.

In some Ascomycetes, syngamy is followed immediately by karyogamy, and the zygote develops directly into a single ascus. In the overwhelming majority of the group, asci are produced indirectly, and there is no fusion of nuclei until this takes place. The zygote sends out hyphae called ascogenous hyphae, recognizably different from the vegetative ones. The cells of the ascogenous hyphae are binucleate; or, arising from a multinucleate zygote, become binucleate by the establishment of crosswalls. The two nuclei of each cell divide concurrently and the cell walls are so placed that each cell receives nuclei of different origin. This effect is achieved in the final cell division before ascus formation by a peculiar process called crozier formation. The terminal cell of the ascogenous hypha becomes bent to the form of a hook; the nuclei divide concurrently, and cell walls appear between the daughter nuclei of each pair; the middle cell of the row of three thus produced remains binucleate and becomes an ascus. The uninucleate terminal and basal cells lie side by side, and may fuse to form a binucleate cell which may become an additional ascus, or else may grow forth and give rise to more asci than one.

The stage consisting of cells with two nuclei of different origin is called the dikaryophase. It is characteristic of Ascomycetes ,and also of Basdiomycetes: among Inophyta, it is a normal and familiar thing. To a concept of cytology founded on studies overlooking the Inophyta, it would appear an extreme anomaly, almost an impossibility. It has the appearance of a rather awkward device for making cells genetically and physiologically diploid while the nuclei remain haploid. In most Ascomycetes it is a brief stage, but there are some, as *Taphrina,* whose mycelium consists prevalently of binucleate cells.

The detailed behavior of nuclei in the ascus was first described by Harper (1895, 1897, 1900) from studies of *Peziza, Sphaerotheca, Erysiphe,* and *Pyronema.* The two nuclei in the primordium of the ascus unite into one. The fusion nucleus divides three times, each time in much the same manner. A centrosome with astral rays is present at the nuclear membrane, apparently outside. It divides, and a spindle forms, inside the intact nuclear membrane, between the daughter centrosomes. The chromosomes appear and divide. As they move toward the poles of the spindle, the nuclear membrane collapses or dissolves, leaving the spindle free in the cytoplasm. The mass of chromatin at each pole of the spindle shreds out into a nuclear network, duly surrounded by a nuclear membrane and usually containing a nucleolus.

Haploid chromosome numbers of Ascomycetes (all of which have been observed in the ascus) include the following:

Ascoidea rubescens, fide Walker (1935)	2
Eremascus albus, fide De Lamater et al. (1953)	6
Glomerella, fide Lucas (1946)	4
Hypomyces Solani var. *Cucurbitae,* fide Hirsch (1949)	4
Lachnea scutellata, fide Brown (1911)	5
Neurospora crassa, fide McClintock (1945)	7
Peziza domiciliana, fide Schultz (1927)	8
Phyllactinia corylea, fide Colson (1938)	10
Pyronema confluens var. *igneum,* fide Brown (1915)	5
Taphrina deformans, fide Martin (1940)	4

According to Harper, when the third division in the ascus is complete, each of the eight nuclei produced by it thrusts forths its centrosome upon a beak. The astral rays of the centrosomes become recurved in the cytoplasm about the nucleus, and grow and multiply until they are converted into a smooth membrane, outside of which a wall is deposited. Most observers have not seen so much detail. Brown (1911) and Dodge (1937) describe the cell membrane of the ascus, apparently under the influence of the centrosome of each nucleus, as cutting into the cytoplasm in an ellipsoid pattern. In *Taphrina* (Martin, 1940), the cytoplasm of the spores is delimited simply by accumulation about the nuclei. By whatever process the ascospores are cut out, some of the cytoplasm of the ascus is excluded and left without nuclei. Harper (1899) proposed to limit the older term free cell formation to processes which have this effect; he observed that the occurrence of such processes distinguishes asci from the sporangia of Oomycetes and Zygomycetes, in which spores are cut out by cleavage.

Harper believed that a fusion of nuclei follows immediately the fusion of gametes; that the karyogamy observed in the ascus is a uniting of diploid nuclei, producing tetraploid nuclei; and that the characteristic three nuclear divisions in the ascus are necessary for reduction of the chromosome number from tetraploid to haploid. These hypotheses, long accepted as possible, were disproved by genetic studies by Betts and Meyer (1939) and Keitt and Langford (1941). In the asci of many species, the spores lie in a single series in which their order is determined by the divisions which produce their nuclei. By refined technique, the spores from a single ascus may be identified, separated, and cultivated. It is then observed that the mycelia grown from the first four spores may differ in some particular character from those grown from the second four spores; those from the first pair of spores may differ from those from the second; but those from two members of any of the pairs, first, second, third or fourth, are always alike. These observations mean that the first two divisions in the ascus constitute the meiotic process, the third being mitotic. Lucas (1946) obtained cytological evidence refined enough to confirm this conclusion.

Asci are almost always produced in fruits, which may be called ascocarps. The ascocarp aside from the asci arises usually from vegetative hyphae; in the Ascomycetes regarded as primitive, it does not begin to develop until after fertilization, but in the higher ones it may develop in advance of fertilization and become the seat of this process.

There are several types of ascocarps, among which three are most familiar. A small ascocarp completely enclosing the asci is a cleistothecium. Cleistothecia were formerly included under the term perithecium; that term will better be limited to small fruits which are globular or vase-like, opening through a single pore, the ostiole, and differing from the pycnidia already described in producing ascospores instead of conidia. A fruit in which the asci form a broad layer which is typically fully exposed at maturity, the whole being ordinarily of the form of a disk or cup, larger than a cleistothecium or perithecium, is an apothecium.

Asci produced in perithecia or apothecia usually discharge the ascospores violently. The mechanism of discharge is apparently simply turgidity. Some asci show no visible adaptations for the discharge of spores; others have lids (opercula) whose position determines the direction of discharge. Certain large apothecia can throw the spores to a distance of 10-20 cm.; the discharge is so governed by temperature and humidity as to occur in gently moving rather than in still air. By blowing across these apothecia one can make them throw out a visible cloud of spores. Heald and Walton (1914) reviewed many older observations of violent discharge by perithecia, the oldest by Pringsheim on *Sphaeria Scirpi,* 1858. Rankin, 1913, found that each ascus in turn breaks loose, comes up to the ostiole, projects through it, throws out its spores, and collapses to make room for another. Weimer (1920) found that the perithecia of *Pleurage curvicolla* bend toward the light and throw the spores to a maximum distance of 45 cm., which is apparently the record.

There is a widely entertained hypothesis that the Ascomycetes evolved from the red algae. It appears to have developed from a piece of classification by Sachs (1874), who proposed a class *Carposporeen,* to consist of the red algae, certain higher green algae, and the Ascomycetes and Basidiomycetes. A number of resemblances support it. Both red algae and Ascomycetes include many parasites; both lack flagellate cells; both have differentiated gametes, the egg bearing a trichogyne; in both, fertilization leads to further development before spores are produced. In addition to these genuine resemblances, an imaginary one was influential, namely the double fertilization ascribed to the red algae by Schmitz and to the Ascomycetes by Harper. Numerous as these resemblances are, they are not now believed to indicate relationship. Atkinson (1915) formulated the counter-argument. The Ascomycetes resemble the Mucorina in nutrition, in producing no flagellate cells, and in multinucleate gametes. The germination of the zygote of the Mucorina, by the production of a hypha bearing a sporangium, resembles the production of ascogenous hyphae by the zygotes of Ascomycetes. Two principal changes would convert Mucorina into Ascomycetes: the zygote should cease to be a resting spore, and cell division within the sporangium should be by free cell formation. This could happen if the centrosomes of the ultimate nuclei of the sporangia were in control of cleavage, and if these nuclei were so far separated that considerable areas of cell membrane would lie beyond the influence of the centrosomes, with the effect that the cell membrane, furrowing in to delimit a spore around each nucleus, would leave some of the cytoplasm outside of all of the spores. The organisms listed below as the first order of Ascomycetes, Endomycetalea, are but poorly known, yet seem genuinely to represent the transition from Mucorina to typical Ascomycetes.

It is not yet possible to formulate a system of orders of Ascomycetes with the expectation that it will not be found to require much amendment[1]. The following will serve tentatively; excellent contemporary authority makes several orders each of the ones listed fourth, fifth, and seventh.

1. Ascus developed directly from the zygote (or apomictically from an unfertilized cell); not producing fruits . . . . Order 1. ENDOMYCETALEA.
1. The zygote giving rise to filaments of cells with more than one nucleus, these producing the asci.
  2. Producing fruits.
    3. The fruits cleistothecia.
      4. Asci scattered in the fruits; mostly saprophytes with branched conidiophores . . . . Order 2. MUCEDINES.
      4. Asci in one cluster, or solitary, in the fruits; mostly parasites with unbranched conidiophores . . . . Order 3. PERISPORIACEA.
    3. The fruits, originally closed, opening by irregular pores or regular or irregular clefts . . . . Order 4. PHACIDIALEA.
    3. The fruits apothecia . . . . Order 5. CUPULATA.
    3. The fruits perithecia.
      4. Producing a normal mycelium . . . . Order 7. SCLEROCARPA.
      4. Parasitic on insects, the mycelium reduced . . . . Order 8. LABOULBENIALEA.
  2. Not producing fruits, the asci arising directly from the mycelium . . . . Order 6. EXOASCALEA.

Order 1. **Endomycetalea** [Endomycetales] Gäumann Vergl. Morph. Pilze 135 (1926).
Subclass *Hemiasci* Engler Syllab. 26 (1892).

---

[1]Luttrell (1951) has presented a complete reorganization of the class. He sets apart as a major subordinate group Bitunicatae five orders in which the ripe ascus exudes a vesicle and discharges the spores from this.

Subclass *Hemiasci* or *Hemiasceae,* with suborder (Unterreihe) *Hemiascineae,* and suborder *Protoascineae* of subclass *Euasci,* Engler in Engler and Prantl Nat. Pflanzelfam. I Teil, Abt. 1: iv (1897), the names not based on those of genera.

Order *Protoascineae* Campbell Univ. Textb. Bot. 165 (1902).

Order *Hemiascales* Engler Syllab. ed. 3: 28 (1903).

Ascomycetes whose asci develop directly from the zygotes. Two families may be recognized.

Family 1. **Endomycetacea** [Endomycetaceae] Schröter in Engler and Prantl Nat. Pflanzenfam. I Teil, Abt. 1: 154 (1894). Family *Ascoideaceae* Schröter op. cit. 145. Mostly saprophytes, the uninucleate or multinucleate cells of the filaments tending to round up, become separate, and function as conidia; the zygotes, produced by syngamy of scarcely differentiated cells, enlarging and becoming asci of 4, 8, or many spores cut out by free cell formation. *Dipodascus, Eremascus, Endomyces, Ascoidea.* The asci of the last are apparently produced asexually (Walker, 1935).

The genus *Protomyces* requires mention. It is a parasite on higher plants, producing walled resting spores which germinate by producing a sporangium of many spores. It is chytrid-like, but its spores are non-motile. Its proper place in classification has for a long time been a puzzle.

Family 2. **Saccharomycetacea** [Saccharomycetaceae] (Rees) Schröter op. cit. 153. Class and family *Saccharomycetes* Winter in Rabenhorst Kryptog.-Fl. Deutschland 1, Abt. 1: (1879). Unicellular, reproducing by budding, i.e., by production upon the cells of outpocketings which are pinched off as additional cells, or by a sexual cycle in which endospores are produced, usually by fours.

These are the organisms which are in English called yeasts. The common bread- and beer-yeast called *Saccharomyces cerevisiae* has a good claim to be considered, economically, the most important of all "fungi." Its metabolism, in which dextrose is converted to alcohol and carbon dioxide, gives a superficial appearance of simplicity, and has attracted much study, contributing much to an understanding of the genuine intricacy of energesis.

In addition to agents of fermentation, this family includes pathogens causing chronic infections of animals. These have been treated as a genus *Torula, Torulopsis, Blastoderma,* or *Cryptococcus.* They have not been observed to produce endospores.

Order 2. **Mucedines** Fries Syst. Myc. 3: 380 (1832).

Order *Gymnoascaceae* Winter in Rabenhorst Kryptog-Fl. Deutschland 1, Abt. 2: 3 (1887).

Suborder *Plectascineae* Engler in Engler and Prantl Nat. Pflanzenfam. I Teil, Abt. 1: v (1897).

Order *Plectascineae* Campbell Univ. Textb. Bot. 169 (1902).

Order *Aspergilliales* Bessey in Univ. Nebraska Studies 7: 304 (1907).

Order *Gymnascales* Clemens Gen. Fung. 93 (1909).

Order *Plectascales* Gäumann Vergl. Morph. Pilze 164 (1926).

Ascomycetes producing cleistothecia in which the asci are scattered; mostly saprophytic and producing branched conidiophores.

The name Mucedines means molds. Under this name Fries listed twelve genera, with *Aspergillus* Link and *Penicillium* Link first. The former is the evident standard genus of the order. Both genera are very common and numerous in species. They are readily recognized under the microscope by the forms of their clusters of conidia.

The conidiophore of *Aspergillus* ends in a globular swelling from which spring many radiating rows of conidia, with the effect that the entire mass, yellow, brown, black, pink, or red in color, is globular. *Penicillium* has a branching conidiophore bearing rows of conidia in a broom-like mass. The masses are usually blue or green, and are familiar on cheese, jam, bread, cardboard, oranges, or almost any organic material.

Particular species of *Penicillium* are involved in the making of genuine Camembert and Roquefort cheeses. The genus has become best known for the production by *P. notatum* of the drug penicillin. In 1929, Dr. Alexander Fleming of London noticed that a mycelium of this species, growing as a contaminant on a plate of bacteria, interfered with the growth of the latter. This observation led to the discovery of a substance clinically useful against actinomycetes, spheres, and Gram positive rods, but not against Gram negative rods. Production was for several years very scant, and the drug expensive accordingly; in the early 1940's, as a war measure, the United States financed large scale production along with the appropriate scientific study (Elder, 1944; Committee on Medical Research, Washington, and the Medical Research Council, London, 1945). Several forms of penicillin have been recognized; they differ in the radicle R in the formula $C_9H_{11}O_4SN_2R$. The structural formula is believed to be as follows (Editorial Board of the Monograph on the Chemistry of Penicillin, 1947):

$$RCONH - CH - CH - S - C \ (CH_3)_2$$

$$OC\text{---} N\text{-------} CH \ COOH.$$

The sexual reproduction of *Aspergillus* and *Penicillium* involves the syngamy of differentiated cells. The zygote sends out ascogenous hyphae which bud off scattered asci; the neighboring cells send out hyphae which become woven into a minute firm-walled cleistothecium enclosing them.

Link, who named *Aspergillus* and *Penicillium,* gave to the ascocarp-producing stage of *Aspergillus* the name *Eurotium.* There is a rule of botanical nomenclature which allows only a tentative status to names given to the conidium-producing stages of inophytes. Thom and his associates (1926, 1945), in presenting a workable system of the species of *Aspergillus,* remarked that "It is better to forget Eurotium along with the technicality."

This order includes a variety of other molds: *Gymnoascus,* producing only a loose weft of hyphae about the asci; *Ctenomyces,* on feathers, recognized by comb-like outgrowths from the loosely woven ascocarps; *Monascus,* its name a misnomer, the minute fruit containing many asci; *Onygena,* saprophytic on horns and hoofs, producing puffball-like fruits as much as 1 cm. high; *Elaphomyces,* forming a mycorrhiza on roots of conifers and producing hypogaeous fruits as large as walnuts.

Order 3. **Perisporiacea** [Perisporiaceae] Fries Syst. Myc. 3: 220 (1829).
Order *Perisporia* Fries op. cit. 1: xlviii (1832).
Suborder *Perisporiaceae* Winter in Rabenhorst Kryptog.-Fl. Deutschland 1, Abt. 2: 21 (1887).
Subsuborder (*Underordnung*) *Perisporiales* Engler in Engler and Prantl Nat. Pflanzenfam. I Teil, 1: v (1897).
Order *Perisporiales* Bessey in Univ. Nebraska Studies 7: 295 (1907).
Ascomycetes producing cleistothecia containing a compact cluster of asci or a solitary ascus; mostly parasites producing unbranched conidiophores.

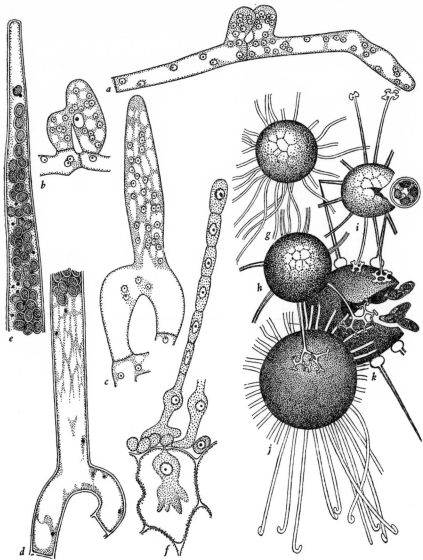

Fig. 25.—Ascomycetes: **a-e,** *Dipodascus albidus* after Juel (1902), x 1,000; **a,** gametes; **b,** syngamy; **c,** development of ascus; **d, e,** lower and upper parts of a mature ascus.   **f,** *Erysiphe graminis,* haustorium penetrating an epidermal cell of a grass and conidiophore bearing a chain of conidia x 500.   **g-k,** Cleistothecia of Perisporiacea x 100: **g,** of *Erysiphe* sp.; **h,** of *Microsphaera* sp.; **i,** of *Podosphaera* sp.; **j,** of *Uncinula* sp.; **k,** of *Phyllactinia* sp.

The more familiar Perisporiacea are those of family **Erysiphea** [Erysipheae] Winter. They are parasites on plants, mostly producing a white mycelium on the surface and sending brief haustoria into the epidermal cells. They produce abundant conidia in erect unbranched chains; this habit explains the common name of powdery mildews. Harper's important studies of the morphology of Ascomycetes were in large part made on powdery mildews. The gametes are uninucleate and unite directly, the egg bearing no trichogyne; the ascogenous hyphae are brief; each minute black globular cleistothecium bears an equatorial whorl of appendages of a form characteristic of the genus. In *Erysiphe* and *Sphaerotheca* (*S. pannosa* is the common rose mildew), the fruits bear unbranched sinuous appendages like vegetative hyphae; the fruit of *Erysiphe* contains several asci, while that of *Sphaerotheca* contains one. In *Microsphaera* (*M. Alni* is the powdery mildew of lilac) and *Podosphaera,* the appendages are dichotomously forked near the tip; the fruit of *Microsphaera* contains several asci, that of *Podosphaera* only one. The appendages of *Uncinula* are hooked at the tip. Those of *Phyllactinia* are like sharp spikes with bulbous bases.

Other Perisporiacea, parasitic or saprophytic on plant material, are comparatively poorly known. The fruits may bear appendages of other characters than those of the Erysiphea, or none, and may be characteristically clustered or borne in stromata. In some examples the fruits have no definite dehiscence mechanism; in others they open by deliquescence or by a separation of plates. Some open by a single pore, and appear transitional to those of order Sclerocarpa; some open by a cleft, or by lobes separated by radiating clefts, and appear transitional to those of order Phacidialea.

Order 4. **Phacidialea** [Phacidiales] Bessey in Univ. Nebraska Studies 7: 298 (1907).

Phacidiacei Fries Syst. Myc. 1: li (1832).

Order *Hysteriaceae* and suborders (of order *Discomycetes*) *Phacidiaceae, Stictideae,* and *Tryblidieae* Rehm in Rabenhorst Kryptog.-Fl. Deutschland 1, Abt. 3: 1, 60, 112, 191 (1896); the ordinal name preoccupied by family HYSTERIACEAE Saccardo.

Suborders *Phacidiineae* and *Hysteriineae* Engler in Engler and Prantl Nat. Pflanzenfam. I Teil, Abt. 1: v (1897).

Orders *Graphidiales* and *Hysteriales* Bessey op. cit. 298, 303.

Order *Hemisphaeriales* Theissen in Ann. Myc. 11: 468 (1913).

Order *Microthyriales* Clements and Shear Gen. Fung. ed. 2: 94 (1931).

Ascomycetes producing fruits which are not typical cleistothecia, apothecia, or perithecia.

This group is here used as a catch-all for three or more distinct groups, which appear to form cross-connections among orders Perisporiacea, Cupulata, and Sclerocarpa. This appearance suggests the probability that the present group, and the usually accepted orders assembled under it, are not natural, but represent parallel developments from several sources. The present groups include moderately numerous ordinary parasites and saprophytes, together with great numbers of lichen-formers. Only the latter are common and familiar in temperate countries. There has been little study of the morphology.

The families which appear tenable are distinguished as follows:

a. Fruits minute and flattened, usually releasing the spores through one or more pores or clefts (Order *Hemisphaeriales* Theissen, *Microthyriales* Clements and Shear).

Family **Microthyriacea** [Microthyriaceae] Lindau (in Engler and Prantl, 1897). Parasitic on plants, surfaces of the fruits marked by radiating ridges.

Family **Micropeltidacea** [Micropeltidaceae] Clements and Shear (1931). Family *Hemisphaeriaceae* Theissen (1913), not based on a generic name. Like the foregoing, but the surface of the fruit not radiate or radiate only at the margin.

Family **Trichothyriacea** [Trichothyriaceae] Theissen and Sydow. Parasitic on inophytes, the mycelium a pseudoparenchymatous layer, asci pendant within the fruits from the apparent summit.

b. Fruits elongate, hard, dark, opening by a narrow cleft (suborder HYSTERIINEAE Engler).

Family **Hysteriacea** [Hysteriaceae] Saccardo Sylloge 2: 721 (1883). Parasitic on higher plants or saprophytic.

Family **Graphidiacea** [Graphidiaceae] Clements (1909). An enormous group of lichens or parasites on lichens, largely tropical and chiefly crustose, the openings of the fruits forming dark lines.

c. Fruits not as above, mostly with a roundish area of asci exposed by the irregular or stellate shattering of a superficial layer; if long and narrow, not hard and dark (Suborder PHACIDIINEAE Engler).

Family **Phacidiea** [Phacidieae] Saccardo Sylloge 8: 705 (1889). *Phacidiaceae* Saccardo (1889). Family *Phacidiaceae* Lindau (in Engler and Prantl, 1896). The dark fruits thin and weak laterally and below.

Family **Tryblidacea** [Tryblidaceae] Rehm (in Rabenhorst, 1896). The dark fruits hard and thick laterally and below.

Family **Stictea** [Sticteae] Saccardo Sylloge 8: 647 (1889). *Stictaceae* Saccardo (1889). Family *Stictidaceae* Lindau (1896). Fruits light-colored or white. Higgins (1914) found that the agents of the shot-hole disease of plums and cherries, which, on the basis of non-fruiting stages, have been called *Cylindrosporium Pruni,* produce on fallen leaves ascocarps distinguishable as three species of the genus *Coccomyces* of the present family.

Order 5. **Cupulata** [Cupulati] Fries Syst. Myc. 1: 2 (1821).
Order *Mitrati* Fries l. c.; order *Uterini* Fries op. cit. 1: liii (1832).
Family *Discomycetes* Fries Epicrisis 1 (1836).
Orders *Discomycetes* and *Tuberaceae* Winter in Rabenhorst Kryptog.-Fl. Deutschland 1, Abt. 2: 3 (1887).
Suborders *Helevellineae, Pezizineae,* and *Tuberineae* Engler in Engler and Prantl Nat. Pflanzenfam. I Teil, Abt. 1: v (1897).
Orders *Helevellineae, Pezizineae,* and *Tuberineae* Campbell Univ. Textb. Bot. 166, 167, 168 (1902).
Orders *Pezizales, Discolichenes, Helvellales,* and *Tuberales* Bessey in Univ. Nebraska Studies 7: 299, 300, 303, 304 (1907).

This order includes primarily the cup fungi, the inophytes which produce cup- or disk-shaped fruits bearing a single layer of closely packed asci on the inner or upper surface. There has been much study of some of them, notably of *Pyronema,* by Harper, Dangeard, Claussen, and Brown. The disk-shaped flesh-colored apothecia of *Pyronema,* 1-3 mm. in diameter, are found particularly on damp charcoal. The mycelium produces differentiated multinucleate antheridia and ascogonia, the latter bearing one-celled multinucleate trichogynes. After syngamy, or sometimes without it, but always to the best of our knowledge without any fusion of nuclei, the ascogonia send out branching filaments which become septate in such fashion that the ultimate cells are binucleate. These cells form croziers and produce asci. During the development of the ascogenous hyphae, other hyphae, more slender, grow up from the vegetative mycelium; these produce a disk of undifferentiated cells below the layer of asci, and send up sterile hairs (paraphyses) among them.

Gäumann (1926) divided the families of this group into two series by the presence or absence of a differentiated operculum at the summit of the ascus. The names being put into neuter form, and family Tuberacea being added, the lists are as follows:

Inoperculata: Patellariacea, Dermateacea, Bulgariacea, Cyttariacea, Mollisi-acea, Helotiacea, Geoglossacea, Tuberacea.

Operculata: Rhizinacea, Pyronemacea, Ascobolacea, Pezizacea, Helvellacea.

Along with these, Clements and Shear (1931) list eight families of lichen-formers, some of them very numerous.

Families Pezizacea and Ascobolacea include the ordinary cup fungi. They are mostly saprophytes in soil or on manure, and do not usually produce conidia. *Peziza* was listed by Fries first in order Cupulata; it is the evident standard genus of the order.

Families Dermateacea and Helotiacea include many parasites on plants. One of the Helotiacea is *Sclerotinia cinerea,* the agent of the brown rot of stone fruits. As an active parasite it produces conidia of a type which, if the fruits were unknown, would place it in the genus *Monilia.* These spread the disease rapidly. The killed fruits fall and the organism lives in them as a saprophyte, replacing their tissues with a hard black mass of hyphae, a sclerotium. This survives the winter and in spring sends up stalked white apothecia.

The Helvellacea have been treated as a separate order, but are not sufficiently numerous and distinct to justify this treatment. They are saprophytes in soil, producing large stalked apothecia bearing an extensive layer of asci which is everted and wrinkled. The most familiar genera are *Elvella* and *Morchella.* The fruits are edible, indeed delicious; they should be boiled briefly, then creamed and served on toast. When found in abundance they should be preserved by drying for use throughout the year.

The Tuberacea, the truffles, also usually treated as a distinct order, produce underground fruits which appear to be apothecia distorted and rolled into balls. They are associated with particular species of trees on which the mycelia are believed to live as mycorhizae (Dangeard, 1894). The asci commonly contain reduced numbers of spores. The fruits are prized by gourmands.

The relationships of the Cupulata are a puzzle. *Pyronema* could be interpreted as representing an evolutionary transition from the order Mucedines to this. Certain parasitic cup fungi produce minute apothecia, hard, dark, and nearly closed, suggesting a transition to order Sclerocarpa. Some species, particularly among the parasites and lichen-formers, seem to intergrade with order Phacidialea, and thence again both to Mucedines and Sclerocarpa. The operculate asci which mark a part of the group occur also in other orders. Thus there is among Ascomycetes an appearance of reticulate relationships, such as reputable naturalists of the past supposed to exist in many groups. The appearance is of course illusory; sufficient study of other groups has made it possible to distinguish the resemblances among them which indicate relationship from those which are results of parallel evolution. The study of the Ascomycetes has not yet been carried this far.

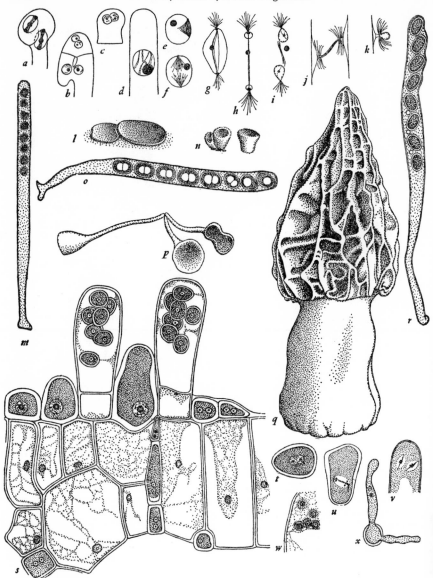

Fig. 26.—Ascomycetes: **a-k**, *Lachnea scutellata* after Brown (1911) x 1,000; **a**, **b**, formation of crozier; **c**, karyogamy; **d**, fusion nucleus; **e-i**, stages of meiosis; **j**, **k**, early stages of free cell formation. **l**, Apothecia x 2, and **m**, ascus x 250, of *Lamprospora leiocarpa*. **n**, Apothecia x 2, and **o**, ascus x 250, of *Aleuria rutilans*. **p**, Apothecia of *Sclerotinia cinerea* x 2. **q**, Fruit x 1, and **r**, ascus x 250, of *Morchella conica*. **s-x**, *Taphrina deformans* after Martin (1940) x 1,000; **s**, growth on surface of an infected leaf; **t**, karyogamy; **u**, mitosis; **v**, homeotypic anaphase in the ascus; **w**, development of ascospores; **x**, germination.

Order 6. **Exoascalea** [Exoascales] Bessey in Univ. Nebraska Studies 7: 305 (1907). Suborder *Protodiscineae* Engler in Engler and Prantl Nat. Pflanzenfam. I Teil, Abt. 1: v (1897), not based on a generic name. Order *Protodiscineae* Campbell Univ. Textb. Bot. 166 (1902). Order *Agyriales* Clements and Shear Gen. Fung. ed. 2: 141 (1931), in part. Ascomycetes parasitic on plants, producing no fruits but a broad layer of asci directly on the mycelium.

The leaves of the hosts of these parasites become swollen and distorted; the diseases recognized by these symptoms are called curly-leaf diseases. The most familiar is the curly-leaf of peaches, caused by *Taphrina (Exoascus) deformans.* Many others are known. The agents of all of these diseases may be regarded as a single family **Exoascacea** [Exoascaceae] Schröter (in Engler and Prantl, 1894), and all are commonly treated as a single genus, *Taphrina* Fries, typified by *T. aurea* on poplar trees; there are differences among them which might well be treated as of generic rank.

Clements and Shear associated the curly-leaf parasites with a collection of saprophytes producing small and undifferentiated disk-like or indefinite fruits, as *Pyronema, Ascocorticium,* and *Agyrium;* and offended against the principles of nomenclature by re-naming the order Agyriales. It is probable that something of the nature of *Agyrium* may represent the transition from order Cupulata to this one.

Martin (1940) described the cytology of *Taphrina deformans.* The mycelium grows between the cells of the host, not penetrating them. It is a dikaryophase mycelium, the cells binucleate, the nuclei dividing concurrently, cell division occurring in such fashion as to separate the daughter nuclei of each pair. In preparation for reproduction, hyphae of short round cells form a single layer between the epidermis and the cuticle of the host. In each cell of these hyphae, the nuclei unite and then divide. The division is mitotic, the fusion nuclei and the daughter nuclei having each eight chromosomes. The cell divides, by a wall parallel to the surface of the leaf, into two. The daughter cell which lies against the tissues of the host dies, and its wall becomes empty; the other cell grows and bursts through the cuticle of the host and becomes an ascus. Its nucleus divides three times; the first two divisions are the meiotic process, and the chromosome number is reduced to four. Cytoplasm accumulates around each of the resulting eight nuclei and is presently cut out by a membrane and a wall. No centrosome is evident at any stage of the process. The spores germinate by sending out buds, as yeasts form buds; sometimes they do this before being discharged from the ascus. So far as Martin could determine, the binucleate condition of the mycelium is established by division of the nucleus of the spore from which it grows.

Order 7. **Sclerocarpa** [Sclerocarpi] Persoon Syst. Meth. Fung. xii (1801). Order *Pyrenomycetes* Fries Syst. Myc. 2:312 (1822); order *Uterini,* suborder *Pyrenomycetes* Fries op. cit. 1: li (1832). Family *Pyrenomycetes* Fries Epicrisis 1 (1836). Order *Pyrenomycetes,* suborders *Hypocreaceae, Sphaeriaceae,* and *Dothideaceae,* Winter in Rabenhorst Kryptog.-Fl. Deutschland 1, Abt. 2: 18, 82, 152, 893 (1887). Suborder *(Unterreihe) Pyrenomycetineae,* sub-suborders *(Unterordnungen) Hypocreales, Dothideales,* and *Sphaeriales* Engler in Engler and Prantl Nat. Pflanzenfam. I Teil, Abt. 1: v, vi (1897).

Order *Pyrenomycetales* Bessey in Univ. Nebraska Studies 7: 295 (1907).
Orders *Hypocreales, Sphaeriales,* and *Dothideales* Gäumann Vergl. Morph.
Pilze 222, 253, 284 (1926).

Ascomycetes producing, from a normal mycelium, perithecia, i. e., small fruits
of the shape of a small globe or flask opening through a single pore, the ostiole.
Sphaeria, which Persoon and Fries listed first under the names which they respectively used, is the evident standard genus; but this genus has become broken up and
lost in the work of subsequent scholars.

This order includes very many species and is by the generality of authority
divided into three. Forms whose perithecia are borne directly on the mycelium,
together with those whose perithecia are borne in or on but distinct from a dark

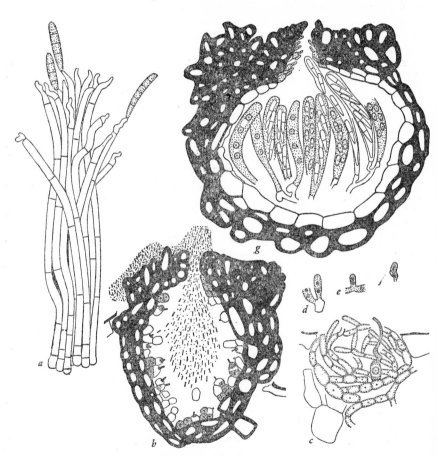

Fig. 27.—*Mycosphaerella personata* after Higgins (1929), x 1,000; **a**, conidiophores and conidia of *Cercospora* type. **b**, longitudinal section of pycnidium;
c, primordium of perithecium with ascogonoium bearing a trichogyne; **d, e**, ascogenous hyphae; **f**, crozier formation; **g**, longitudinal section of mature perithecium.

stroma, are assigned to order Sphaeriales. Forms with perithecia in or on a brightly colored stroma are Hypocreales. Those whose perithecia are cavities with a wall indistinguishable from a dark stroma are Dothideales. These groups are not confidently acceptable as natural: the stromatic Sphaeriales (Wehmeyer, 1926), the Hypocreales, and the Dothideales appear each to include more than one line of descent from Sphaeriales with solitary perithecia.

As a general rule, each perithecium develops in consequence of a separate act of fertilization, of a differentiated ascogonium, either by an antheridium, a spermatium, or otherwise.

Gäumann recognized fourteen families in the present group or groups. To these are to be added a great number of lichen-formers, properly Sphaeriales and Hypocreales, but construed as a single family Verrucariacea; and a smaller number, representing the Dothideales, and called Mycoporacea.

Exmples include the following:

Among Sphaeriales with solitary perithecia, *Mycosphaerella* is a genus of more than one thousand parasites on plants, mostly inconspicuous, causing leaf spots. Their conidia are of various types, *Septoria, Phleospora, Ramularia, Cercospora. Venturia,* another numerous genus, includes *V. inaequalis,* causing apple scab; its conidia are of a type called *Fusicladium.*

Four species of *Neurospora* were discovered by Shear and Dodge (1927) as the fruiting stages of a red mold on bread called *Monilia sitophila.* Genetic study of this genus particularly by Tatum, Beadle, and their associates (Ryan, Beadle, and Tatum, 1943; McClintock, 1945; Beadle and Tatum, 1945; Tatum and Bell, 1946; Mitchell and Houlahan, 1946; Tatum, Barratt, Fries, and Bonner, 1950) has yielded results of the highest theoretical significance. Normal cultures require no other food than minerals, carbohydrate, and a single vitamin, biotin (Butler, Robbins, and Dodge, 1941). Either spontaneously or under violent treatment (with x-rays, ultra-violet radiations, or mustard gas) the cultures give rise to many mutations, behaving as Mendelian recessives, each consisting of the inability to synthesize some one vitamin or amino acid. These observations mean that life in its aspect of metabolism consists of unit chemical processes, each controlled by a specific enzyme, each enzyme being dependent upon a specific area in a specific chromosome.

Among stromatic Sphaeriales, *Glomerella,* with conidial stages identified as *Gloeosporium* or *Colletotrichum,* attacks many plants; *G. cingulata* causes the bitter rot of apples. *Valsa, Diatrype,* and *Diaporthe* are numerous in species. *Endothia parasitica* causes the chestnut blight, destructive in the eastern United States. *Xylaria, Daldinia,* and other genera are saprophytic on wood; the former produces black fruits, club-shaped or branched; the latter, fruits of the form of black knobs which may reach the size of golf balls.

Among Hypocreales, *Nectria cinnabarina* is common as a saprophyte on dead twigs of poplar. It produces small wart-like red stromata which bear first conidia, then perithecia. *Claviceps purpurea* causes a disease of rye; it produces conidia of various types, and converts the grains of rye into sclerotia. These bodies are called ergot; they are extremely poisonous, sometimes dangerously so, because they may be ground with the grain. They are used in medicine. After lying in the earth through the winter, the sclerotia send up fruits of the form of a stalk bearing a knob consisting of radiating perithecia. *Cordyceps* kills subterranean larvae or pupae of insects and then sends up a stalk bearing an elongate head of many perithecia.

The Dothideales include *Plowrightia morbosa,* the agent of the black knot of plums. Diseased twigs become swollen and covered with a black stroma which bears, according to the season, conidia of various types or else perithecia.

Order 8. **Laboulbenialea** [Laboulbeniales] Engler Syllab. ed. 3: 42 (1903).
> Order *Laboulbeniaceae* Thaxter, the name (ascribed to Peyritsch) preoccupied
>   by family Laboulbeniaceae Berlese in Saccardo Sylloge 8: 909 (1889).
> Suborder *Laboulbeniineae* Engler in Engler and Prantl Nat. Pflanzenfam.
>   I Teil, Abt. 1: vi (1897).
> Class *Laboulbeniomycetes* Engler Syllab. l. c.
> Class *Laboulbenieae* Schaffner in Ohio Naturalist 9: 450 (1909).

Parasites on insects, the mycelium scant or reduced to a single cell, producing antheridia which discharge spermatia into the air and small numbers of perithecia.

These organisms have the appearance of exceptional setae on their hosts, which are not usually seriously injured by them. They were first mentioned in a note by the entomologist Rouget, 1850; Montagne and Robin, in Robin's book on parasitic plants, 1853, gave the first names, *Laboulbenia Rougetii* and *L. Guerinii,* the generic name honoring the entomologist Laboulbène. Only a few scholars, notably Thaxter (1896, 1908, 1924, 1926, 1931) have given much attention to this group; they have distinguished well over a thousand species, forming three families and about fifty genera.

Many Laboulbenialea occur as two forms, male and hermaphrodite. A male individual produces a series of flask-shaped antheridia, each of which discharges into the air, one at a time, a series of globular naked sperms. A hermaphrodite individual produces first a series of antheridia as described and then one or more perithecia. A perithecium consists of a wall, of a definite number of cells produced in definite order and pattern, surrounding an egg which bears a trichogyne; the trichogyne protrudes from the perithecium and receives the sperms. The zygote gives rise to a fascicle of asci which crowd aside and destroy the inner cells of the wall and discharge the ascospores (usually eight in the ascus, and divided into two cells) through the ostiole.

Those who would link the Ascomycetes with the red algae entertain the hypothesis that the Laboulbenialea represent the transition. This hypothesis is surely mistaken. The Laboulbenialea are a highly specialized group, not a link between others. They appear to have evolved from Sphaeriales with solitary perithecia.

Class 3. HYPHOMYCETES Fries

Classes Hyphomycetes and *Coniomycetes* Fries Syst. Myc. 3: 261, 455 (1832).
Families *Hyphomycetes* and *Coniomycetes* Fries Epicrisis 1 (1836).
Fungi imperfecti or *Deuteromycetes* Auctt.
Inophyta of which the structures involved in sexual reproduction are unknown.

It has been noted that a particular genus of Ascomycetes may produce conidia of more types than one, as *Sclerotinia* produces types called *Monilia* and *Botrytis,* and *Glomerella* produces types called *Gloeosporium* and *Colletotrichum.* The same type may be produced by many genera; the *Monilia* type recurs in *Neurospora,* which does not belong to the same order as *Sclerotinia.* Collecting naturalists, and plant pathologists in the pursuit of their duties, are constantly encountering conidial stages whose assignment to an order of Ascomycetes is impossible. It is an obvious practical necessity that a register of these observations be kept. The register is provided by the present group, one which is named, defined, and assigned to the category of classes, and divided into named orders, families, and genera under which specimens may be identified as of species old or new. Class, orders, families, and genera are known not to -be valid taxonomic groups; many of the ostensible species are known, and most of the rest are believed, to be stages of organisms which would in other stages have other names. Almost all of them are Ascomycetes; Zygomycetes and Basidiomycetes do not usually occur in unidentifiable stages.

The ascus-bearing stages are constantly being discovered. When this happens, the species is re-named in its proper place among Ascomycetes. Theoretically, it loses its place in the list of imperfect fungi; practically, it retains it, because the next collector or plant pathologst is most likely to try to find it there.

The system of Hyphomycetes is as follows:

Order 1. **Phomatalea** [Phomatales] Clements Gen. Fung. 121 (1909).
Sphaeropsideae Saccardo Sylloge 8: xvi (1889).
Order *Sphaeropsidales* Engler in Engler and Prantl Nat. Pflanzenfam. I Teil
Abt. 1**: v (1900), not based on a generic name.
Order *Phomales* Clements and Shear Gen. Fung. ed. 2: 175 (1931).
Producing pycnidia. The four families correspond with as many groups of Ascomycetes.

Family 1. **Phomatacea** [Phomataceae] Clements Gen. Fung. 121 (1909). Family *Sphaerioideae* or *Sphaerioidaceae* Saccardo; but *Sphaeria* belongs to order Sclerocarpa. Family *Phomaceae* Clements and Shear (1931). Pycnidia hard and black as in Sphaeriales and Dothideales. *Phoma, Ascochyta, Diplodia, Septoria,* each of many species.

Family 2. **Zythiacea** [Zythiaceae] Clements Gen. Fung. 128 (1909). Family *Nectrioideae* or *Nectrioidaceae* Saccardo; but *Nectria* belongs to order Sclerocarpa. Pycnidia in brightly colored stromata as of Hysteriales.

Family 3. **Leptostromatacea** [Leptostromataceae] Saccardo Sylloge 3: 625 (1884). Pycnidia in shield-like stromata, like the fruits of Microthyriacea.

Family 4. **Discellacea** [Discellaceae] Clements and Shear Gen. Fung. ed. 2: 192 (1931). Family *Excipulaceae* Saccardo; but *Excipula* is a cup fungus. Pycnidia wide open like the fruits of Phacidiea.

Order 2. **Melanconialea** [Melanconiales] Engler in Engler and Prantl Nat. Pflanzenfam. I Teil, Abt. 1**: v (1900).
The conidia borne on a stroma but not in pycnidia.
Family **Melanconiacea** [Melanconiaceae] (Saccardo, without category) Lindau in Engler and Prantl op. cit. 398, the single very numerous family: *Gloeosporium; Coryneum, C. Beijerinckii,* the shot-hole of almonds; *Pestallozia.*

Order 3. **Nematothecia** [Nematothecii] Persoon Synops. Meth. Fung. xix (1801). Orders *Dematiei, Sepedoniei, Tubercularini,* and *Stilbosporei* Fries Syst. Myc. Order *Hyphomycetes* (Fries) Auctt. Order *Moniliales* Clements Gen. Fung. 138 (1909). Conidia directly on the mycelium, or none.
Family 1. **Tuberculariea** [Tubercularieae] Saccardo Sylloge 4: 635 (1886). *Tuberculariaceae* Saccardo (1889). Family *Tuberculariaceae* Lindau (1900). Scarcely distinct from Melanconiacea, the conidia on a mass of interwoven hyphae

less compact than a stroma. *Fusarium,* an enormous number of species producing as conidia crescent-shaped rows of cells. Snyder and Hansen (1941, 1945) find that the fruiting stages are species of *Hypomyces, Nectria, Gibberella,* or *Calonectria,* all Hypocreales.

Family 2. **Stilbellacea** [Stilbellaceae] Bessey Morph. and Tax. Fungi 584 (1950). Family *Stilbeae* Saccardo Sylloge 4: 563 (1886). *Stilbaceae* Saccardo (1889). Family *Stilbaceae* Lindau (1900); Bessey observed that the type of the genus *Stilbum* does not belong to this family. Mostly molds producing coremia.

Family 3. **Dematiea** [Dematieae] Saccardo Sylloge 4: 235 (1886). *Dematiaceae* Saccardo (1889). Family *Dematiaceae* Lindau (1900). Dark-colored parasites, as *Helminthosporium, Cladosporium,* and *Cercospora,* or molds, as *Alternaria.*

Family 4. **Moniliacea** [Moniliaceae] Clements Gen. Fung. 138 (1909). *Mucedineae* Persoon, family *Mucedineae* or *Mucedinaceae* Saccardo, not based on a generic name. White or brightly colored parasites or molds, as *Oidium,* with colorless spores in chains, *Monilia, Botrytis,* etc. The parasites on animals which have been referred to *Monilia* are currently called *Candida.*

Family (?) 5. Sterile mycelia. Many mycorhizae must be left here. *Rhizoctonia,* dark net-like masses of hyphae occurring as parasites or saprophytes. *Trichophyton,* parasitic on the skins of man and animals, causing ringworm, athlete's foot, etc.

Class 4. BASIDIOMYCETES (Sachs ex Bennett and Thistleton-Dyer) Winter

Order *Basidiosporeae* and subordinate group *Basidiomycetae* Cohn in Hedwigia 11: 17 (1872).

Basidiomyceten Sachs Lehrb. Bot. ed. 4: 249 (1874).

BASIDIOMYCETES Bennett and Thistleton-Dyer in Sachs Textb. Bot. English ed. 847 (1875).

Class BASIDIOMYCETES Winter in Rabenhorst Kryptog.-Fl. Deutschland 1, Abt. 1: 72 (1884).

Classes *Teliosporeae* and *Basidiosporeae* Bessey in Univ. Nebraska Studies 7: 305, 306 (1907).

Classes *Teliosporeae* and *Basidiomycetae* Schaffner in Ohio Naturalist 9: 450 (1909).

Inophyta which produce, as a feature of the sexual cycle, conidiophores called basidia, each producing typically four conidia called basidiospores.

Germinating basidiospores give rise to mycelia of cells with solitary haploid nuclei. Syngamy occurs among cells of these mycelia, usually simply by contact of undifferentiated cells; the rusts produce differentiated sperms in spermagonia resembling the pycnidia of Ascomycetes. In some species any haploid hypha may conjugate with any; in some there are two mating types, and in some four. Raper (1953) has studied the interesting genetics of the mating types.

The cell produced by syngamy remains undifferentiated, but gives rise, by concurrent division of its nuclei, to a dikaryote mycelium. The nuclei are minute, and mitosis has rarely been seen. The nuclear divisions are often followed by a peculiar manner of cell division, comparable to the crozier formation of Ascomycetes, and producing structures called clamp connections.

Either the original haploid mycelium or the dikaryophase may produce conidia without nuclear change. Such reproduction is familiar among the rusts, rather unfamiliar among other Basidiomycetes.

Only the dikaryophase produces the specialized conidiophores called basidia, which are regularly the seat of karyogamy and meiosis. There is a considerable variety of types of basidia. Van Tieghem (1893) originated the terminology applicable to these; Martin (1938) has attempted to refine it, and Linder (1940) to simplify it.

Frequently, the seat of meiosis is a thick-walled resting spore or an otherwise differentiated cell called a probasidium, upon which the proper basidium develops, after meiosis, as an outgrowth. A basidium arising in this fashion is commonly elongate and divided into four cells each of which produces a basidiospore. Such a hypha-like basidium may be called a promycelium or a phragmobasidium; the latter term is applicable also to an elongate four-celled basidium which does not arise from a probasidium. In a few Basidiomycetes, the basidium is divided into four cells by longitudinal walls; such basidia are called cruciate basidia. In the familiar Basidiomycetes the basidium does not become divided by walls and is called a holobasidium or autobasidium. Gäumann (1926) distinguished two types of holobasidia: the stichobasidium, in which the spindles of the dividing nuclei lie at various levels and in various directions, and which frequently produces more than four nuclei; and the chiastobasidium, in which the spindles lie transversely near the summit, and which regularly produces just four nuclei. Dodge, translating Gäumann (1928), denies much importance to this distinction.

The meiotic divisions have repeatedly been studied. Apparent centrosomes have been seen at the poles of the spindles (Lewis, 1906; Lander, 1933), but not by most microtechnical methods (Savile, 1939; Ritchie, 1941). The chromosomes gather as usual at the middle of the spindle and divide. The nuclear membrane becomes indistinct, but the nuclear sap remains distinct from the cytoplasm nearly until the completion of division; it then disappears, leaving the groups of daughter chromosomes connected by a spindle of the appearance of a dark streak in the cytoplasm.

Observed haploid chromosome numbers include the following:

Coleosporium, fide Moreau (1914)	2
Coleosporium Vernoniae, fide Olive (1949)	8
Coprinus, fide Vokes (1931)	4
Eocronartium, fide Fitzpatrick (1918)	4
Exidia, fide Whelden (1935)	4
Gymnosporangium, fide Stevens (1930)	2
Melampsora, fide Savile (1939)	4
Myxomycidium flavum, fide Martin (1938)	8
Puccinia, fide Savile (1939)	4
Transchelia, fide Savile (1939)	4
Russula, fide Ritchie (1941)	4
Scleroderma, fide Lander (1933)	2
Uromyces, fide Savile (1939)	4

Savile suggests that some at least of the reports of a chromosome number of 2 may have resulted from misinterpreted observations of one pair of choromosomes behind another.

Normally, only the two meiotic divisions, producing four nuclei, occur in the basidium; exceptionally, there are further, mitotic, divisions, resulting in more than four spores on the basidium. The basidiospores are usually borne on slender stalks called sterigmata. Sterigmata and spores are formed by evagination of the wall of the basidium; the nuclei migrate through the sterigmata into the spores.

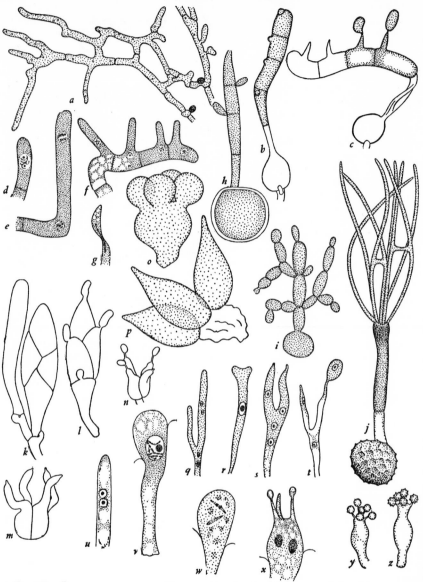

Fig. 28.—Basidiomycetes: **a**, Two germinating basidiospores of *Agaricus campes-tris* produce mycelia which anastomose freely, the cells becoming plurinucleate, after Hein (1930), x 500.   **b, c**, Young and older basidia of *Cystobasidium sebaceum,* after Martin (1939).   **d-g**, *Eocronartium muscicola* after Fitzpatrick (1918); **d**, fus-ion nucleus; **e**, homeotypic division in the basidium; **f**, four-celled basidium; **g**, pro-duction of basidiospore.   **h, i, j**, Basidia of *Ustilago Heufleri, U. Hordei,* and *Tille-*

(Continued bottom p. 145)

Most basidia discharge the spores actively, to a distance of a fraction of a milli-meter. Buller (1929) observed that just before a spore is cast off a minute droplet of liquid appears at the summit of the sterigma. This occurs in precisely the same fashion in mushrooms, rusts, certain smuts, and the yeast-like organism *Sporobolo-myces.* Buller inferred that the force which discharges the spore is surface tension in the droplet. The fruits of Basidiomycetes are evidently adapted to the feebleness of the mechanism by which the spores are discharged. If the fruits are cup-like, they open laterally or downward. The basidia of mushrooms stand horizontally on gills which are commonly less than one millimeter apart, allowing the spores to fall from between them without touching them.

The groups of Ascomycetes and Basidiomycetes are evidently related. Morels and mushrooms, truffles and puffballs, taste alike. The technical scholar will be con-vinced that the groups are related by the occurrence in both of a dikaryophase stage, a character too strongly in contrast with those of the generality of organisms to be a probable product of parallel evolution. Gäumann quotes an old opinion of Vuillemin (1893), "qu'une baside est un asque dont chaque cellule-fille avant de passer à l'état de spore, fait saillie au dehors et se transforme en une sorte de conidie pour mieux s'adapter au transport par la vent." In dealing with the Zygomycetes, Gäumann emphasized the apparent evolution of conidia from endospores by evagina-tion of the walls of the sporangia. Largely, as it seems, by Gäumann's influence, Vuillemin's hypothesis has become generally accepted.

Gäumann was disposed to derive the Basidiomycetes from something like *Asco-corticium,* and began his account of several of the groups of Basidiomycetes with forms having scant flat fruits, or having basidia which spring directly from the substratum or host. Linder (1940) suggested a derivation from Cupulata or Sclero-carpa having operculate asci. He took note that many such asci open by producing a vescicle, bounded by the stretched inner wall of the ascus, into which the asco-spores pass. This led to the conclusion that the Basidiomycetes producing probasidia are the lowest, and to this extent his reasoning appears cogent. He went on to identify the rusts as the lowest Basidiomycetes, which seems far-fetched, the rusts being distinctly a specialized group.

The generally accepted groups of Basidiomycetes are those which were set forth by Engler (1897, 1900), as follows:

Subclass HEMIBASIDII, having basidia bearing indefinite numbers of spores; the smuts.

Subclass EUBASIDII, the basidia bearing definite numbers of spores.

Order (*Reihe*) PROTOBASIDIOMYCETES, the basidia divided into cells.

Suborder (*Unterreihe* or *Ordnung*) AURICULARIINEAE, the basidia divided by transverse walls.

Sub-suborder (*Unterordnung*) UREDINALES, the rusts.

---

tia Tritici, after Sartoris (1924). **k, l,** Basidia of *Patouillardina cinerea* after Martin (1935). **m,** Basidium of *Sebacina sublilacina* after Martin (1934). **n,** Basidium of *Protodontia Uda* after Martin (1932). **o, p,** younger and older basidia of *Tulas-nella phaerospora,* after Martin (1939). **q-t,** Development of the basidium of *Guepinia Spathularia,* after Bodman (1938). **u-x,** *Russula emetica* after Ritchie (1941); binucleate primordium of basidium, fusion nucleus, homeotypic division, development of basidiospores. **y, z,** Basidia of *Lycogalopsis Solmsii* after Martin (1939). x 1,000 except as noted.

Sub-suborder Auriculariales.

Suborder Tremellineae, the basidia divided by longitudinal walls.
(At this point should appear *Reihe* Autobasidiomycetes, to include eight *Unterreihen* of ordinary Basidiomycetes. The name Autobasidiomycetes does not appear in the table of contents, the text, or the index of the *Natürlichen Pflanzenfamilien;* it was published in Engler's Syllabus, 1892).

Rearranging these groups according to current opinion, and suppressing the subsidiary categories, one arrives at the following system of orders:

1. Producing probasidia or transversely divided
basidia, usually both.
2. Probasidia, if formed, terminal on the
hyphae.
3. Mostly saprophytic and producing
gelatinous fruits..................Order 1. Protobasidiomycetes.
3. Parasitic, mostly not producing
fruits; the rusts.................. Order 2. Hypodermia.
2. Probasidia produced by rounding up and
deposition of thick walls by the generality of the cells of the mycelium........Order 3. Ustilaginea.
1. Without probasidia, the basidia divided longitudinally.............................. Order 4. Tremellina.
1. Without probasidia, the basidia undivided.
2. Fruits gelatinous, basidia producing only
two spores on stout sterigmata...........Order 5. Dacryomycetalea.
2. Not as above.
3. Basidia in a layer which forms without protection or becomes exposed...Order 6. Fungi.
3. Basidia formed in closed fruits
which do not open to expose them
as a single layer..................Order 7. Dermatocarpa.

Order 1. **Protobasidiomycetes** Engler in Engler and Prantl Nat. Pflanzenfam. I Teil, Abt 1**: iii (1900).
Suborder *Auriculariineae* and sub-suborder *Auriculariales* Engler l. c.
Order *Auricularineae* Campbell Univ. Textb. Bot. 175 (1902).
Order *Auriculariales* Bessey in Univ. Nebraska Studies 7: 309 (1907).
Basidiomycetes mostly producing probasidia, the basidia divided by transverse walls, mostly saprophytic and producing gelatinous fruits.

This order includes the family **Auriculariacea** [Auriculariaceae] Lindau in Engler and Prantl Nat. Pflanzenfam. I Teil, Abt. 1**: 83 (1900), from which two or three others have been segregated; about fifteen genera and about 125 species.

Martin (1943) has discussed the name of the genus *Auricularia* and of its type species. The organism in question is surely the Jew's ear, *Tremella Auricula* L.; the genus *Auricularia* Bulliard 1795 can have nothing else as a type. The right name of the species is *Auricularia Auricula* (L.) Underwood 1902. It is a saprophyte on logs and sticks, producing flattened brown gelatinous fruits a few centimeters in diameter, vaguely resembling human ears. There are no probasidia. Hyphae growing toward the surfaces of the fruits produce a palisade of elongate basidia. Each basidium becomes divided by transverse walls into four cells, and each of these sends out to the surface an elongate sterigma which bears a curved basidiospore. The organism produces also conidia, either from the mycelium, the fruits, or directly from the basidiospores.

A series of unfamiliar other genera, *Platygloea, Cystobasidium, Septobasidium,* etc., have been studied notably by Martin (1934, 1937, 1939, 1942). *Jola* and *Eocronartium* are parasites on mosses. All of these genera produce probasidia, from which four-celled phragmobasidia arise, as a layer near the surfaces of the fruits. Most of them produce also conidia.

Order 2. **Hypodermia** [Hypodermii] Fries Syst. Myc. 3: 460 (1832).
> *Urédinées* Brongniart in Bory de Saint Vincent Dict. Class. Hist. Nat. 16: 471 (1830).
> Order *Uredineae* Winter in Rabenhorst Kryptog.-Fl. Deutschland 1, Abt. 1: 74 (1884).
> Sub-suborder *Uredinales* Engler in Engler and Prantl Nat. Pflanzenfam. I Teil, Abt. 1**: iii (1900).
> Order *Uredinales* Bessey in Univ. Nebraska Studies 7: 306 (1907).
> Order *Pucciniales* Clements and Shear Gen. Fung. ed. 2: 147 (1931).

The rusts: parasitic Basidiomycetes, the haploid and dikaryote mycelia usually attacking different hosts; the dikaryote mycelium producing probasidia, these not usually compacted into fruits, usually heavily walled and serving as resting spores, becoming or giving rise to four-celled phragmobasidia.

The typical reproductive structure of the haploid stage is the aecium, a cup-shaped structure which releases spores called aeciospores; this stage is accordingly called the aecial stage, and its host the aecial host. In addition to aecia, this stage usually produces pycnidia or spermagonia. The typical reproductive structures of the dikaryote mycelium are clusters (telia) of spores called teliospores or teleutospores; this stage, then, is the telial stage, and its host the telial host. The telial stage usually produces, beside the teliospores, others called uredospores. The teliospore, or rather (since the teliospore commonly consists of two or more cells) each cell of the teliospore, is a probasidium, producing a promycelium which bears four basidiospores. These statements mean that a normal rust produces spores of five kinds. Rusts producing different kinds of spores were formerly supposed to be different genera; such were the *Aecidium, Uredo,* and *Puccinia* of Persoon, who, however, remarked of *Uredo linearis,* "vereor, ne junior plantula Pucciniae graminis modo sit." De Bary first proved that *Aecidium Berberis* is yet another stage of *Puccinia graminis.*

The dikaryophase is initiated, of course, by syngamy among cells of the aecial stage. In *Phragmidium violaceum,* Blackman observed this to take place between different cells of the same hypha. Christman (1905) and Moreau (1914), studying other species of *Phragmidium,* observed fusion to take place between tips of different hyphae. Craigie, 1927, showed that *Puccinia graminis* occurs in two mating types, and that the fertilizing elements are pycniospores or spermatia. De Bary (1884) had suggested that this is the truth; his suggestion waited some forty years to be confirmed. Allen (1930) has described much of the detail. The pycniospores are carried out of the pycnidium in exuding fluid, and are carried by insects; they make protoplasmic connection with paraphyses growing from pycnidia of the opposte mating type. The binucleate uredospores arise from a dikaryote mycelium, but the cup-shaped wall of the aecium is produced by the haploid mycelium.

The first-formed reproductive structures of the dikaryote mycelium on the telial host are usually uredospores, which remain binucleate and have the function of spreading the infection of the telial host.

Teliospores may be compacted into palisade-like masses which break through the epidermis of the host; the masses may be gelatinous and yellow, like fruits of Auriculariacea. In other genera, the teliospores are gathered into hard, microscopically stout columns, and in yet others they break through the epidermis in masses not compacted, each teliospore on a separate stalk. The teliospores of *Phragmidium* are chains of several probasidia; those of the many species of *Puccinia* are chains reduced to two probasidia; those of *Ravenelia* are globular clusters of probasidia. Almost always, the teliospores are thick-walled; outside of the tropics, they have the function of overwintering. Each probasidium contains two nuclei. These unite as a preliminary to germination: this was first observed by Sappin-Trouffy (in Dangeard and Sappin-Trouffy, 1893). Thereafter the probasidium gives rise to the four-celled promycelium.

The life cycle thus described is not perfectly stable. Aeciospores, uredospores, and young teliospores are alike dikaryote, and are genetically identical. Spores of the structure and behavior of any of these types may be produced by processes which normally lead to another. Thus in *Puccinia Malvacearum,* the hollyhock rust, syngamy leads directly to the production of teliospores on the host of the haploid mycelium; spermagonia, aecia, and uredosori are not produced.

Four families of rusts may be recognized (various authorities make fewer or more). There are about five thousand species.

Family 1. **Melampsoracea** [Melampsoraceae] Dietel in Engler and Prantl Nat. Pflanzenfam. I Teil, Abt. 1**: 38 (1900). Teliospores forming a single compact layer and germinating by producing promycelia. The aecial stages are mostly on conifers. Some have telial stages on ferns, and Faull (1929) regards these as most primitive; others attack a variety of flowering plants.

Family 2. **Coleosporiacea** [Coleosporiaceae] Auctt. The teliospores themselves becoming basidia by transverse division. In some examples, as *Gallowaya,* they are thin-walled.

Family 3. **Cronartiacea** [Cronartiaceae] Auctt. The teliospores compacted into columns. *Cronartium,* with aecial stages on pines; *C. ribicola,* the important white pine blister rust, its telial stage on gooseberries and currants.

Family 4. **Uredinacea** [Uredinaceae] Cohn in Hedwigia 11: 17 (1872). Family *Pucciniaceae* Dietel op. cit. 48. The bulk of the rusts, producing teliospores on individual stalks. *Hemileia vastatrix,* the coffee rust; *Phragmidium* spp., autoecious (attacking a single host) on Rosaceae; *Gymnosporangium,* the aecial stage on junipers, the telial (with no uredospores) on plants of the apple tribe; *Puccinia,* a great number of species. The races which attack barberry and grasses are all called *Puccinia graminis*; but there are morphologically distinguishable strains on wheat, rye, oats, timothy, *Agrostis,* and blue grass. Leading an active sexual life and capable of mutation, these strains are subdivisible into large numbers of races distinguished by capacity to attack different races of hosts. Given a specimen of rust on wheat, one determines by trial upon seedlings of ten varieties of wheat to which of 189 numbered races it belongs. The races occur characteristically in different wheat-growing areas. If one breeds wheat for resistance to rust, there is good probability of success against the races occurring locally; but some other race is likely to move into the area (Stakman, 1947).

Order 3. **Ustilaginea** [Ustilagineae] (Tulasne and Tulasne) Winter in Rabenhorst Kryptog.-Fl. Deutschland 1, Abt. 1: 73 (1884).
USTILAGINEAE Tulasne and Tulasne in Ann. Sci. Nat. Bot. sér. 3, 7: 73 (1847).
Subclass *Hemibasidii* Engler Syllab. 26 (1892).
Order *Ustilaginales* Bessey in Univ. Nebraska Studies 7: 306 (1907).

The smuts: parasitic Basidiomycetes completing their development on a single host, the dikaryophase mycelium breaking up into thick-walled black spores, these functioning as probasidia, the basidia usually bearing more than four basidiospores.

In the apparently more primitive smuts, the promycelia are four-celled phragmobasidia. The haploid nuclei divide before passing into the basidiospores, with the effect that each cell of the promycelium buds off a series of basidiospores. In other examples the promycelia do not become divided by walls, but are of the character of holobasidia. The basidiospores of some species are capable of budding like yeasts. In some species, they are capable of syngamy with each other, and in some they send out hyphae which bear conidia of characteristic form. In many species, syngamy has not been observed, but is believed to take place between vegetative hyphae. Hybridization, and mutation, particularly in the capacity to attack particular races of hosts, take place freely in smuts, which are accordingly well fitted to cope with the efforts of plant breeders.

The smuts are believed to be somewhat degenerate descendants of the rusts.

There are two families, about thirty genera, about six hundred species.

Family 1. **Ustilaginacea** [Ustilaginaceae] Cohn in Hedwigia 11: 17 (1872). The basidia divided by transverse walls. *Ustilago*, on grasses and other plants.

Family 2. **Tilletiacea** [Tilletiaceae] Dietel in Engler and Prantl Nat. Pflanzenfam. I Teil, Abt. 1**: 15 (1900). The basidia not divided by walls. *Tilletia*, on grains, etc. *Tuburcinia, Doassansia,* the resting spores produced in globular masses.

Order 4. **Tremellina** [Tremellinae] Fries Syst. Myc. 1: 2 (1821); 2: 207 (1822).
Order *Tremellinei* Fries Hymen. Eur. 1 (1874).
Order *Tremellineae* Winter in Rabenhorst Kryptog.-Fl. Deutschland 1, Abt. 1: 74 (1884).
Suborder *Tremellineae* Engler in Engler and Prantl Nat. Pflanzenfam. I Teil, Abt. 1**: iii (1900).
Order *Tremellales* Bessey in Univ. Nebraska Studies 7: 309 (1907).
Order *Tulasnellales* Gäumann Vergl. Morph. Pilze 487 (1926).

Saprophytic Basidiomycetes producing gelatinous fruits bearing a layer of basidia which typically become divided into four cells by longitudinal walls. Each cell produces a long stout sterigma which reaches the surface of the fruit and bears a spore. The mycelia, the young fruits, or the basidiospores may bear conidia.

The number of species is perhaps one hundred. Nearly all belong to family **Tremellacea** [Tremellaceae] Cohn in Hedwigia 11: 17 (1872). Martin (1935, 1937, 1939) has given much study to this group. It is clearly related to the Protobasidiomycetes; *Patouillardina*, having basidia divided by oblique walls, is clearly transitional. *Tremella, Sebacina, Tremellodendron, Hyaloria.*

Tulasnella differs from the generality of Tremellina in producing holobasidia of a peculiar type, with bulbous sterigmata (Lindau interpreted the sterigmata as basidiospores borne without sterigmata and not released, but producing conidia; it may be that this interpretation is more sound than the obvious one). It is supposed that the holobasidia of this genus are derived from the cruciate basidia of proper Tremellina by a line of descent separate from those which have produced the holobasidia of other groups. By leaving *Tulasnella* in order Tremellina, we spare ourselves the recognition of one more insignificant order.

Order 5. **Dacryomycetalea** [Dacryomycetales] Gäumann Vergl. Morph. Pilze 490 (1926).

Suborder *Dacryomycetineae* Engler in Engler and Prantl Nat. Pflanzenfam. I Teil, Abt. 1**: iv (1900).

Saprophytic Basidiomycetes producing small gelatinous fruits bearing holobasidia in which two of the nuclei produced by meiosis undergo degeneration, while two pass into the basidiospores by way of stout sterigmata which give the basidium the form of a Y. Conidia are produced either from the mycelium, from the young fruits, or from the basidiospores.

There is a single family **Dacryomycetacea** [Dacryomycetaceae] Hennings in Engler and Prantl Nat. Pflanzenfam. I Teil, Abt. 1**: 96 (1900). *Dacryomyces, Dacryomitra, Guepinia.* Bodman (1938) observed the details of the cytological processes in the basidia.

This insignificant order, like *Tulasnella* and the two great orders next to be considered, is evidently derived from Protobasidiomycetes, through Tremellina, by loss of septa in the basidia; the peculiarities of its basidia suggest an independent origin.

Order 6. **Fungi** L. Sp. Pl. 1171 (1753).

Order *Hymenothecii* Persoon Syst. Meth. Fung. xvi (1801).

Class *Hymenomycetes* and orders *Pileati* and *Clavati* Fries Syst. Myc. 1: 1, 2 (1821).

Family *Hymenomycetes* Fries Espicrisis 1 (1836).

Family AGARICACEAE Cohn in Hedwigia 11: 17 (1872).

Order *Hymenomycetes* Winter in Rabenhorst Kryptog.-Fl. Deutschland 1, Abt. 1: 74 (1884).

Suborders *Exobasidiineae* and *Hymenomycetineae* Engler in Engler and Prantl Nat. Pflanzenfam. I Teil, Abt. 1**: iv (1900).

Orders *Hymenomycetales* and *Exobasidiales* Bessey in Univ. Nebraska Studies 7: 307, 308 (1907).

Order *Agaricales* Clements Gen. Fung. 102 (1909).

Orders *Cantharellales, Polyporales,* and *Agaricales* Gäumann Vergl. Morph. Pilze 495, 503, 519 (1926).

Basidiomycetes producing holobasidia in a layer which is or becomes exposed to the air, usually on fruits which are woody, leathery, or fleshy, rather than waxy or gelatinous.

The layer of basidia is called the hymenium. In the lowest members of the group, the hymenium is formed directly on the mycelium, on the surface of the host or substratum; in higher examples, it is formed on the surface of more or less complicated fruits; in the highest, it is formed in closed fruits which open to expose it. The area of the hymenium, and the number of basidia it can bear, is increased when it is not smooth, but thrown into teeth, ridges, plates, or other projections. Families have been distinguished chiefly on the basis of the form of the hymenium. The system is not reliably entirely natural; Overholts (1929) pointed out various microscopic details which promise to contribute to a more natural system. Among these are cystidia, swollen cells imbedded in the hymenium and projecting from it; in some examples at least, they are sterile basidia and serve to hold apart the ridges bearing the hymenium. Other microscopic features are setae, similar to cystidia but hard, dark, and pointed; slender hairs called paraphyses; latex ducts; and crystalline inclusions.

There are some fifteen thousand species. The following families are for the most part the conventionally accepted ones.

Family 1. **Exobasidiacea** [Exobasidiaceae] Hennings in Engler and Prantl Nat. Pflanzenfam. I Teil, Abt. 1**: 103 (1900). The basidia directly on the mycelium. A small group, mostly parasitic on plants. *Exobasidium.*

Family 2. **Thelephoracea** [Thelephoraceae] (Saccardo) Hennings (1900). Order *Thelephorei* Fries Hymen. Eur. 1 (1874). Family *Thelephorei* Winter (1884). *Thelephoraceae* Saccardo Sylloge 8: xiii (1889). Fruits of various form, gelatinous, fleshy or leathery, the hymenium covering the surface generally except where it faces upward. *Corticium,* saprophytic, the fruit a mere appressed layer; *Stereum,* leathery shelf-like extensions from decaying sticks and logs: these genera seem to lead into family Polyporacea. *Cora,* a tropical variant of *Stereum,* is the only lichen-forming basidiomycete. *Thelephora, Craterellus,* the fruits club-, funnel-, or cup-like.

Family 3. **Clavariacea** [Clavariaceae] (Saccardo) Hennings (1900). Order *Clavariei* Fries (1874). Family *Clavariei* Winter (1884). *Clavariaceae* Saccardo (1889). Fruits fleshy, club-like or branched; stag-horn fungi. *Clavaria,* generally edible.

Family 4. **Hydnacea** [Hydnaceae] (Saccardo) Hennings (1900). Order *Hydnei* Fries (1874). Family *Hydnei* Winter (1884). *Hydnaceae* Saccardo (1889). Hymenium on the surface of downward-pointing teeth. Fruits assigned to the genus *Hydnum* may be massive or variously branched or mushroom-shaped, leathery or fleshy; the fleshy examples are edible. Fruits of *Irpex* are little leathery brackets projecting from sticks and logs, distinguished from *Stereum* or *Polystictus* by the masses of fine teeth projecting below.

Family 5. **Polyporacea** [Polyporaceae] (Saccardo) Hennings (1900). Order *Polyporei* Fries (1874). Family *Polyporei* Winter (1884). *Polyporaceae* Saccardo (1889). The hymenium lining vertical tubes open below. These are mostly woody or leathery shelf fungi, mostly saprophytic on wood, numerous and varied in detail. Cooke (1940) recognized forty-six genera in North America. *Polyporus, Fomes, Polystictus.* In *Daedalea,* the pores are not cylinders but slits; this genus leads into *Lenzites,* in which the hymenium is borne on radiating plates, and which is conventionally stationed in Agaricacea. *Boletus* has stout fleshy mushroom-shaped fruits, yellow to brown, turning green when bruised. These fruits are unattractive, but some species are eaten; others are supposed to be poisonous.

Family 6. **Agaricacea** [Agaricaceae] Cohn in Hedwigia 11: 17 (1872). Order *Agaricini* Fries (1874). Family *Agaricini* Winter (1884). The hymenium on vertical plates, radiating from a center, called gills.

These are the Fungi whose fruits are called mushrooms or toadstools. The fruits are mostly mushroom-shaped, sometimes shelf-like; the texture is usually fleshy, varying to leathery on the one hand, and on the other to deliquescent, i.e., becoming converted after maturity into black fluid. There has been much study of the development of the fruits (Levine (1922) and Hein (1930) give extensive bibliographies). This occurs in any of several different fashions, leading to recognizable differences in the mature structure. For the identification of agarics, many mushroom books are available. Any interested person, noting the details of structure which result from the different courses of development, together with the color of the spores (of one of five classes, white, pink to red, light brown to rust color, dark brown or purple, or black), will find identification reasonably easy. Popular interest in agarics is concerned, of course, with the edible and poisonous. Many amateur mycophagists need to be convinced that there is no single test for poisonous agarics except the final one. One who encounters an unfamiliar species may chew and eat a small scrap of it; if it is tasty and without bad after-effects, one may collect and eat the same species when one again recognizes it by its technical characters. At the present point, it is expedient to mention only a few examples.

Deliquescent agarics with black spores are called inky caps and constitute the genus *Coprinus*. All are edible; they should be fried in butter and served on toast.

Fruits of *Agaricus campestris,* the field mushroom, are rather large, white or gray on top, the stalk marked by a ring but no cup, the gills pink when young, dark brown to nearly black when mature. Anything of this character is safely edible.

Fruits of *Pleurotus* have an excentric or lateral stalk, or none, being shelf- or bracket-like, fleshy, with white spores. All species are edible. The most familiar is the oyster mushroom, *P. ostreatus,* producing large white to gray fruits on dead trees, commonly on poplars.

Fruits of *Amanita* are marked by cup and ring, and bear white spores. Some species are known to be edible; others, as the fly agaric, *A. muscaria,* recognized by a red cap flecked with white, are extremely poisonous.

Family 7. **Podaxacea** [Podaxaceae] Fischer in Engler and Prantl Nat. Pflanzenfam. I Teil, Abt. 1**: 332 (1900). *Gyrophragmium* produces fruits much like those of *Agaricus,* but coming up only to ground level, and drying and shattering irregularly instead of opening like mushrooms. The gills are quite evident in immature fruits. *Podaxon* is similar, but does not form definite gills. These organisms are conventionally stationed in the next order, but their obvious natural position is next to Agaricacea.

Order 7. **Dermatocarpa** [Dermatocarpi] Persoon Syst. Meth. Fung. xiii (1801).
  Order *Lytothecii* Persoon op. cit. xv.
  Class *Gasteromycetes* and orders *Angiogastres* and *Trichospermi* Fries Syst. Myc. 2: 275, 276 (1822).
  Family *Gasteromycetes* Fries Epicrisis 1 (1836)
  Order *Gasteromycetes* Winter in Rabenhorst Kryptog.-Fl. Deutschland 1, Abt. 1: 864 (1884).
  Suborders *Phallineae, Hymenogastrineae, Lycoperdineae, Nidulariineae,* and *Plectobasidiineae* Engler in Engler and Prantl Nat. Pflanznfam. I Teil, Abt. 1**: iv (1900).
  Orders *Phallineae, Lycoperdineae,* and *Nidularineae* Campbell Univ. Textb. Bot. 186, 187, 188 (1902).
  Orders *Hymenogastrales, Phallales, Lycoperdales, Nidulariales,* and *Sclerodermatales* Bessey in Univ. Nebraska Studies 7: 306-307 (1907).
  Orders *Plectobasidiales* and *Gasteromycetes* Gäumann Vergl. Morph. Pilze 537, 544 (1926).

Basidiomycetes producing holobasidia enclosed in fruits, not forming a continuous layer or not exposed as such, not discharging the spores directly into the air, sterigmata more or less suppressed.

Distinguished by negative characters, this order may be suspected of being artificial; but Engler's attempt to correct this produced orders which were small and numerous

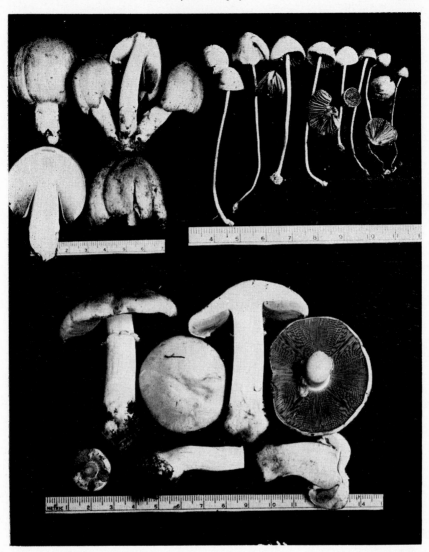

FIG. 29.—Fruits of Agaricacea: **upper left**, *Coprinus atramentarius;* **upper right**, *Galera tenera;* **below**, *Agaricus campestris*. Photographs by the late Dr. J. J. McCabe, by courtesy of the Department of Botany, University of California.

to an unsatisfactory degree, and to some of which the suspicion of artificiality continued to attach.

Dodge, translating Gäumann (1928), took account of the course of development of the fruits in rearranging those families whose fruits are characteristically produced underground. The roll of families which appear tenable is as follows.

A. Fruits typically formed underground.

Family 1. **Rhizopogonacea** [Rhizopogonaceae] Dodge in Gäumann Comp. Morph. Fungi 469 (1928).

Family 2. **Sclerodermea** [Sclerodermei] Winter in Rabenhorst Kryptog.-Fl. Deutschland 1, Abt. 1: 865 (1884). Family *Sclerodermataceae* Fischer in Engler and Prantl Nat. Pflanzenfam. I Teil, Abt. 1**: 334 (1900).

Family 3. **Hydnangiacea** [Hydnangiaceae] Dodge in Gäumann op. cit. 485.

Family 4. **Hymenogastrea** [Hymenogastrei] Winter in Rabenhorst op. cit. 865. Family *Hymenogastraceae* de Toni in Saccardo Sylloge 7: 154 (1888).

Family 5. **Hysterangiacea** [Hysterangiaceae] Fischer in Engler and Prantl op. cit. 304.

B. Fruits appearing on the surface of the ground.

Family 6. **Lycoperdacea** [Lycoperdaceae] Cohn in Hedwigia 11: 17 (1872). These are the common puffballs, *Lycoperdon, Bovista, Calvatia, Lycogalopsis,* etc. The contents of the more or less globular fruits become disorganized, leaving a mass of spores mixed with fibers (modified hyphae constituting a capillitium), enclosed in one or more continuous layers of tissue (peridia) which open usually through one stellate pore at the summit. *Geaster* has a double peridium. The outer peridium becomes split by meridional clefts from the apex nearly to the base, and the lobes curl back in damp weather, exposing the inner peridium with its terminal pore. The appearance of the fruit in the damp condition explains the common name, earth star, and the scentific name of the same meaning.

Family 7. **Tulostomea** [Tulostomei] Winter in Rabenhorst op. cit. 866. Family *Tulostomataceae* Fischer in Engler and Prantl op. cit. 342. *Tulostoma* produces at ground level puffball-like fruits which are found to stand upon buried stalks some centimeters long. The basidia bear the spores scattered along the sides instead of in a crown at the summit. This is probably a minor deviation from the condition in ordinary puffballs, and not a token of independent origin.

Family 8. **Nidulariea** [Nidulariei] Winter in Rabenhorst 1. c. Family *Nidulariaceae* de Toni in Saccardo Sylloge 7: 28 (1888). The bird's nest fungi, *Nidularia, Cyathus,* etc., with small fruits growing on sticks or earth, the outer peridium opening and exposing several peridioles.

Family 9. **Sphaerobolacea** [Sphaerobolaceae] Fischer in Engler and Prantl op. cit. 346. *Sphaerobolus,* a saprophyte on wood, produces minute puffball-like fruits which discharge mechanically a globular mass of spores.

Family 10. **Clathracea** [Clathraceae] Fischer in Engler and Prantl op. cit. 280. Closely related and transitional to the following family.

Family 11. **Phalloidea** [Phalloidei] Winter in Rabenhorst 1. c. Family *Phallaceae* Fischer in Engler and Prantl op. cit. 289. The stinkhorns, *Phallus, Dictyophora, Mutinus,* etc. These organisms produce highly specialized fruits. A fruit is first seen as a white globe, as large as a marble or a golf-ball, at ground level. It has a leathery peridium containing certain structures imbedded in gelatinous matter: there is a firm thimble-shaped structure upon whose surface the basidia develop; below or within this there is a body of the form of a hollow cylinder of spongy structure. When the spores are ripe, the spongy body grows, so to speak, by unfolding, and becomes, it may be within an hour, a stalk as much as 15 cm. tall. This happens usually during the night or at dawn, and is not commonly observed. The growing stalk carries the basidium-bearing structure into the air, bursting the peridium, which remains as a cup about the base, and exposing the spores in a mass of jelly which is of an odor repulsive to man but attractive to carrion-seeking insects. The latter are used as agents of dissemination.

Chapter X
PHYLUM PROTOPLASTA

Phylum 6. PROTOPLASTA Haeckel

Stämme PROTOPLASTA and *Myxomycetes* Haeckel Gen. Morph. 2: xxiv, xxvi (1866).

Subphylum *Plasmodroma* Doflein Protozoen 13 (1901), in part.

Subphylum *Rhizoflagellata* Grassé Traité Zool. 1, fasc. 1: 133 (1952), not order RHIZOFLAGELLATA Kent (1880).

Further names for the myxomycetes as a phylum are cited below under class Mycetozoa.

Organisms without photosynthetic pigments, mostly with flagellate stages, the flagella simple or acroneme, not paired and equal nor solitary and posterior; commonly occurring also in amoeboid stages. By Haeckel's original publication, the type or standard is *Amoeba,* i.e., *Amiba diffluens.*

Amoeboid organisms are those whose protoplasts lack walls or shells, or are only incompletely covered by them, and which thrust forth temporary bodies of protoplasm, called pseudopodia, functional in motion and in predatory nutrition. Pseudopodia are of several types. If massive and blunt they are lobopodia. If fine and straight, not anastomosing and usually not branching, they are filopodia; or, if they contain inner filaments, axopodia. If fine, branching, and anastomosing, they are rhizopodia.

The characters of the pseudopodia distinguish the accepted primary groups of amoeboid organisms. Variations in this character tend to run parallel to variations in the structure and composition of shells and skeletons: to a considerable extent, the accepted groups appear natural. This applies to the second, third, and fourth among the classes treated below. The phylum, on the other hand, is acknowledgedly artificial. Some of its groups appear to have had their origins (presumably more origins than one) among the chrysomonads; others are of unguessed origin.

1. Flagellate in the vegetative condition................Class 1. ZOOMASTIGODA.
1. Amoeboid in the vegetative condition.
   2. Producing rhizopodia; with shells, these
     usually calcareous............................Class 3. RHIZOPODA.
   2. Producing filopodia or axopodia; mostly
     with skeletons, these usually siliceous..........Class 4. HELIOZOA.
   2. Producing lobopodia.
     3. Producing flagellate reproductive
      cells; mostly macroscopic, subaerial..........Class 2. MYCETOZOA.
     3. Not as above; without flagellate
      stages...................................Class 5. SARKODINA.

Class 1. ZOOMASTIGODA Calkins

Subclass *Zoomastigina* Doflein Lehrb. Prot. ed. 4: 462 (1916).
Class ZOOMASTIGODA Calkins Biol. Prot. 285 (1926).
Class *Zooflagellata* Grassé Traité Zool. 1, fasc. 1: 574 (1952).
Class *Zoomastigophorea* Hall Protozoology 170 (1953).

Non-pigmented flagellates having acroneme or simple flagella; amoeboid stages, if they occur, having lobopodia. The standard is *Bodo.* Four orders are to be recognized.

1. Flagella one or two.......................... Order 1. Rhizoflagellata.
1. Flagella four to eight (in each neuromotor system, if these are more than one).
   2. Axostyles, if present, homologous with flagella; parabasal body commonly absent.................................... Order 2. Polymastigida.
   2. Axostyles present, not homologous with flagella; parabasal body present, disappearing during mitosis.....................Order 3. Trichomonadina.
1. Flagella of indefinite large numbers............ Order 4. Hypermastigina.

Order 1. **Rhizoflagellata** [Rhizo-Flagellata] Kent Man. Inf. 1: 220 (1880).
    Orders *Trypanosomata* (the mere plural of a generic name) and *Flagellato-Pantostomata* in part Kent op. cit. 218, 229.
    Suborders *Monadina* in part and *Heteromastigoda* Bütschli in Bronn Kl. u. Ord. Thierreichs 1: 810, 827 (1884).
    Protomastigina Klebs in Zeit. wiss. Zool. 55: 293 (1893).
    Order *Protomonadina* Blochmann Mikr. Tierwelt ed. 2, 1: 39 (1895).
    Subclasses *Pantostomatineae* and *Protomastigineae* Engler in Engler and Prantl Nat. Pflanzenfam. I Teil, Abt. 1a: iv (1900).
    Orders *Pantostomatales* and *Protomastigales* Engler Syllab. ed. 3: 7 (1903).
    Orders *Cercomonadinea* and *Monadidea* in part Poche in Arch. Prot. 30: 139, 140 (1913).
    Orders *Pantostomatineae* and *Protomastigineae* Lemmermann in Pascher Süsswasserfl. Deutschland 1: 30, 52 (1914).
    Order *Rhizomastigina* Doflein Lehrb. Prot. ed. 4: 704 (1916).
    Orders *Pantostomatida* and *Protomastigida* Calkins Biol. Prot. 286, 288 (1926).
    Orders *Trypanosomidea* Grassé, *Bodonidea* Hollande, and *Proteromonadina* Grassé in Grassé Traité Zool. 1, fasc. 1: 602, 669, 694 (1952).
    Orders *Rhizomastigida* and *Protomastigida* Hall Protozoology 171, 173 (1953).
    Non-pigmented flagellates with one flagellum or two unequal flagella, these simple or acroneme; commonly with amoeboid stages, or amoeboid while bearing flagella. The type, being the sole genus of Rhizo-Flagellata as originally published, is *Mastigamoeba,* i. e., *Chaetoproteus* Stein.
    As the synonymy shows, most authorities have made these organisms two orders, Pantostomatales (or some such name), amoeboid in the vegetative condition, and Protomastigina (or the like), not definitely so. *Monas,* and the choanoflagellates and Amphimonadaceae, usually included in the latter order, have in the present work been given places elsewhere. The residue of the Protomastigina are not sharply different in character from the original Rhizoflagellata, and are accordingly placed in the same order. The resulting group is not a very numerous one. Some examples appear to occur naturally as predators in uncontaminated waters; the majority have been found in foul or contaminated waters, or in feces, and are believed to be naturally entozoic, either commensal or parasitic. Further examples are parasites in blood. A cytological character marking the majority of the goup, but not confined to it, is the parabasal body (better, perhaps, the kinetoplast; Kirby, 1944). This is a

rather massive extranuclear body regularly present in the cell and distinct both from the centrosome and the blepharoplast. In the present group, it divides when the nucleus does. Thus this group, although marked chiefly by characters which are negative or derived, appears possibly to be natural.

1. Flagella two.
    2. Cells not notably slender.................Family 1. CERCOMONADIDA.
    2. Cells notably slender....................Family 2. TRYPANOPLASMIDA.
1. Flagellum one.
    2. Not regularly markedly amoeboid..........Family 3. OICOMONADACEA.
    2. Conspicuously amoeboid..................Family 4. CHAETOPROTEIDA.

Family 1. **Cercomonadida** [Cercomonadidae] Kent Man. Inf. 1: 249 (1880). Family *Bodonina* Bütschli in Bronn Kl. u. Ord. Thierreichs 1: 827 (1884). Family *Bodonaceae* Senn in Engler and Prantl Nat. Pflanzenfam. I Teil, Abt. 1a: 133 (1900). Family *Bodonidae* Doflein Protozoen 73 (1901). Family *Cercobodonidae* Hollande 1942. Family *Proteromonadidae* Grassé Traité Zool. 1, fasc. 1: 694 (1952). Non-pigmented flagellates, the bodies not notably slender, with two flagella, one directed anteriorly, the other trailing. Fischer (1894) found both of the flagella of *Bodo* to be acroneme.

In *Bodo* both flagella are free of the body. There are numerous species, in infusions or foul or polluted waters, or entozoic in a wide variety of animals, from insects to men. *Prowazekia, Proteromonas,* and *Pleuromonas* are doubtfully distinct. *Rhynchomonas,* from fresh or foul waters, is distingished by a protoplasmic beak in which the anterior flagellum is imbedded. *Cercomonas,* of like habitats, has the trailing flagellum grown fast to the cell membrane; the cells exhibit a considerable capacity to send out lobopodia.

Biflagellate organisms which can lose their flagella and take on the appearance of ordinary amoebas have repeatedly been discovered and variously named. So far as the pseudopodia are lobopodia and the flagella are unequal, these organisms belong in this family; but many accounts fail to establish the equality or inequality of the flagella, with the result that the names used in them cannot be applied with confidence. This is true of various organisms originally named under *Pseudospora, Dimastigamoeba,* and *Naegleria.* The earliest generic name definitely applicaple to organisms as described in *Cercobodo* Senn, 1910.

Bělǎr (1914, 1916, 1920, 1921), Kühn (1915), and others have described mitosis in various examples of this family; the most detailed account is of *Bodo Lacertae* in Bělǎr's paper of 1921. The flagella spring from a blepharoplast from which a rhizoplast extends into the nucleus. The chromatin is reticulate, not massed in a karyosome, but no centrosome has been recognized in it when it is not dividing. The rhizoplast, where it passes through the cytoplasm, is surrounded by stainable *Ringkörper.* The parabasal body, located on the posterior side of the nucleus, is massive and often irregular. In division, the blepharoplast divides, each part retaining one flagellum and generating an additional one. The rhizoplast appears to begin to split, but presently it and the *Ringkörper* become invisible. Within the intact nuclear membrane there appears a spindle with evident centrosomes at the poles. The centrosomes come presently to the inner surface of the nuclear membrane, while the blepharoplasts move to adjacent positions on the outside. Chromosomes duly assemble at the equator of the spindle and undergo division. Division of the nucleus is completed by constriction of the nuclear membrane; the parabasal body undergoes constriction; the cell divides by constriction lengthwise. The *Ringkörper* and the rhizoplast are apparently regenerated by the blepharoplast.

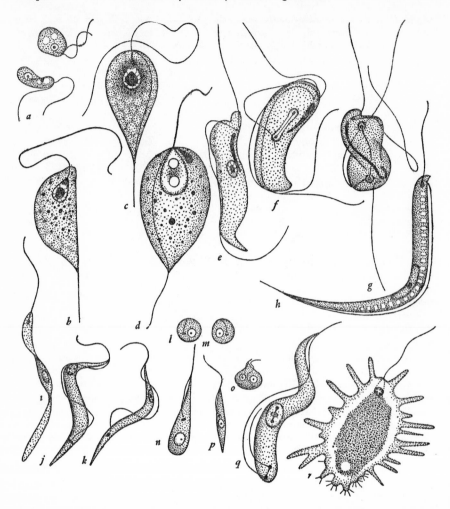

Fig. 30.—Rhizoflagellata: **a**, *Bodo* sp. x 1,000. **b**, **c**, *Cercomonas longicauda*
as identified by Wenyon (1910) in material from a cholera patient; **d**, the same as
identified by Hovasse (1937) in swamp water. **e-h**, *Cryptobia* spp.; **e-g**, cell and
division stages of a species from the conger eel after Martin (1910); **h**, a species
from siphonophores after Keysselitz (1904) x 1,000. **i**, *Phytomonas Donovani* after
França (1914). **j-p**, *Trypanosoma Lewisi*; **j**, **k**, forms from the rat after Minchin
(1909); **l-p**, forms from the flea *Ceratophyllus fasciatus* after Minchin & Thompson (1915). **q**, Division stage of *Trypanosoma Brucii* after Kühn & Schuckmann
(1911). **r**, *Chaetoproteus* (*Mastigamoeba aspera*) after Schulze (1875) x 100.
x 2,000 except as noted.

Alexeieff (1924) described fusions of pairs of cells of *Bodo edax.*

Family 2. **Trypanoplasmida** [Trypanoplasmidae] Hartmann and Jollos 1910. Family *Cryptobiidae* Poche in Arch. Prot. 30: 148 (1913). Family *Trypanophidae* Hollande in Grassé Traité Zool. 1, fasc. 1: 680 (1952). Organisms of essentially the structure of *Cercomonas,* but notably slender in adaptation to parasitic life, the trailing flagellum forming the margin of an undulating membrane on the body. Parasitic in various invertebrates and in the gut and blood of fishes.

The numerous species may be included in a single genus *Cryptobia* Leidy (*Trypanoplasma* Laveran and Mesnil; *Trypanophis* Keysselitz).

According to Martin's (1910) description of a species from the eel *Conger niger,* both flagella spring from a blepharoplast ("basal granule") at the anterior end. As preliminary to division, the blepharoplast and flagella divide, and one blepharoplast migrates to the posterior end of the cell. The nucleus divides by constriction of the nuclear membrane. There is a prominent parabasal body ("kinetonucleus") which divides by constriction, as does the cell, transversely.

Bělǎr (1916) described sexual fusions of differentiated individuals of a species parasitic in snails.

Family 3. **Oicomonadacea** [Oicomonadaceae] Senn in Engler and Prantl Nat. Pflanzenfam. I Teil, Abt. 1a: 118 (1900). Family *Trypanosomidae* Doflein Protozoen 55 (1901). Family *Trypanosomatidae* Grobben 1904. Family *Oicomonadidae* Hartog. Non-pigmented anteriorly uniflagellate organisms, not markedly amoeboid while in the flagellate condition.

Oikomonas includes organisms of the character of the family without particular specialization, occurring in contaminated water or soil, and as commensals in the intestine of animals.

The bulk of the family consists of the slender-celled parasites which may be celled trypanosomes in the broad sense of the word. From the viewpoint of man, these are the most important flagellates, and they have been the most intensely studied. Some are known only from the guts of insects; some occur alternatively in insects and plants; some in insects and vertebrates; and some in vertebrates and in invertebrates other than insects, as ticks and leeches. The range of parasitization is as though the group had evolved as parasites in insects, and had been carried to other hosts by the activity of insects and other biting or sucking invertebrates.

Most trypanosomes occur in varied forms. The forms are designated by words which originated as names of genera and remain in use as such. (1) The leptomonas form has an anterior flagellum but no undulating membrane; it resembles a cell of *Oikomonas* but is notably slender. (2) The leishmania form has no flagellum; the cell is rounded up and lives attached to, or inside of, cells of the host. (3) In the crithidia form, the base of the flagellum is continued as an undulating membrane more or less to the middle of the cell. (4) In the trypanosoma form, the base of the flagellum is continued as an undulating membrane to the posterior end of the cell.

The accepted genera are distinguished (artificially, as one may suspect) by stages produced and groups of hosts attacked, as follows:

1. With leptomonas stages in insects and in
   Euphorbiaceae, Ascelepiadaceae, and other
   plants with milky juice....................................*Phytomonas.*
1. Confined to invertebrate animals.
   2. Trypanosoma stage known...........................*Herpetomonas.*
   2. Trypanosoma stage unknown; crithidia

stage known.......................................*Crithidia.*
  2. Trypanosoma and crithidia stages un-
    known..*Leptomonas.*
1. Attacking vertebrate animals.
  2. Trypanosoma stage known...........................*Trypanosoma.*
  2. Trypanosoma stage unknown........................*Leishmania.*

Man has been concerned particularly with *Trypanosoma gambiense,* the agent of African sleeping sickness; *T. Cruzi,* the cause of Chagas' disease; *T. Brucii, T. Evansi, T. equinum,* and *T. equiperdum,* which cause in domestic animals the diseases, respectively, nagana, surra, mal de caderas, and dourine; *Leishmania Donovani* and *L. tropica,* causing kala azar and oriental sore; and *L. brasiliensis,* causing espundia, ferida brava, or chicleros' ulcer, usually appearing as a grievous disfigurement of the features.

Schaudinn (1903), having studied a trypanosome occurring in mosquitoes and in the owl *Athene noctua,* described the nucleus as undergoing repeated unequal divisions. It appeared to him that when a cell is to produce a flagellum, one of the minor nuclei produced by unequal division generates it. Prowazek (1903) described similar phenomena in a *Herpetomonas* occurring in flies. These reports led Woodcock (1906) to apply to the proper nucleus of trypanosomes the term trophonucleus, and to the large granule near the base of the flagellum the term kinetonucleus.

There has been much other study of the cytology of trypanosomes (as by Minchin, 1908, 1909; Robertson, 1909; Woodcock, 1910; Minchin and Woodcock, 1910, 1911; Kühn and Schuckmann, 1911; Minchin and Thomson, 1915; Schuurmans Stekhoven, 1919). This has not confirmed the foregoing accounts and conclusions, but appears to have established the following points.

The base of the flagellum is slightly swollen and may be construed as a blepharoplast. Separated from the blepharoplast by a distance of one or two microns there is a conspicuous parabasal body (the kinetonucleus of Woodcock). Fine strands connecting the blepharoplast, parabasal body, and nucleus, have been observed. Most of the stainable material in the resting nucleus is aggregated in a globular karyosome. In mitosis, the karyosome breaks up to form a moderate number of chromosomes and a central granule, evidently a centrosome, which stains more heavily than the chromosomes. It divides before the chromosomes, the daughter centrosomes remaining connected by a fine fiber, the centrodesmose. An obscure spindle forms about the centrodesmose; the chromosomes undergo division within the spindle, and the daughter chromosomes assemble about the centrosomes. Mitosis is completed by constriction of the nuclear membrane.

The blepharoplast divides at the same time as the nucleus. The flagellum splits to a short distance and one of the branches breaks loose; one daughter blepharoplast retains essentially the whole of the original flagellum while the other generates one which is almost entirely new. The parabasal body undergoes constriction. The cell membrane cuts in in such fashion as to divide the cell longitudinally. The blepharoplast and the parabasal body persist through the non-flagellate leishmania stage. Reports that the nucleus may generate these structures, or that one of them may generate another, were apparently mistaken.

Schaudinn described complicated processes by which a trypanosome generates differentiated male and female gametes which duly undergo syngamy. His account is believed to have resulted from mistaking stages of a sporozoan for those of a trypanosome. Still, the occurrence of syngamy among trypanosomes is inherently probable.

Family 4. **Chaetoproteida** [Chaetoproteidae] Poche in Arch. Prot. 30: 172 (1913). Family *Rhizomastigina* Bütschli in Bronn Kl. u. Ord. Thierreichs 1: 810 (1884). Family *Rhizomastigaceae* Senn in Engler and Prantl Nat. Pflanzenfam. I Teil, Abt. 1a: 113 (1900). Family *Mastigamoebidae* Kudo Protozoology ed. 3: 263 (1946). Amoeboid organisms bearing one anterior flagellum, either permanently or temporarily. In polluted soil or water, or commensal or pathogenic in animals.

The oldest genus, *Chaetoproteus* Stein (*Mastigamoeba* F. E. Schulze, 1875; *Dinamoeba* Leidy ?) remains poorly known. This organism and *Mastigella* are described as fairly large; *Craigia* is much smaller. *Rhizomastix* is doubtfully distinct from *Craigia*. Early names of this family appear to refer to *Rhizomastix* as the type, but the family is much older than the genus, and the names are not valid.

Order 2. **Polymastigida** Calkins Biol. Prot. 292 (1926).
> Family *Polymastigina* Bütschli in Bronn Kl. u. Ord. Thierreichs 1: 842 (1884).
> Order *Polymastigina* Blochmann Mikr. Tierwelt ed. 2, 1: 47 (1895).
> Subclass *Distomatineae* Engler in Engler and Prantl Nat. Pflanzenfam. I Teil, Abt. 1a: iv (1900).
> Order *Distomatinales* Engler Syllab. ed. 3: 7 (1903), not based on a generic name.
> Orders *Pyrsonymphina, Oxymonadina, Retortomonadina,* and *Distomata* Grassé Traité Zool. 1: fasc. 1: 788, 801, 824, 963 (1952).

Non-pigmented flagellates with simple or acroneme flagella of definite number, from four to eight (two in *Retortomonas*), in the individual neuromotor system, and accordingly on the individual cell, except when the neuromotor systems are multiplied; not of the definite characters of the following order. Free-living, chiefly in foul waters, or commensal or parasitic in animals. *Polymastix* is presumably the type of the group. It was listed with a query in Bütschli's original publication of family Polymastigina.

In the generality of Polymastigida, the cells are dorsiventral and have single nuclei and neuromotor systems. There are derived examples in which the cells are spirally twisted. There is a group in which the cells are double, having two nuclei and neuromotor systems. In another group there are two or more neuromotor systems, usually with more than one nucleus; the cells consist of units in a whorled or spiral arrangement, so that as wholes they are of radial symmetry.

The neuromotor system consists primarily of (1) the flagella; (2) one or more blepharoplasts from which the flagella spring; (3) one or more rhizoplasts linking together the parts of the system; and (4) a centrosome located just outside the nuclear membrane. Furthermore, (5) a parabasal body may be present. (6) An axostyle is a rod imbedded in the cytoplasm. In *Hexamita* the axostyles are the proximal ends of backwardly directed flagella; axostyles occurring in various other genera of the order appear also to be homologous with flagella.

Nuclear and cell division have been observed in various genera, as in *Hexamita* by Swezy (1915); in *Streblomastix* by Kidder (1929); in *Giardia* by Kofoid and Christianson (1915) and Kofoid and Swezy (1922); and in *Oxymonas* by Connell (1930).

Cleveland (1947) observed in *Saccinobaculus* a multiplication of nuclei followed by their fusion in pairs, and by meiosis in the fusion nuclei: thus there is a sexual cycle without fusion of cells. It is not probable that sexual reproduction does not occur in the generality of the group, but it has not been observed in any others.

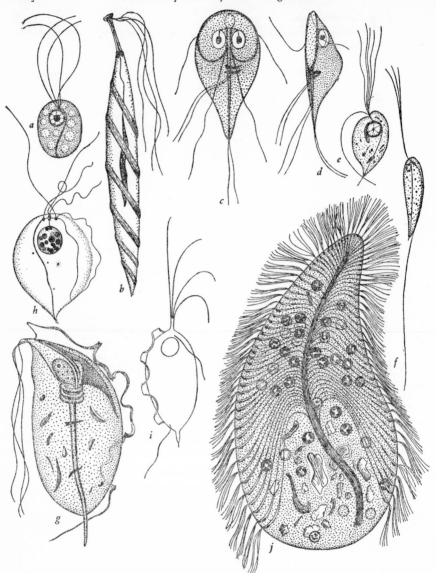

Fig. 31.—Polymastagida: **a**, *Polymastix Melolonthae* after Swezy (1916). **b**, *Streblomastix Strix* x 1,000 after Kidder (1929).   **c, d**, *Giardia enterica* after Kofoid & Swezy (1922).   Trichomonadina: **e**, *Hexamastix Termopsidis* after Kirby (1930).   **f.** *Tricercomitus Termopsidis* after Kirby (1930).   **g**, *Macrotrichomonas pulchra* after Kirby (1938).   **h**, *Trichomonas tenax* x 4,000 after Hinshaw (1926). **i**, *Pentatrichomonas obliqua* after Kirby (1943).   **j**, *Snyderella Tabogae* x 500 after Kirby (1929).   x 2,000 except as noted.

In making the clearly natural group of trichomonads a separate order, Kirby (1947) removed the majority of the species formerly assigned to this order, and left a miscellany of small isolated families. It seems not expedient to make them several small orders, as Grassé has done; rather they are to be held together until their respective relationships become evident. A hint of Hall has led in the present work to the transfer of family Trimastigida to order Ochromonadalea.

1. With a single nucleus and neuromotor system.
    2. Cells not spirally twisted, at least not as wholes and not conspicuously.............Family 1. Tetramitida.
    2. Entire cells conspicuously spirally twisted.
        3. With four free flagella...............Family 2. Streblomastigida.
        3. With four or eight flagella whose proximal ends are grown fast to the cell membrane.....................Family 3. Dinenymphida.
1. With one or several nuclei and two or more neuromotor systems........................Family 4. Oxymonadida.
1. With two nuclei and neuromotor systems.......Family 5. Trepomonadida.

Family 1. **Tetramitida** [Tetramitidae] Kent Man. Inf. 1: 312 (1880). Families *Tetramitina* and *Polymastigina* Bütschli in Bronn Kl. u. Ord. Thierreichs 1: 841, 842 (1884). Family *Tetramitaceae* Senn in Engler and Prantl Nat. Pflanzenfam. I Teil, Abt. la: 143 (1900). Family *Polymastigidae* Doflein Protozoen 83 (1901). Family *Chilomastigidae* Wenyon (1926). Family *Costiidae* Kudo Handb. Prot. 153 (1931). Family *Retortomonadidae* Wenrich 1932. Cells mostly dorsiventral and with four flagella; these uniform or differentiated; when differentiated, one or two may trail behind the cell. Axostyles present or absent, parabasal bodies not reported. Like the order, the family is a miscellany; good authority has made as many as four families of the few genera. *Tetramitus*, free-living, unfamiliar. *Costia*, occurring usually as sessile parasites on fishes. *Polymastix*, in insects. *Monocercomonoides*, in insects and vertebrates. *Chilomastix*, in insects and vertebrates, cells marked by a cytostomal groove into which one of the flagella, shorter than the others, is recurved. The species which occurs in man (usually, as it appears, as a harmless commensal) is in most works called *C. Mesnili;* the correct name is apparently **Chilomastix Hominis** (Davaine) n. comb[1]. Current authority places next to *Chilomonas* the biflagellate *Retortomonas*, also in insects and vertebrates, and having cells of essentially the same structure.

---

[1]Kofoid (1920) gave the history involved in this combination. Davaine, 1860, described the flagellates *Cercomonas Hominis* var. A and var. B. The two forms are not of the same species, and Moquin-Tandon, in the same year, re-named them respectively *C. Davainei* and *C. obliqua*. They are not of the same genus, being respectively a *Chilomastix* and a *Pentatrichomonas*, under which genera they have various names. Kofoid named them respectively *Chilomastix davainei* and *Trichomonas hominis*. In so doing, he may be held to have exercised his right to choose a type in a group in which no type has been designated; but it is arguable on the contrary that an author who designates a var. A designates the type in doing so. It is on the basis of this argument that the new combination here published is applied to the *Cercomonas Hominis* var. A of Davaine.

Family 2. **Streblomastigida** [Streblomastigidae] Kofoid and Swezy in Univ. California Publ. Zool. 20: 15 (1919). The only known species is *Streblomastix Strix,* a slender spirally twisted organism with four anterior flagella, free-swimming or attached in the gut of the termite *Termopsis.* The significance of the epithet *Strix* (a Greek noun meaning screech owl) as applied to this species is not clear.

Family 3. **Dinenymphida** [Dinenymphidae] Grassi in Atti Accad. Lincei ser. 5. Rendiconti Cl. Sci. 20, 1° Semestre: 730 (1911). Elongate flagellates, the four or eight anterior flagella adherent to the body and spirally twisted with it, free at their distal ends. Often beset with spirochaets, which have been mistaken for additional flagella; the family has been misplaced in order Hypermastigina. *Dinenympha* and *Pyrsonympha* in termites; *Saccinobaculus* in the wood roach *Cryptocercus.*

Family 4. **Oxymonadida** [Oxymonadidae] Kirby in Quart Jour. Micr. Sci. n. s. 72: 380 (1928). Flagellates with radially symmetrical bodies including two or more neuromotor systems, entozoic in termites of subfamily Kalotermitinae. Each pear-shaped cell of *Oxymonas* has one nucleus and two neuromotor systems (Kofoid and Swezy, 1926). In *Microrhopalodina* (*Proboscoidella*) each cell contains a whorl of nuclei, each with its separate neuromotor system (Kofoid and Swezy, 1926; Kirby, 1928). These organisms are superficially closely similar to the Calonymphida, from which Kirby distinguished them.

Family 5. **Trepomonadida** [Trepomonadidae] Kent Man. Inf. 1: 300 (1880). Family *Hexamitidae* Kent op. cit. 318. *Distomata* Klebs in Zeit. wiss. Zool. 55: 329 (1893). Family *Distomataceae* Senn in Engler and Prantl Nat. Pflanzenfam. I Teil, Abt. 1a: 148 (1900). Flagellates each with two nuclei and two neuromotor systems. In most examples, each half-cell is dorsiventral, and the whole isobilateral, with two cytostomes. Most of the genera, *Trepomonas, Gyromonas, Trigonomonas,* are free-living in fresh or foul waters and have been little studied. *Hexamita* occurs both free-living and entozoic, in roaches and in all classes of vertebrates; the cells have eight flagella (*Octomitus* Prowazek and *Urophagus* Moroff are synonyms). In *Giardia* the half-cells are asymmetric, and the whole cells dorsiventral, with one cytostome. There are several species, serious pathogens in mammals. The valid name of the species in man, usually known as *G. Lamblia,* appears to be *G. enterica* (Grassi) Kofoid (1920).

Order 3. **Trichomonadina** Grassé Traité Zool. 1, fasc. 1: 704 (1952).
    Order *Trichomonadida* Kirby in Jour. Parasitol. 33: 215, 224 (1947), preoccupied by family Trichomonadidae Wenyon (1926).
Flagellates of the general nature of the Polymastigida having in each neuromotor system one trailing flagellum; axostyle present, rigid, apparently not homologous with the flagella; parabasal body present, disappearing during mitosis. Entozoic, the majority of the species, to the number of fully 150, occurring in termites.

The base of the trailing flagellum may be underlain by a cresta, a more or less prominent body distinct both from parabasal body and from axostyle. The trailing flagellum may be grown fast to the cell membrane and converted into an undulating membrane; in this case it is underlain by a rod called the costa, apparently homologous with the cresta (Kirby, 1931).

Nuclear and cell division have been described in *Trichomonas* by Kuczynski (1914), Kofoid and Swezy (1915, 1919; the *Trichomitus* described in the latter year is a *Trichomonas*) and Hinshaw (1926). The centrosome (or a combined centrosome and blepharoplast, the centroblepharoplast of Kofoid and Swezy, 1919) lies outside the nuclear membrane. This structure divides and the daughter structures move apart along the nuclear membrane. They remain connected, usually until mitosis is complete, by a stainable strand, the paradesmose. Definite chromosomes, usually few in number, and an intranuclear spindle, are formed. Mitosis is completed by constriction of the nuclear membrane. In what appears to be the typical course of cell division, the rhizoplast and blepharoplast divide when the centrosome does. Of other parts of the neuromotor system, some may remain connected to one blepharoplast and some to the other; some may disappear. The parts needed to complete a neuromotor system are regenerated in each daughter cell.

1. With a single nucleus and neuromotor system.
  2. Lacking a cresta, costa, or undulating membrane.........................Family 1. Monocercomonadida.
  2. With a trailing flagellum whose base is underlain by a cresta.................Family 2. Devescovinida.
  2. With a trailing flagellum grown fast to the cell membrane, forming an undulating membrane underlain by a costa.....Family 3. Trichomonadida.
1. With several nuclei and neuromotor systems.. Family 4. Calonymphida.

Family 1. **Monocercomonadida** [Monocercomonadidae] Kirby in Jour. Parasitol. 33: 225 (1947). Minute flagellates of the appearance of certain Tetramitida, but having a firm axostyle, the parabasal body disappearing and a paradesmose forming between the daughter centrosomes during mitosis; lacking a cresta, costa, or undulating membrane; entozoic in termites and other insects, and in all classes of vertebrates. *Monocercomonas, Hexamastix, Tricercomitus.*

Family 2. **Devescovinida** [Devescovinidae] Doflein Lehrb. Prot. ed. 3: 537 (1911). Subfamily *Devescovininae* Kirby in Univ. California Publ. Zool. 36: 215 (1931). Organisms with three anterior flagella and a larger trailing flagellum underlain by a cresta; confined to termites of the families Mastotermitidae, Hodotermitidae, and Kalotermitidae, being most abundant in the last. The cells, usually fairly large, ingest scraps of wood and are presumed to contribute to the lives of their hosts by digesting it. *Devescovina, Gigantomonas, Macrotrichomonas, Foaina, Parajoenia, Metadevescovina.* Spirochaets which share the habitat of these organisms are commonly found adhering to their cell membranes, and were mistaken for additional flagella in the original descriptions of some of the genera.

Family 3. **Trichomonadida** [Trichomonadidae] Wenyon Protozoology 1: 646 (1926). Flagellates with three or more flagella directed forward and one trailing, the proximal part of the latter grown fast to the cell membrane and forming an undulating membrane underlain by a costa. Entozoic in a wide variety of animals. *Trichomonas,* normally with four anterior flagella, is the most numerous genus. It occurs in termites, including those of the advanced family Termitidae, in which scarcely any other flagellates occur; it does not ingest wood, and is not believed to be beneficial to its hosts. It occurs also in all classes of vertebrates. Man harbors *Trichomonas tenax* as a commensal in the mouth. *T. vaginalis* may be a serious pathogen. **Pentatrichomonas obliqua** (Moquin-Tandon) comb. nov.,[1] commensal (or pathogenic?) in the gut has at the anterior end a fifth flagellum separate from the other four (Kirby, 1943).

---

[1] cf. footnote, p. 165.

Family 4. **Calonymphida** [Calonymphidae] Grassi in Atti Accad. Lincei ser. 5, Rendiconti Cl. Sci. 20, 1° Semestre: 730 (1911). Flagellates with radially symmetrical bodies including more than two nuclei and neuromotor systems, the latter of trichomonad type; entozoic in termites of subfamily Kalotermitinae. These flagellates ingest scraps of wood and are believed to contribute to the nutrition of their hosts. In *Coronympha* each cell contains one whorl of nuclei each with its separate neuromotor system (Kirby, 1929). In *Stephanonympha,* the nuclei and neuromotor systems are so numerous as to form a spiral band of several cycles in the anterior part of the cell. In *Calonympha,* besides numerous neuromotor systems associated with nuclei, there are others free of any nucleus; in *Snyderella,* the two types of structures are independently multiplied.

Order 4. **Hypermastigina** Grassi in Atti Accad. Lincei ser. 5, Rendiconti Cl. Sci. 20, 1° Semestre: 727 (1911).
Order *Trichonymphidea* Poche in Arch. Prot. 30: 149 (1913).
Order *Hypermastigida* Calkins Biol. Prot. 295 (1926).
Order *Lophomonadida* Light in Univ. California Publ. Zool. 29: 486 (1927).
Orders *Joeniidea, Lophomonadina, Trichonymphina,* and *Spiratrichonymphina,* Grassé Traité Zool. 1, fasc. 1: 837, 851, 862, 916 (1952).

Flagellates, mostly large and of radial symmetry, with single nuclei and indefinitely numerous flagella. Entozoic in roaches and in termites excluding those of family Termitidae. *Lophomonas* is to be regarded as the type.

Cleveland (1925, 1926) found it possible, by starvation or by exposure to high pressures of oxygen or high temperatures, to rid insects of all of their intestinal flagellates or of some of the kinds. When completely freed of flagellates, wood roaches and termites of the lower families are able to remain alive only for a few weeks. The life of *Termopsis* is not prolonged by the presence of *Streblomastix,* and it is prolonged only moderately by the presence of *Trichomonas Termopsidis.* But if infested with either *Trichonympha Campanula* or *T. sphaerica,* it can survive indefinitely on a diet of pure cellulose. Both species ingest the ground scraps of wood which reach the part of the intestine in which they occur; it is evident that they serve their hosts as agents of digestion. Cleveland's observations raise unanswered questions as to the occurrence of fixation of nitrogen; it is known only that termites are quite economical in their use of nitrogenous compounds available to them.

The Hypermastigina have elaborate neuromotor systems. There is regularly a large centroblepharoplast. In what appears to be the relatively primitive type of cell division, as in *Trichonympha* (Kofoid and Swezy, 1919), the neuromotor system of the mother cell is divided between the daughter cells. In *Spirotrichonympha* (Cupp, 1930), only the centroblepharoplast divides; the neuromotor system of the mother cell remains attached to one of the daughter centroblepharoplasts, while the other generates the remaining parts of a complete system. In *Lophomonas* (Kudo, 1926), and *Kofoidia* (Light, 1927), the neuromotor system of a dividing cell is absorbed or discarded, with the exception of the centroblepharoplasts, from which new systems develop.

In *Trichonympha* and *Spirotrichonympha* the details of nuclear division have much the appearance of meiosis. A double set of chromosomes appears, and the chromosomes form pairs which are divided in the spindle. It is supposed that this appearance is produced by a precocious splitting of the chromosomes.

In species of *Trichonympha, Leptospironympha,* and *Eucomonympha* from the wood roach *Cryptocercus,* Cleveland (1947, 1948) observed the syngamy of undifferentiated or differentiated gametes; the appearance of the process is as though the egg ingested the sperm. Syngamy is followed immediately by meiosis. This means that vegetative individuals are haploid. *Barbulanympha* achieves without syngamy an alternation of haploid and diploid stages. Diploid cells are produced when a centroblepharoplast fails to divide, with the result that the nucleus remains intact, while chromosomes appear and divide. Reduction division, by the separation of undivided chromosomes, occurs when a centroblepharoplast divides at an exceptionally early stage. Cleveland concluded that the early division of the central body is the event which primarily distinguishes meiosis from mitosis. It is possible that he has recognized an essential feature of the evolution of the sexual cycle. His words suggest the idea that the sexual cycle may have originated within the present group. This is an impossibility; the sexual cycle is a normal character of nucleate organisms, and is fully established in nucleate organisms far more primitive than these.

There are fewer than one hundred known species of Hypermastigina. They are treated as seven families.

1. Body without segmented appearance.
   2. Flagella distributed generally over the
      surface of the body or its anterior part. . . . Family 1. TRICHONYMPHIDA.
   2. Flagella in spiral bands. . . . . Family 2. HOLOMASTIGOTOIDIDA.
   2. Flagella in tufts.
      3. Flagella in a single tuft. . . . . Family 3. LOPHOMONADIDA.
      3. Flagella in two tufts. . . . . Family 4. HOPLONYMPHIDA.
      3. Flagella in four tufts. . . . . Family 5. STAUROJOENIIDA.
      3. Flagella in many tufts. . . . . Family 6. KOFOIDIIDA.
1. Body with segmented appearance. . . . . Family 7. TERATONYMPHIDA.

Family 1. **Trichonymphida** [Trichonymphidae] Leidy ex Doflein Lehrb. Prot. ed. 3: 537 (1911). The numerous flagella distributed generally over the surface of the body or its anterior part. *Trichonympha (Leidyopsis), Eucomonympha,* etc.

Family 2. **Holomastigotoidida** [Holomastigotoididae] Janicki in Zeit. wiss. Zool. 112: 644 (1915). Family *Spirotrichonymphidae* Grassi in Mem. Accad. Lincei Cl. Sci. ser. 5, 12: 333 (1917). The numerous flagella arranged in spiral bands. *Holomastigotoides, Spirotrichonympha,* etc.

Family 3. **Lophomonadida** [Lophomonadidae] Kent Man. Inf. 1: 321 (1880). Family *Joeniidae* Janicki in Zeit wiss. Zool. 112: 644 (1915). The numerous flagella assembled in a single anterior tuft. *Lophomonas,* in cockroaches, all of the flagella directed forward. *Joenia, Joenina, Joenopsis,* etc., in termites, the outer flagella directed backward.

Family 4. **Hoplonymphida** [Hoplonymphidae] Light in Univ. California Publ. Zool. 29: 138 (1926). The flagella assembled in two anterior tufts. *Hoplonympha, Barbulanympha,* etc.

Family 5. **Staurojoeninda** [Staurojoenindae] Grassi in Mem. Accad. Lincei Cl. Sci. ser. 5, 12: 333 (1917). The flagella assembled in four anterior tufts. *Staurojoenina.*

Family 6. **Kofoidiida** [Kofoidiidae] Light in Univ. California Publ. Zool. 29: 485 (1927). The flagella fused at their bases into several bundles. *Kofoidia,* a single known species in *Kalotermes.*

Family 7. **Teratonymphida** [Teratonymphidae] Koidzumi in Parasitology 13: 303 (1921). Family *Cyclonymphidae* Reichenow. Elongate and segmented, with a single

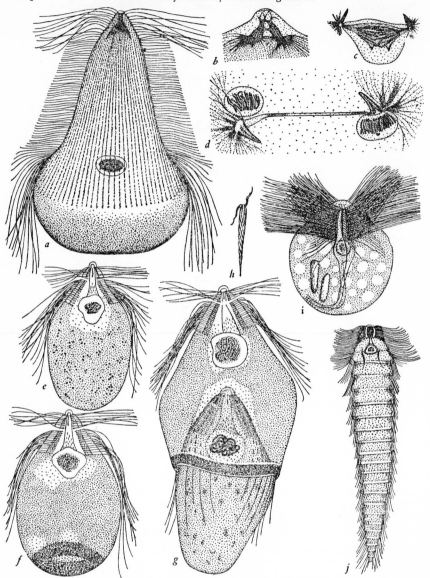

FIG. 32.—HYPERMASTIGINA: **a-d,** *Trichonympha Campanula* after Kofoid & Swezy (1919); **a,** cell x 250; **b,** division of centroblepharoplast and formation of paradesmose, and **c** and **d,** earlier and later stages of mitosis x 500. **e, f, g,** Sperm, egg, and fertilization of *Trichonympha* sp. from the roach *Cryptocercus* after Cleveland (1948). **h,** *Hoplonympha Natator* x 250 after Light (1926). **i,** *Staurojoenina assimilis* x 250 after Kirby (1926). **j,** *Teratonympha mirabilis* after Koidzumi (1921).

nucleus in the anterior segment; flagella distributed generally on the surface, most abundant on an anterior beak. *Teratonympha* Koidzumi (*Cyclonympha* Dogiel), a single known species in *Reticulitermes*.

Class 2. **MYCETOZOA** de Bary

Order *Dermatocarpi* Persoon Syst. Meth. Fung. xiii (1801), in part.

Suborder *Myxogastres* Fries Syst. Myc. 3: 3 (1829); suborder *Trichospermi* Fries op. cit. 1: xlix (1832), in part.

Suborder *Myxomycetes* Link 1833.

Mycetozoen de Bary in Bot. Zeit. 16: 369 (1858); Zeit. wiss. Zool. 10: 88 (1859).

Stamm Myxomycetes Haeckel Gen. Morph. 2: xxvi (1866).

Class MYCETOZOA de Bary ex Rostafinski Versuch Systems Mycetozoen 1 (1873).

Division *Mycetozoa* and classes *Myxogasteres* and *Phytomyxini* Engler and Prantl Nat. Pflanzenfam. II Teil: 1 (1888).

Division *Myxothallophyta* Engler in Engler and Prantl Nat. Pflanzenfam. I Teil, Abt. 1: iii (1897).

Stamm *Myxophyta* Wettstein Handb. syst. Bot. 1: 49 (1901).

Division *Phytosarcodina, Myxothallophyta,* or *Myxomycetes* Engler Syllab. ed. 3: 1 (1903).

Division *Myxomycophyta* Tippo in Chron. Bot. 7: 205 (1942).

Order *Mycetozoida* Hall Protozoology 227 (1953).

Organisms whose walled resting cells produce in germination anteriorly unequally biflagellate cells; these giving rise to bodies called plasmodia, being multinucleate bodies of amoeboid character.

1. Predatory, subaerial, producing macroscopic
   spore-bearing fruits.
   2. Spores produced within the fruits..............Order 1. ENTERIDIEA.
   2. Spores produced on the surfaces of the
      fruits......................................Order 2. EXOSPOREA.
1. Parasitic, not producing definite fruits..............Order 3. PHYTOMYXIDA.

Order 1. **Enteridiea** [Enteridieae] Rostafinski Vers. 3 (1873).

Cohort *Endosporeae* and orders *Anemeae, Heterodermeae, Reticularieae, Amaurochaeteae, Calcareae,* and *Calonemeae* Rostafinski op. cit.

Order *Endosporea* Lankester in Enc. Brit. ed. 9, 19: 840 (1885).

Orders *Protodermieae* and *Columniferae* Rostafinski ex Berlese in Saccardo Sylloge 7: 328, 417 (1888).

Cohorts *Amaurosporales* and *Lamprosporales,* with numerous orders with names in -*aceae,* Lister Monog. Mycetozoa 21-23 (1894).

Subclass *Myxogastres* and orders *Physaraceae, Stemonitaceae, Cribrariaceae, Lycogalaceae,* and *Trichiaceae* Macbride North American Slime Molds 20 (1899).

Subsuborder (!) *Endosporinei* Poche in Arch. Prot. 30: 200 (1913).

Orders *Physarales, Stemonitales, Cribrariales, Lycogalales,* and *Trichiales* Macbride op. cit. ed. 2 (1922).

Order *Liceales* Macbride and Martin (1934).

Suborder *Eumycetozoina* Hall Protozoology 230 (1953).

Predatory Mycetozoa producing macroscopic fruits, these producing internal uninucleate spores. The type is *Lycogala,* the sole genus of the order as originally published.

The fruits of many examples are of the appearance of minute puffballs, and Persoon and Fries classified them as puffballs; Fries took note that they are *primitus mucilaginosi* and made them a suborder distinct from the proper puffballs. De Bary studied the non-reproductive stages; concluded "dass die Myxomyceten nicht dem Pflanzenreiche angehören, sondern dass sie *Thiere,* und zwar der Abtheilung der Rhizopoden angehörig, sind"; and renamed the group Mycetozoen. This name was apparently first published in Latin form, in the category of classes, by de Bary's student Rostafinski. Conventional botany continues to list Myxomycetes as a class of Fungi; conventional zoology makes the group an order of Rhizopoda or Sarcodina.

The spores germinate readily in water or appropriate solutions (Jahn, 1905; Gilbert, 1929; Smith, 1929). Their nuclei usually divide once or twice, during or just after germination; thus each spore produces from one to four naked cells.

It is in germinating spores that mitosis is most easily observed. Mitosis takes place in a clear area, about which some observers have found a persistent nuclear membrane. The spindle is sharp-pointed. Only a few observers (as Skupienski, 1927) have discerned definite centrosomes. When the one or two divisions associated with germination are complete, the flagella grow forth from the areas of the poles of the mitotic spindle. All earlier observers described the spores as uniflagellate, but Ellison (1945) and Elliott (1949) found them biflagellate. The flagella may be apparently equal or moderately unequal; or one of them may be very brief. Each nucleus remains connected to the base of the flagella by a conical body of clear protoplasm, the *Geisselglocke* of Jahn (Jahn, 1904; Howard, 1931).

The flagellate cells are not spores, but gametes; they fuse with each other. Skupienski (1917) affirms that they are of two mating types. Fusion is at first by pairs, and Howard (1931) found that each zygote develops into a plasmodium by itself. All other observers (de Bary, 1858, 1859; Cienkowski, 1863; Skupienski, 1917, 1927; Schünemann, 1930) have found the zygotes to fuse with each other and with further gametes. The flagella are lost. The nuclei fuse in pairs; those which fail to find partners are digested.

The cell formed by the fusion of zygotes and gametes is a young plasmodium. The term was coined by Cienkowski (1863, p. 326): "Das Protoplasmanetz der Myxomyceten werde ich mit den Namen Plasmodium bezeichnen." The plasmodium nourishes itself in predatory fashion, on fungus spores, bacteria, and other digestible objects, and grows accordingly. Mitosis occurs simultaneously in all nuclei of the plasmodium, and takes 20 to 40 minutes; it has accordingly only rarely been observed (Lister, 1893; Howard, 1932). Plasmodia do not ordinarily divide, but grow to great sizes. They are not very familiar objects because during most of their life they keep to dark and moist places, chiefly among vegetable remains. Drouth does not kill them; they can become dry and hard while retaining the capacity to resume activity upon the return of moisture. When an active plasmodium reaches a certain stage, its reactions change; it moves out into the light and to dry places. A plasmodium in this stage is conspicuous, being of the form of a network which may be many centimeters in diameter, in some species brilliantly colored. The whole is a single naked protoplast.

Each plasmodium proceeds to produce a fruit or fruits. The entire mass may heap itself up, or it may break up into portions, large or minute. In species whose plasmodia break up into small fragments, each of these may secrete a column of lifeless material, a millimeter or more in height, and ascend upon it. Each separate body of protoplasm secretes an external wall and begins to undergo cleavage within it. Harper (1900) described the details of the process. All authorities agree that the nuclei undergo a flare of divisions at this time (Strasburger, 1884; Harper, 1900, 1914; Bisby, 1914). It is almost certain that there are two flares of division, constituting the meiotic process, but few authorities have positively affirmed this (Schünemann, 1930)[1]. Cleavage is carried to the point of producing uninucleate protoplasts. While this is taking place, many species secrete a network of hollow tubes or a system of hollow fibers, called the capillitium, by deposition of lifeless material outside the cell membranes. In species which produce a true capillitium, all of the uninucleate protoplasts secrete walls and become spores. Strasburger found the capillitium and the walls of the spores to consist of impure cellulose; others have found no cellulose. In many species which do not produce a true capillitium, an analogous structure called a pseudocapillitium, consisting of solid bodies of various forms, is modelled from a part of the nucleate protoplasm which is deprived of its reproductive function and killed. In many species, much calcium carbonate is deposited in the wall, or both in the wall and in the capillitium or pseudocapillitium.

A small separate fruit is called a sporangium. A fruit of the form of a large mass, or of many sporangia not completely separate, is an aethalium. The spores are released by collapse of the outer wall.

These organisms are of no known economic importance. There are some forty genera, between four and five hundred species. As Lister remarked, the same species occur everywhere: collections from Colombia (Martin, 1938) and from Mount Shasta (Cooke, 1949) consist entirely of familiar species.

Rostafinski (1873) arranged the genera in two cohorts, seven orders, and nineteen tribes, the last with names in *-aceae*. His subsequent monograph of the group (1875) was regrettably published in a barbarous language, and is for nomenclatural purposes a nullity. All later systems are based on Rostafinski's original system. The group being essentially uniform, it is properly treated as a single order.

Definite families were first established by Lankester, mostly under names which Rostafinski had applied to tribes. Berlese (in Saccardo, 1888) provided a complete set of names in *-aceae*, valid under botanical rules; Poche provided a complete set in *-idae*, valid under zoological rules. Authorities have differed moderately as to the list of families; here, somewhat arbitrarily, fourteen are maintained.

   1. Capillitium none (order *Cribrariales* Mac-
     bride).
      2. Producing separate sporangia, pseudo-
       capillitium none.
         3. Sporangia shattering irregularly or
           opening through a terminal oper-
           culum........................Family 1. Liceacea.
         3. Sporangia opening through numer-
           ous pores, the walls becoming sieve-
           like...........................Family 2. Cribrariacea.
      2. Fruits aethalioid, pseudocapillitium
       present.

---

[1]While the present work was in proof, Wilson and Ross (1955) established the point that meiosis occurs immediately before the formation of spores.

3. Aethalia consisting of more or less separate sporangia.
    4. Sporangia tubular, opening through terminal pores........ Family 3. TUBIFERIDA.
    4. Sporangia indistinct, their walls becoming freely punctured and converted into a reticulate pseudocapillitium............. Family 4. RETICULARIACEA.
3. Aethalia not consisting of distinguishable sporangia................Family 5. LYCOGALACTIDA.
1. Capillitium present.
  2. Fruits without considerable deposits of calcium carbonate.
    3. Spores black or dark, capillitial hairs smooth (order *Stemonitales* Macbride).
      4. Fruits aethalioid, capillitium poorly defined, without a central axis..................... Family 6. AMAUROCHAETACEA.
      4. Fruits of separate sporangia with a definite capillitium including a central axis (columella).
        5. Capillitium spreading horizontally from the columella...................Family 7. STEMONITEA.
        5. Capillitium spreading chiefly from the summit of the columella.............Family 8. ENERTHENEMEA.
    3. Spores pallid or yellow (order *Trichiales* Macbride).
      4. Capillitial hairs smooth, unbranched or sparsely branched.
        5. Capillitial threads horizontal, attached at both ends.....................Family 9. MARGARITIDA.
        5. Capillitial threads running at random, not attached at the ends......... Family 10. PERICHAENACEA.
      4. Capillitium reticulate, sculptured, but not with spiral bands.. .Family 11. ARCYRIACEA.
      4. Capillitial threads unbranched or sparsely branched, sculptured with spiral bands.........Family 12. TRICHIACEA.
  2. Fruits containing considerable deposits of calcium carbonate (order *Physarales* Macbride).
    3. Calcium carbonate both in walls and in capillitium.................Family 13. PHYSAREA.

3. Calcium carbonate in walls but not
   in capillitium...................Family 14. DIDYMIACEA.

Family 1. **Liceacea** [Liceaceae] (Rostafinski) Lankester in Enc. Brit. ed. 9, 19: 841 (1885). Tribe *Liceaceae* Rostafinski Vers. 4 (1873). Order *Liceaceae* Lister Monog. Mycetozoa 149 (1894). Family *Liceidae* Doflein 1909. Family *Orcadellidae* Poche in Arch. Prot. 30: 200 (1913). Family *Orcadellaceae* Macbride N. Am. Slime Molds ed. 2: 203 (1922). Sporangia separate, sessile or stalked, without capillitium or pseudocapillitium, the walls shattering irregularly or opening by means of a terminal operculum. *Licea, Orcadella.*

Family 2. **Cribrariacea** [Cribrariaceae] (Rostafinski) Lankester l. c. Tribe *Cribrariaceae* Rostafinski op. cit. 5. Order *Cribrariaceae* Macbride N. Am. Slime Molds 20 (1899). Order *Heterodermaceae* Lister op. cit. 136. Family *Cribrariidae* Poche l. c. The wall of the stalked fruit becoming sieve-like. *Cribraria. Dictydium.*

Family 3. **Tubiferida** [Tubiferidae] Poche in Arch. Prot. 30: 200 (1913). Order *Tubulinaceae* Lister op. cit. 152 (1894). Family *Tubulinidae* Doflein 1909. Family *Tubiferaceae* Macbride in N. Am. Slime Molds ed. 2: 203 (1922). Aethalia consisting of tubular sporangia opening through terminal pores. *Tubifer* (its older name *Tubulina* preoccupied), *Lindbladia, Alwisia.*

Family 4. **Reticulariacea** [Reticulariaceae] (Rostafinski) Lankester l. c. Tribes *Dictydiaethaliaceae* and *Reticulariaceae* Rostafinski op. cit. 5, 6. Order *Reticulariaceae* Lister op. cit. 156. Family *Dictydiaethaliidae* Poche l.c. Aethalia of indistinct sporangia whose walls become porous and are converted into a reticulate pseudocapillitium. *Reticularia, Dictydiaethallium,* etc.

Family 5. **Lycogalactida** [Lycogalactidae] Poche in Arch. Prot. 30: 201 (1913). Tribe *Lycogalaceae* de Bary. Order *Lycogalaceae* Macbride N. Am. Slime Molds 20 (1899). Family *Lycogalaceae* Macbride and Martin Myxomycetes (1934). Aethalia with a pseudocapillitium, not divided into sporangia. *Lycogala,* the brownish fruits a few millimeters in diameter clustered on wood, of much the appearance of small puffballs.

Family 6. **Amaurochaetacea** [Amaurochaetaceae] (Rostafinski) Berlese in Saccardo Sylloge 7: 401 (1888). Tribe *Amaurochaetaceae* Rostafinski op. cit. 8. Order *Amaurochaetaceae* Lister op. cit. 134. Family *Amaurochaetidae* Doflein 1909. Fruits aethalioid with dark spores and a poorly defined capillitium without a central axis. *Amaurochaete.*

Family 7. **Stemonitea** Lankester in Enc. Brit. ed. 9, 19: 841 (1885). Tribes *Stemonitaceae* and *Brefeldiaceae* Rostafinski op. cit. 6, 8. Families *Stemonitaceae* and *Brefeldiaceae* Berlese in Saccardo op. cit. 390, 402. Order *Stemonitaceae* Macbride N. Am. Slime Molds 20 (1899). Family *Stemonitidae* Doflein 1909. Families *Brefeldiidae* and *Stemonitidae* Poche op. cit. 202. Sporangia with dark spores and a capillitium of smooth threads spreading from a central axis, the columella. *Stemonitis,* common, the clustered stalked fruits of the appearance of minuscule dark bottle-brushes. *Brefeldia, Comatricha; Diachea,* exceptional in containing much lime in the stalk and wall.

Family 8. **Enerthenemea** Lankester l. c. Tribes *Echinosteliaceae* and *Enerthenemaceae* Rostafinski op. cit. 7, 8. Families *Echinosteliaceae* and *Enerthenemaceae* Berlese in Saccardo op. cit. 389, 402. Family *Lamprodermaceae* Macbride N. Am. Slime Molds ed. 2: 189 (1922). Like Stemonitea, in which this family has usually been included, but the capillitium attached chiefly at the summit of the columella. *Enerthenema, Clastoderma, Lamproderma, Echinostelium.*

Family 9. **Margaritida** [Margaritidae] Doflein 1909. Order *Margaritaceae* Lister op. cit. 202. Family *Dianemaceae* Macbride N. Am. Slime Molds ed. 2: 237 (1922). Sporangia with pale or yellow spores and a capillitium of smooth threads attached at both ends. *Dianema, Margarita.*

Family 10. **Perichaenacea** [Perichaenaceae] (Rostafinski) Lankester l. c. Tribe *Perichaenaceae* Rostafinski op. cit. 15. Sporangia with pale or yellow spores and a capillitium of unattached smooth threads. *Perichaena, Ophiotheca.*

Family 11. **Arcyriacea** [Arcyriaceae] (Rostafinski) Lankester l. c. Tribe *Arcyriaceae* Rostafinski op. cit. 15. Order *Arcyriaceae* Lister op. cit. 182. Family *Arcyriidae* Doflein 1909. Sporangia with pale or yellow spores and a reticulate capillitium, usually sculptured, but not with spiral bands. *Arcyria, Lachnobolus.*

Family 12. **Trichiacea** [Trichiaceae] (Rostafinski) Berlese in Saccardo Sylloge 7: 437 (1888). Tribe *Trichiaceae* Rostafinski op. cit. 14. Family *Trichinaceae* Lankes-

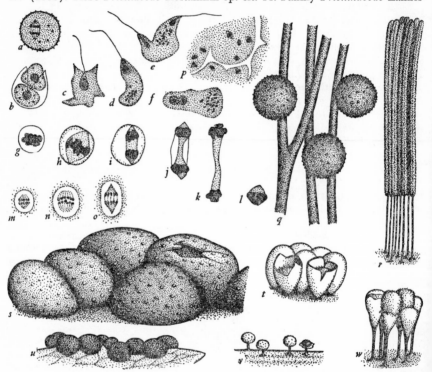

Fig. 33.—Mycetozoa. **a-f**, Spore, germination, gametes, syngamy, and zygote of *Physarum polycephalum* after Howard (1931) x 1,000. **g-l**, Stages of mitosis in the plasmodium of *Physarum polycephalum* after Howard (1932) x 2,000. **m-o**, Stages of mitosis in the plasmodium of *Trichia* after Lister (1893) x 1,000. **p**, Cleavage in the developing fruit of *Physarum polycephalum* after Howard (1931) x 1,000. **q**, Capillitium and spores of *Lepidoderma Chailletii* x 1,000. **r-w**, fruits of Mycetoza x 5; **r**, *Stemonitis splendens;* **s**, *Lycogala epidendrum;* **t**, *Leocarpus fragilis;* **u**, *Lepidoderma Chaillettii;* **v**, *Physarum notabile;* **w**, *Hemitrichia intorta.*

ter l. c.; the genus *Trichina* does not belong to this family! Order *Trichiaceae* Macbride N. Am. Slime Molds 20 (1899). Family *Trichiidae* Doflein 1909. Sporangia with pale or yellow spores, the capillitium of free threads, unbranched or sparsely branched, marked with spiral bands. *Trichia, Hemitrichia, Oligonema, Calonema.*

Family 13. **Physarea** Lankester l. c. Tribes *Cienkowskiaceae, Physaraceae,* and *Spumariaceae* Rostafinski op. cit. 9, 13. Families *Cienkowskiaceae, Physaraceae,* and *Spumariaceae* Berlese in Saccardo op. cit. 328, 329, 387. Order *Physaraceae* Macbride N. Am. Slime Molds 20 (1899). Family *Physaridae* Doflein 1909. Fruits sporangial or aethalioid, with capillitium, both wall and capillitium containing considerable deposits of calcium carbonate. *Physarum,* with some seventy-five species, is the most numerous genus of Mycetozoa; the little gray sporangia may be spherical or irregular, sessile or stalked. *Fuligo septica* produces dirty yellow aethalia reaching several centimeters in diameter on vegetable trash; observed on spent tan bark, it has the common name of flowers of tan. *Badhamia, Craterium, Leocarpus, Chondrioderma, Spumaria,* etc.

Family 14. **Didymiacea** [Didymiaceae] (Rostafinski) Lankester l. c. Tribe *Didymiaceae* Rostafinski op cit. 12. Order *Didymiaceae* Lister op. cit. 93. Family *Didymidae* Doflein 1909. Family *Didymiidae* Poche op. cit. 202. Family *Collodermataceae* Macbride and Martin Myxomycetes 145 (1934). Sporangia with deposits of calcium carbonate in the wall and a simple capillitium free of mineral deposits. *Didymium, Leangium, Lepidoderma, Colloderma.*

Order 2. **Exosporea** (Rostafinski) Lankester in Enc. Brit. ed. 9, 19: 841 (1885).
Cohors *Exosporeae* Rostafinski Vers. 2 (1873).
Order *Ectosporeae* Engler Syllab. 2 (1892).
Order *Ceratiomyxaceae* (Schröter) Lister Monog. Mycetozoa 25 (1894).
Subsuborder (!) *Exosporinei* Poche in Arch. Prot. 30: 200 (1913).
Organisms of much the character of the Enteridiea, but the spores forming a single layer on the surface of the fruits. There is a single family with only one well-marked species.

Family **Ceratiomyxacea** [Ceratiomyxaceae] Schröter (in Engler and Prantl, 1889). *Ceratiomyxa* Schröter (*Ceratium* Albertini and Schweinitz, 1805, non Schrank, 1793); *C. fruticulosa* (O. F. Müller) Macbride. The fruits are white pillars, sometimes branched, 1-2 mm. tall, of secreted material. Each spore of the single superficial layer generates a microscopic stalk and ascends upon it before becoming walled. Meiosis then takes place, making the spores 4-nucleate; the chromosome number is cut from 16 to 8 (Gilbert, 1935). In germination, the contents of the spore are released as a single amoeboid protoplast, whose nuclei divide once; the cell then divides into eight, and these generate flagella (Rostafinski, 1873; Jahn, 1905; Gilbert, 1935).

Order 3. **Phytomyxida** Calkins Biol. Prot. 330 (1926).
Class *Phytomyxini* Engler and Prantl Nat. Pflanzenfam. II Teil: 1 (1889); class *Phytomyxinae* op. cit. I Teil, Abt. 1: iii (1897).
Order *Phytomyxinae* Campbell Univ. Textb. Bot. 71 (1902).
Class *Plasmodiophorales* Engler Syllab. ed. 3: 1 (1903).
Order *Plasmodiophorales* Sparrow in Mycologia 34: 115 (1942).
Suborder *Plasmodiophorina* Hall Protozoology 228 (1953).
Intracellular parasites chiefly of higher plants, attacking also algae, Oomycetes, and beetles, being naked multicellular plasmodia producing walled resting cells,

the walls containing no cellulose; these releasing naked infective cells with paired unequal simple flagella.

This inconsiderable group was made known by the discovery of *Plasmodiophora Brassicae,* the agent of the clubroot disease of cabbage, by Woronin (1878). The proper place of the group in classification has been a puzzle; some students treat it as a class of myxomycetes, others as an order of chytrids. The known characters— paired unequal simple flagella; cells naked in the vegetative condition; and non-production of cellulose—assure us that this group has nothing to do with proper chytrids, nor with Oomycetes of chytrid body type. The traditional association with myxomycetes is tenable. Alternatively, the group would not be out of place next to order Rhizoflagellata (anyone who chooses to put it there should take note that the class name Phytomyxini is older than Zoomastigoda).

The plasmodium causes often much hypertrophy of the host tissue. In some forms the mature plasmodium becomes walled; the protoplast undergoes cleavage into uninucleate portions; these become swimming cells and are released through a discharge tube. These forms are of much the appearance of Lagenidialea. In the majority of the group the naked plasmodium undergoes cleavage; the resulting protoplasts become walled; the resulting spores or cysts, released by decay of the host, discharge their contents as one or two swimming cells. Ledingham (1939) and Sparrow (1947) report both types of development as occurring in *Polymyxa.* Karling (1944) found the walls to contain no cellulose. Ellison (1945) found the flagella to be simple.

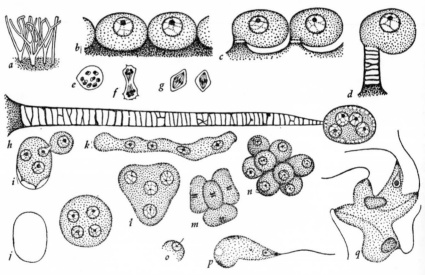

Fig. 34.—*Ceratiomyxa fruticulosa.* **a,** Fruits x 5. **b-q,** reproductive processes after Gilbert (1935); **b,** young spores on the surfaces of the fruit; **c, d,** the same raised on stalks; **e, f,** heterotypic division; **g,** homeotypic division; **h,** the mature spore on its stalk; **i-n,** germination and subsequent processes: the amoeboid protoplast passes through a "thread stage" before rounding up and dividing into four and then into eight; **o,** production of flagellum; **p,** "zoospore" (gamete); **q,** gametes fusing to initiate the plasmodium. All x 1,000 except Fig. **a.**

In the growing plasmodium, a nucleus which is not dividing contains an endosome ("nucleolus"). During mitosis, which occurs within the intact nuclear membrane, the endosome becomes elongate, and a ring of chromatin, within which separate chromosomes have not been distinguished, forms about its middle. The resulting "cruciform" figure resembles some which have been seen in trypanosomes. The nuclear divisions which occur immediately before cleavage are of a different character: no endosome is seen, but there is a spindle with centrosomes at the poles, and definite chromosomes are present. The occurrence of these two types of nuclear division has been noted by every careful observer, Schwartz (1914), Horne (1930), Cook (1933), Ledingham (1939), and Karling (1944). Horne was probably correct in supposing the divisions which precede cleavage to be meiotic. Conjugation of the flagellate cells of *Spongospora* has been observed.

There are monographic accounts of the Phytomyxida by Cook (1933) and Karling (1942). The group may be treated as a single family with a dozen genera and about twenty-five species.

Family **Plasmodiophorea** [Plasmodiophoreae] Berlese in Saccardo Sylloge 7: 464 (1888). Family *Plasmodiophoreen* Zopf Pilzthiere 129 (1885). Family *Plasmodiophoraceae* Engler Syllab. 1 (1892). Family *Woroninaceae* Minden 1911. Families *Phytomyxidae* and *Woroninidae* Poche in Arch Prot. 30: 198 (1913). *Plasmodiophora, Polymyxa, Spongospora,* and *Sorosphaera* attack land plants; *Tetramyxa, Ligniera,* and *Sorodiscus,* chiefly aquatic seed plants; *Woronina* and *Octomyxa,* Oomycetes; *Phagomyxa,* brown algae; *Sporomyxa* (Léger, 1908) and *Mycetosporidium,* beetles.

Class 3. **RHIZOPODA** Siebold

Order *Foraminifères* d' Orbigny in Ann. Sci. Nat. 7: 128, 245 (1826).
Order FORAMINIFERA Zborewski 1834.
Rhizopodes Dujardin in Compt. Rend. 1: 338 (1835).
Class *Foraminifera* d'Orbigny in de la Sagra Hist. Cuba vol. 8 (1839).
Order *Polythalamia* Ehrenberg in Abh. Akad. Wiss. Berlin (1838): table 1 (1839).
Class RHIZOPODA and orders MONOSOMATIA and *Polysomata* Siebold in Siebold and Stannius Lehrb. vergl. Anat. 1: 3, 11 (1848).
Reticulosa Carpenter 1862.
Stamm Rhizopoda and Class *Acyttaria* Haeckel Gen. Morph. 2: xxvii (1866).
Thalamophora R. Hertwig Hist. Radiolar. 82 (1876).
Class *Reticularia* Lankester in Enc. Brit. ed. 9, 19: 845 (1885).
Order *Reticulosa* Poche in Arch. Prot. 30: 203 (1913).
Order *Granuloreticulosa* de Saedeleer in Mem. Mus. Roy. Hist. Nat. Belgique 60: 7 (1934).
Order *Foraminiferida* Hall Protozoology 250 (1953).
Amoeboid organisms, the pseudopodia of the character of rhizopodia, i.e., fine, freely branching and anastomosing; producing shells, these usually calcareous; commonly reaching macroscopic dimensions; mostly marine.

The first examples of *rhizopodes* mentioned by Dujardin were *milioles, vorticiales,* and *le gromia:* the genus *Miliola* is to be construed as the type. These organisms, the proper rhizopods, are in general usage called Foraminifera, but that name was originally applied in the category of orders.

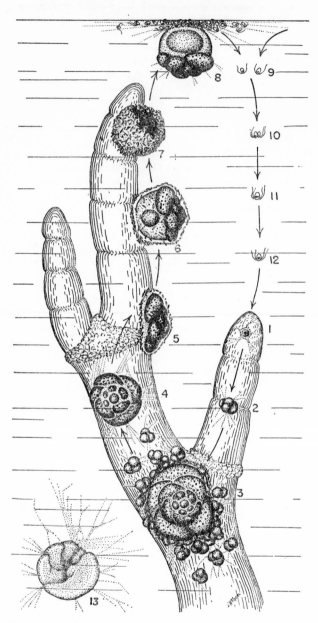

Fig. 35.—Life cycle of *"Tretomphalus,"* i.e., *Discorbis* or *Cymbalopora,* from Myers (1943); **1-3,** microspheric individuals, in **3** releasing young megalospheric individuals; **4-8** megalospheric individuals; **9-12,** gametes and syngamy.

The individual rhizopod originates as a minute amoeboid cell which secretes a shell from which the pseudopodia project. In the fresh-water forms, each protoplast, after moderate growth, divides into two, one of which retains the original shell while the other secretes a new one. In some of the marine forms, the original protoplast, having a cylindrical or irregular shell, enlarges this as it grows. In the great majority of the group, the original shell, called the proloculus, is of definite size and form and has a constricted orifice. When the protoplast reaches a certain stage, it expands, protrudes from the orifice, and secretes an extension of the shell in the form of a second chamber. In some few examples, the second chamber is the final one, being capable of indefinite extension. But again in the great majority, the second chamber, although different from the proloculus, resembles it in being definite in form and in having a constricted orifice. After further development, the protoplast again protrudes through the orifice and secretes a third chamber, generally of the same form as the second, though often larger. Repetition of this process produces macroscopically visible bodies. Even though becoming a centimeter or more in diameter, the individuals continue to be single cells.

As a result of different patterns of growth, the developed shells are of highly varied forms, linear, globular, or coiled in one plane; trochoid or rotaloid, that is, helical, of the form of a low cone; of the form of high cones; or screw-like, with the chambers in fixed longitudinal rows. The growth pattern may change during the life of the individual. There are apparently degenerate forms, simple or irregular. It is highly probable that some of the forms have evolved repeatedly.

The shells may be of gelatinous material or of chitin, without or with imbedded grains of sand. Exceptionally, they are siliceous. They are sometimes of crystallized calcium carbonate with imbedded grains of sand. In the bulk of the group they consist of crystallized calcium carbonate without foreign matter, and are of either of two types of texture: vitreous, that is, hyaline, and punctured by numerous pores a few microns in diameter; or porcellanous, white by reflected light and amber by transmitted light, and with no perforations except the proper orifices. In fossil shells, other textures than these may occur; it is supposed that these are products of modification during preservation. Some of the textures, like some of the forms, are believed to have evolved repeatedly.

Most rhizopods occur in two forms which are most readily distinguished by the size of the proloculi. This was first pointed out by Munier-Calmas, 1880; who, jointly with Schlumberger, 1885, designated the smaller and larger proloculi respectively *microspères* and *mégasphères*. Lister (1895), by study in culture of *Elphidium crispum* (*Polystomella crispa* Lamarck), showed that the two forms are alternate generations. He observed that the microspheric cells become multinucleate during growth, while the megalospheric cells remain uninucleate until just before reproduction. The reproduction of the megalospheric cells is by release of numerous minute biflagellate cells.

Schaudinn (1902) confirmed much of what Lister had observed. He was mistaken in describing nuclear division (except just before the production of the swimming cells) as non-mitotic; and correct in identifying the swimming cells as gametes. Winter (1907) observed a similar life cycle in *Peneroplis,* but described the gametes as having solitary flagella.

Myers (1934, 1935, 1936), dealing with *Patellina* and *Spirillina,* described the details of mitosis. This takes place within an intact nuclear membrane, and is completed by its constriction. The spindle is blunt-ended; there is no evidence of centrosomes. The chromosomes are numerous, long, and slender; the mitotic figures resemble those of Pyrrhophyta. Reduction of the chromosome number is said to be effected by a single nuclear division, the last one before the formation of gametes, which cuts the chromosome number of *Patellina* from 24 to 12, and that of *Spirillina* from 12 to 6. Before they reach this stage, the megalospheric individuals have gathered themselves in clusters of two or more within cyst walls consisting of secreted gelatinous matter and scraps from the neighborhood. Gametes from one individual are unable to unite with each other. The gametes are amoeboid, positively without flagella. In *Discorbis* and *Cymbalopora,* however, Myers (1943) observed the production of biflagellate gametes.

Le Calvez (1950) has cleared up various questions raised by earlier studies. Some forms, as *Discorbis orbicularis,* appear to lack a sexual cycle. *Patellina* and *Spirillina* produce amoeboid gametes 40-50μ in diameter. Most rhizopods produce biflagellate gametes 1.5-4μ long. Le Calvez found the flagella definitely unequal. In *Discorbis mediterranensis* he showed that the megalospheric individuals are of two mating types. Earlier zoologists, apparently misled by familiarity with the normal life cycle of animals, had identified meiosis as occurring at the time of gametogenesis; it is the fact, on the contrary, that it occurs in the last two nuclear divisions in the microspheric individuals. The megalospheric and microspheric stages of rhizopods are respectively haploid and diploid, like the gametophytes and sporophytes of plants.

With the possible exception of some of the one-chambered fresh water forms, the rhizopods are clearly a natural group. The fresh water forms appear to intergrade with organisms which Pascher identified as chrysomonads.

The shells of dead rhizopods may under appropriate conditions be preserved through geologic ages. Natural chalk consists of shells of *Textularia* mixed with coccoliths. Certain forms of limestone consist chiefly of shells of *Miliola.* Certain fossil rhizopods have long been known as indicators of division of geologic time. Since about 1917, it has been found that the whole group offers one of the beautiful illustrations of evolution as related to geologic time: the shells of rhizopods found under magnification in a particular stratum serve promptly and precisely to identify it. The services of experts on "Foraminifera" have acquired a high economic value in the petroleum industry: these experts have found themselves promoted from the status of pure biologists to that of economic geologists.

Among some eleven hundred genera which have been published, Galloway (1933) maintains 542. Of the number of species one can only say that it is a matter of thousands, but probably not many tens of thousands. Economic micropaleontologists find themselves dealing with great numbers of forms which are slightly, yet significantly, distinct. They find it expedient not to name these, but to identify them by comparison with available collections.

Some of the marine and fossil forms are similar, on a small scale, to the animal *Nautilus,* and Linnaeus placed some of them in that genus. Montfort and Lamarck treated them as several genera of mollusks. In first distinguishing these organisms as the order *Foraminifères* of class *Céphalopodes,* d'Orbigny intended to contrast them with *Nautilus,* in whose shells a series of chambers are connected, not by holes (*foramina*) but by cylindrical tubes. Dujardin (1835) found that his *Rhizopodes* are without definite organs. Their shells enclose a clear semiliquid substance; their apparent tentacles are merely temporary structures, formed of this substance, thrust forward in the direction of the movement of the shell and withdrawn as it advances. Dujardin named this substance *sarcode;* it is, of course, the same which has since been called protoplasm. The effect of his discoveries was to show that the rhizopods or Foraminifera are not mollusks, but one-celled organisms.

Very much taxonomic study has been given to this interesting group. The standard system, in the modern period of practical concern with the group, has been that of Cushman (1928).

Galloway (1933), attempting to recognize phylogeny and concluding that certain types of form and texture of shells have evolved repeatedly, has radically revised Cushman's system and set up a system of thirty-five families. The following survey of the group is based on Galloway's system. The names applied to the families are those which he has cited as the oldest, and the groups treated as orders are the blocks of families which to him appeared natural.

1. Shell one-chambered, or of a proloculus followed by one other chamber, not of a series
of similar chambers........................... Order 1. MONOSOMATIA.
1. Shell a series of similar chambers.
   2. Shell porcellanous, imperforate.............Order 2. MILIOLIDEA.
   2. Not as above.
      3. Not specialized as in the following
      orders.............................Order 3. FORAMINIFERA.
      3. Shell hyaline, perforate, typically
      trochoid, i.e., having the succes
      sively larger chambers helically ar
      ranged so that all may be seen from
      one side and only the last whorl
      from the other.......................Order 4. GLOBIGERINIDEA.
      3. Chambers of the fundamentally
      planispiral shell with specialized
      walls containing channels or pro
      ducing chamberlets...................Order 5. NUMMULITINIDEA.

Order 1. **Monosomatia** (Ehrenberg) Siebold in Siebold and Stannius Lehrb. vergl. Anat. 1: 11 (1848).

MONOSOMATIA Ehrenberg in Abh. Akad. Wiss. Berlin (1838): table 1 (1839).
Order *Astrorhizidea* Lankester in Enc. Brit. ed. 9. 19: 846 (1885).
Order *Imperforida* Delage and Hérouard Traité Zool. 1: 107 (1896).
Order *Archi-Monothalamia* Calkins Biol. Prot. 354 (1926).

Rhizopoda consisting of a single chamber, or of a proloculus followed by one other chamber; exceptionally, after passing through a stage of this character, producing a series of similar chambers.

Family 1. **Allogromiida** [Allogromiidae] Cash and Wailes. Minute, with one-chambered chitinous or gelatinous shells, usually subglobular; large in fresh water. *Allogromia* Rhumbler; *Mikrogromia* Hertwig, the pseudopods of sister cells retaining contact so that small colonies are formed; etc.

Family 2. **Astrorhizida** [Astorhizidae] Brady (1881). Family *Astrorhizina* Lankester (1885). Family *Astrorhizidaceae* Lister. Families *Rhizamminidae, Saccamminidae,* and *Hyperamminidae* Cushman. Shell of agglutinated foreign material, usually elongate, often branched, but not coiled. In *Astrorhiza* there is a central chamber from which grow elongate arms. In *Rhizammina,* the shell is tubular, open at both ends; in *Bathysiphon* it is a tube closed at one end; in *Hyperammina* a proloculus is formed before the extended tube.

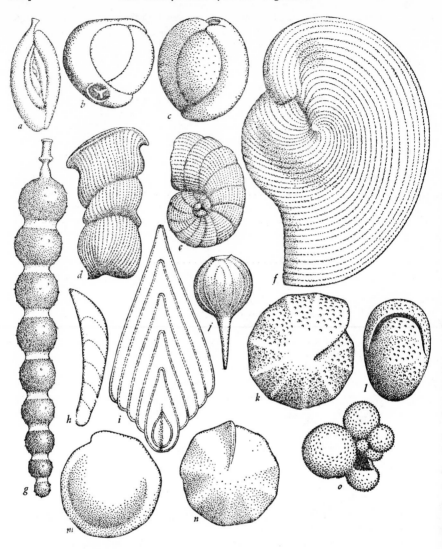

Fɪɢ. 36—Shells of Rhizopoda. **a,** *Ophthalimidium.* **b, c,** *Triloculina.* **d,** *Vertebralina.* **e,** *Peneroplis.* **f,** *Archaias* x 25. **g,** *Nodosaria.* **h,** *Dentilina.* **i,** *Flabellina.* **j,** *Lagena.* **k, l,** *Nonion.* **m, n,** *Rotalia.* **o,** *Globigerina.* x 50 except as noted.

Family 3. **Spirillinidea** Reuss 1861. Family *Spirillinina* Lankester (1885). Family *Silicinidae* Cushman. In *Spirillina*, the perforate hyaline one-chambered shell is planispirally coiled; the family is distinguished by a shell of this form in the young stages if not throughout life. *Silicina* is a Jurassic fossil whose shell is silicified. In *Patellina* the spirally coiled first chamber is followed by others arranged in a helix.

Family 4. **Ammodiscida** [Ammodiscidae] Rhumbler 1895. Like the preceding family, but the shell consisting of agglutinated foreign matter. *Ammodiscus* etc.

Order 2. **Miliolidea** Lankester in Enc. Brit. ed. 9, 19: 846 (1885).

Order *Flexostylida* Calkins Biol. Prot. 355 (1926).

Rhizopoda with imperforate porcellanous shells, a numerous and important group.

Family 1. **Miliolida** [Miliolidae] d'Orbigny (1839). Families *Nubecularina, Miliolina*, and *Hauerinina* Lankester (1885). *Fisherinidae* Cushman. The genus *Cornuspira*, known from the carboniferous, differs from *Spirillina* only in the texture. Evidently evolved from this are genera of planispirally coiled tubes divided into chambers, and from these others in which the series of chambers becomes straight or irregular, as in *Vertebralina* and *Tubinella*. There is an important block of genera in which each cycle of chambers is of two members, the second opening at the opposite end of the body from the first. In *Ophthalmidium* and *Pyrgo* alternate chambers lie regularly on opposite sides of a body whose form is that of an elliptic flake. In other genera of this group successive chambers are not opposite each other, but separated by less than 180°, so that more than two appear on the outside. In *Triloculina* three chambers are externally visible. In *Miliola* Lamarck (*Miliolina* Lamarck, the latter name applied to fossil representatives of the same genus) the chambers are 144° apart, so that five appear on the outside. In many members of the family the apertures are partially blocked by teeth, single, double, or multiple, or extended as bars clear across.

Family 2. **Soritina** Ehrenberg (1839). Family *Helicosorina* Ehrenberg op. cit. Family *Peneroplidea* Reuss 1861. Family *Peneroplidina* Lankester (1885). Family *Peneroplidae* Cushman. Family *Soritidae* Galloway (1933). Specialized derivatives of the lower Miliolida: planispiral shells in which the chambers become successively larger, as in *Peneroplis*, and, by a further development, divided into large numbers of secondary chambers, as in *Archaias, Sorites,* and *Orbitolites. Spirolina,* the shell coiled in the oldest part, straight in the remainder.

Family 3. **Alveolinea** Ehrenberg (1839). Family *Alveolinida* Schultze 1854. Families *Alveolinina* and *Keramosphaerina* Lankester, *Alveolinellidae* and *Keramosphaeridae* Cushman. Another group of specialized derivatives, the planispirally coiling chambers broadened and divided into many chamberlets with separate apertures, the entire body more or less globular. *Borelis; Fasciolites* Parkinson 1811 (*Alveolina* d'Orbigny 1826); *Alveolinella; Keramosphaera* Brody, a rare antarctic form. The organisms of these last two families resemble, as a parallel development, those of order Nummulitinidea, from which they are distinguished by the texture.

Order 3. **Foraminifera** Zborewski 1834.

Order *Polythalamia* and subordinate group *Polysomatia* Ehrenberg in Abh. Akad. Wiss. Berlin (1838): table 1 (1839).

Order *Polysomatia* Siebold in Siebold and Stannius Lehrb. vergl. Anat. 1: 11 (1848).

Orders *Lituolidea, Textularidea,* and *Lagenidea* Lankester in Enc. Brit. ed. 9, 19: 847 (1885).

Order *Perforida* Delage and Hérouard Traité Zool. 1: 107 (1896).
Orders *Nodosalida* and *Textulinida* Calkins Biol. Prot. 355, 356 (1926).
Comparatively unspecialized Rhizopoda, the shells of various textures, not porcellanous; not usually of trochoid form, and if so, usually not vitreous.

Early students, Montfort, Lamarck, and d'Orbigny, were much concerned with organisms which they called *Geophonus, Vorticialis,* or *Polystomella.* These names represent organisms of much the appearance of *Nautilus;* all are synonyms of *Elphidium* Montfort, which is to be considered the type or standard genus of Foraminifera.

Family 1. **Endothyrina** Lankester (1885). Family *Endothyridae* Rhumbler 1895. Fossils, pre-Cambrian to Carboniferous, the calcareous shells granular or fibrous, not porcellanous or vitreous. *Cayeuxina* Galloway (1933) includes minute globular shells solitary or irregularly clustered, described by Cayeux, 1894, from the pre-Cambrian of Brittany; *Matthewina* Galloway includes Cambrian fossils of similar character. *Endothyra* and *Cribrospira* are Carboniferous forms, planispirally coiled; *Tetrataxis* produced trochoid shells.

Family 2. **Nodosinellida** [Nodosinellidae] Rhumbler 1895. Shells like those of the Endothyrina or containing imbedded grains of sand, one-chambered or forming straight or curved, not coiled, rows. Mostly Carboniferous, rare as late as the Eocene. *Archaelagena, Nodosinella, Nodosaroum, Pedangia,* etc.

Family 3. **Reophacida** [Reophacidae] Cushman 1827. A small group of forms apparently degenerate from the foregoing, the chambers in straight, curved, or irregular series, walls chitinous or sandy; sometimes parasitic in other rhizopods. *Reophax,* etc., surviving to the present in cold deep water.

Family 4. **Trochamminida** [Trochamminidae] Schwager 1877. Family *Trochamminina* Lankester (1885). Family *Plocapsilinidae* Cushman. Cells planispiral or trochoid, becoming evolute or irregular; walls with imbedded grains of sand. Pennsylvanian to recent, abundant only in the Cretaceous. *Trochammina, Plocapsilina,* etc.

Family 5. **Lituolidea** Reuss 1861. Family *Lituolidae* Brady (1881). Families *Lituolina* and *Loftusiina* Lankester. Family *Lituolidaceae* Lister. Families *Loftusiidae* and *Neusinidae* Cushman. Shells spiral or becoming evolute or irregular, with walls of agglutinated siliceous or calcareous matter, the chambers subdivided as in order Nummulitinidea. *Cyclammina, Lituola, Loftusia, Neusina,* etc.; Mississippian to recent, most abundant in the Cretaceous and at present.

Family 6. **Orbitolinida** [Orbitolinidae] Martin 1890. Specialized derivatives of the preceding family, walls agglutinated as in that group, the numerous chambers forming a conical or nearly circular body. *Dictyoconus, Orbitolina,* etc. Mesozoic and Eocene.

Family 7. **Ataxophragmidea** Schwager 1877. Families *Valvulinidae* and *Verneulinidae* Cushman 1927. Family *Ataxophragmidae* Galloway (1933). Having walls of agglutinated material and allied to the preceding families; chambers of the shell tending to form an elongate, screw-like spiral. *Valvulina, Ataxophragmium, Verneulina,* etc.; since early Mesozoic, abundant in the present.

Family 8. **Textularina** Ehrenberg (1839). Family *Textularidae* d'Orbigny (1839). Family *Textulariaceae* Lister. Walls more or less agglutinated, the chambers usually in an elongate spiral with two members to a cycle, so that they form two series, the body as a whole tending to be wedge-shaped. *Textularia, Cuneolina, Vulvulina,* etc.; Ordovician to the present.

Family 9. **Nodosarina** Ehrenberg (1839). Family *Nodosarida* Schultze 1854. Family *Lagenidae* Brady (1881). Family *Lagenina* Lankester (1885). Family *Nodo-*

saridae Rhumbler. Family *Lagenaceae* Lister. Walls calcareous, hyaline, perforate; chambers planispiral in the earliest forms, becoming curved or straight in the majority; orifice ordinarily of radiating slits, becoming reduced to a single slit. A numerous group, Triassic to the present. *Lenticulina* Lamarck (*Lenticulites* Lamarck and *Cristellaria* Lamarck are synonyms) is *Nautilus*-like. *Hemicristellaria* and *Vaginulina* resemble the sheath of a dagger; *Flabellina* and *Frondicularia* resemble fans; *Glandulina* is shaped like a jug. *Nodosaria* is like a row of enlarging beads. *Lagena* is a one-chambered form connected to *Nodosaria* by transitions, and evidently reduced, not primitive.

Family 10. **Polymorphinida** [Polymorphinidae] d'Orbigny. Families *Polymorphinina* and *Ramulinina* Lankester (1885). Specialized irregular forms related to the preceding, as indicated by orifices of the same character. *Polymorphina,* etc., present in the Mesozoic, abundant in the Cenozoic to the present.

Family 11. **Nonionidea** [Noninideae] Reuss 1860. Family *Polystomellina* Lankester (1885). Family *Hantkeninidae* Cushman. Shells mostly nautiloid, that is, planispiral with successively larger chambers, a few of the highest trochoid; walls hyaline, perforate; aperture generally a transverse slit. *Nonion* Montfort (*Nonionina* d'Orbigny) and *Elphidium* Montfort (*Geophonus* Montfort, *Vorticialis* Lamarck, *Polystomella* Lamarck, the apparent type of Foraminifera) are simply nautiloid; *Hantkenina* is ornamented with spines. Jurassic to the present.

Order 4. **Globigerinidea** Lankester in Enc. Brit. ed. 9, 19: 847 (1885).

Orders *Rotalidea* and *Chilostomellida* Lankester l. c., both names having previous use in the category of families.

Order *Rotalida* Calkins Biol. Prot. 356 (1926).

The main body of Rhizopoda with perforate hyaline shells, many-chambered, the chambers primitively of the trochoid arrangement.

Family 1. **Rotalina** Ehrenberg (1839). Family *Rotalidea* Reuss 1861. Family *Rotalidae* Brady (1861). Family *Rotalina* Lankester. Family *Rotaliaceae* Lister. Families *Globorotaliidae, Anomalinidae,* and *Planorbulinidae* Cushman. A numerous family, including unspecialized forms, *Globorotalia, Rotalia,* etc., as well as degenerate and irregular forms, *Planopulvinulina,* etc., and moderately specialized ones with conical or disk-shaped bodies of numerous chambers, *Cymbalopora, Planorbulina,* etc. Triassic, rare; Jurassic to the present, common.

Family 2. **Acervulinida** Schultze 1854. Family *Rupertiidae* Cushman. A small group of degenerate derivatives of the foregoing, the bodies attached, irregular, sometimes reduced to one chamber. *Rupertia, Acervulina,* etc., Cretaceous to the present.

Family 3. **Tinoporidea** Schwager 1877. Family *Calcarinidae* Cushman. Another small group derived from Rotalina, the disk-shaped cells with a whorl of prominent spines. *Calcarina, Tinoporus,* etc. Cenozoic, to the present.

Family 4. **Asterigerinida** [Asterigerinidae] d'Orbigny (1839). Two genera, *Asterigerina* and *Amphistegina,* diverging from Rotalina in having each chamber divided into two by an oblique wall. Doubtfully in the Cretaceous; Eocene to the present.

Family 5. **Chapmaniida** [Chapmaniidae] Galloway (1933). The numerous chambers arranged in a low cone whose inside is filled with deposited solid material. *Chapmania, Halkyardia, Dictyoconoides.* Eocene and Oligocene.

Family 6. **Chilostomellida** [Chilostomellidae] Brady (1881). Family *Chilostomellaceae* Lister. A few genera of reduced derivatives of Rotalina with few chambers. *Allomorphina, Chilostomella, Sphaeroidina,* etc. Jurassic to the present.

Family 7. **Orbulinida** Schultze 1854. Family *Globigerinida* Carpenter 1862. A few genera with the chambers mostly few, subglobular, clustered rather than arranged in a definite pattern. *Orbulina. Globigerina,* abundant, pelagic in all oceans, the shells abundant in the ooze on the bottom. Pennsylvanian, doubtful; Jurassic, rare; Cretaceous to the present, common.

Family 8. **Pegidiida** [Pegidiidae] Heron-Allen and Earland 1928. A few genera much like the Orbulinida but with thinner walls. *Pegidia,* etc. Oligocene to the present.

Family 9. **Heterohelicida** [Heterohelicidae] Cushman 1927. A numerous group, the shells screw-like, biseriate, uniseriate, sheath-like or fan-like, the walls often with exterior ornamentation; paralleling the Nodosarina, but without the radiate orifices. *Heterohelix, Sagrina, Eouvigerina, Pavonina, Plectofrondicularia, Bolivina, Mucronina.* Common, Jurassic to the present.

Family 10. **Buliminida** Jones 1876. Family *Uvellina* Ehrenberg (1839), not based on a generic name. Family *Buliminina* Lankester (1885). Shells mostly high spirals, screw-like, often with spines or other external ornamentation, the orifices various, commonly comma-shaped. *Turrilina, Bulimina, Virgulina,* etc. Triassic to the present.

Family 11. **Cassidulinida** [Cassidulinidae] d'Orbigny (1839). A small group with high-spiralled shells and comma-shaped orifices, evidently derived from the foregoing family. *Cassidulina,* etc. Eocene to the present.

Family 12 **Uvigerinida** [Uvigerinidae] Galloway and Wissler, 1927. Further variants from Heterohelicida, the high-spiralled shells with chambers in three rows at first, varying to biseriate and uniseriate. *Uvigerina, Siphonogenerina,* etc. Jurassic to the present, common since the Miocene.

Family 13. **Pleurostomellida** [Pleurostomellidae] Reuss 1860. An additional rather small family of the same general character as the few preceding. *Pleurostomella, Nodosarella, Daucina, Ellipsoidina,* etc. Cretaceous to the present, commonest in upper Cretaceous and Eocene.

Order 5. **Nummulitinidea** Lankester in Enc. Brit. ed. 9, 19: 848 (1885).

Rhizopoda with large specialized shells, the walls hyaline, perforate, generally thickened and traversed by channels and thrown into internal ridges which subdivide the chambers.

Family 1. **Fusulinida** [Fusulinidae] Möller 1878. Carboniferous fossils, the chambers short and broad, numerous, in a planispiral coil, forming bodies which are usually fusiform or globular. *Orobias, Fusulina, Triticina, Verbeekina,* etc.

Family 2. **Nummulitida** [Nummulitidae] Reuss 1861. Family *Camerinidae* Meek and Hayden 1865. Family *Nummulinidae* Brady (1881). Family *Nummulitina* Lankester (1885). Family *Nummulitaceae* Lister. Mostly disk-shaped, planispiral, the walls not highly specialized. *Camerina* Bruguière 1792 (*Nummulites* Lamarck 1801), *Operculina, Heterostegina,* etc. Jurassic to the present, most abundant in the Eocene.

Family 3. **Orbitoidida** [Orbitoididae] Schubert 1920. Similar to the foregoing, the numerous chambers divided into numerous chamberlets. A considerable group of Mesozoic and Cenozoic fossils. *Orbitoides, Cyclosiphon,* etc.

Family 4. **Cycloclypeina** Lankester (1885). Family *Cycloclypeidae* Galloway (1933). Similar to the preceding. A number of Mesozoic and Cenozoic genera, most numerous in the Eocene. *Asterocyclina.* The only living species is *Cycloclypeus Carpenteri* Brady.

Class 4. **HELIOZOA** Haeckel

Family *Polycystina* Ehrenberg in Abh. Akad. Wiss. Berlin 1838: 128 (1839).
Rhizopoda radiaria seu RADIOLARIA J. Müller in Abh. Akad. Wiss. Berlin (1858):
   16 (1859).
Echinocystida Claparède.
Order RADIOLARIA Haeckel Radiolarien 243 (1862).
Stamm Moneres for the most part, and classes HELIOZOA and *Radiolaria,* Haeckel
   Gen. Morph. 2: xxii, xxviii, xxix (1866).
Subclasses *Heliozoa* and *Radiolaria* Bütschli in Brown Kl. u. Ord. Thierreichs 1, 1
   Teil: *Inhalt* (1882).
Class *Proteomyxa* Lankester in Enc. Brit. ed. 9, 19: 839 (1885).
Subclasses *Proteomyxiae, Heliozoariae,* and *Radiolariae* Delage and Hérouard
   Traité Zool. 1: 66, 156, 169 (1896).
Class *Actinopoda* Calkins Biol. Prot. 318 (1926).
Class *Actinopodea* and orders *Helioflagellida, Heliozoida, Radiolarida,* and *Proteo-*
   myxida Hall Protozoology 202, 203, 212, 220 (1953).
Subphylum *Actinopoda* Grassé and Deflandre, and classes *Acantharia, Radiolaria,*
   and HELIOZOA Trégouboff in Grassé Traité Zool. 1, fasc. 2: 267, 270, 321, 437
   (1953).
Organisms having pseudopodia of the character of filopodia, stiffly radiating, or of
axopodia, stiffly radiating and having inner fibers; often with siliceous skeletons.

Here, not without authority, one combines in one class the three groups which have
been treated as the classes Proteomyxa, Heliozoa, and Radiolaria; and adds further
two families of shelled amoebas.

Cienkowski (1865) listed as "Monaden" the new species or genera *Monas amyli,*
Colpodella (apparently a chytrid), *Pseudospora,* and *Vampyrella.* They are minute
fresh-water amoeboid organisms, in part having flagellate stages. Haeckel (1866)
placed most of them (the *Monas* under the new generic name *Protomonas*), together
with his own discoveries *Protamoeba* and *Protogenes,* and also the bacteria, in his
Stamm Moneres, i.e., his group of Protista without nuclei. Later (1868) he omitted
the bacteria. Zopf (1885) found several of Haeckel's Moneres to possess nuclei, and
Lankester renamed the group Proteomyxa. Publication of subsequent original observations of these organisms has been scant and scattered; they remain poorly known.

The Heliozoa as conventionally construed are also mostly inhabitants of fresh
water. Ehrenberg observed some of them and took them for Infusoria with immobile
cilia. There are only a few dozen species of Heliozoa *sensu stricto* (Schaudinn, 1896):
the whole group is no more than a reasonable order.

The Radiolaria (this name also used at this point in its conventional sense) are
marine. Examples were first observed as floating gelatinous bodies. These were taken
for fragments and remained unnamed until 1834, when Mayen named *Physematium*
and *Sphaerozoum.* Fossil skeletons of many examples were described by Ehrenberg
(1839). Huxley (1851) named *Thalassicolla* and gave an accurate account of its
structure. It was by work on organisms of this group that Haeckel first distinguished
himself (1862).

Haeckel dealt further with this group in four important papers (1879, 1882, 1887,
1887-1888). In the last of these, the Radiolaria are a class of four legions, eight sublegions, twenty orders, 85 families, 739 genera, and more than four thousand species.
The categories, Haeckel explained, are purely relative: Radiolaria would as well be

a phylum, the legions classes, and so forth. This idea served him as license for confounding the application of many names, by shifting them among the categories, or by substituting new names for old. All subsequent authors have followed Haeckel's system of Radiolaria, applying names as best they might.

The class Heliozoa in the extended sense here proposed may be organized as five orders distinguished as follows:

1. Cells without a central capsule, i.e., without
a firm membrane surrounding the inner part
of the protoplast.............................Order 1. RADIOFLAGELLATA.
1. Cells with a central capsule.
    2. Central capsule of spherical symmetry
    or with three planes of symmetry at
    right angles, punctured by many pores.
        3. Pores of the central capsule evenly
        distributed; skeleton absent or present, without spicules which cross the
        central capsule or meet in its center......Order 2. RADIOLARIA.
        3. Pores of the central capsule clustered; skeleton including spicules
        which cross the central capsule or
        meet in its middle....................Order 3. ACANTHARIA.
    2. Central capsule of radial symmetry, with
    one opening.............................Order 4. MONOPYLARIA.
    2. Central capsule of isobilateral symmetry,
    with one main opening and two minor
    ones...................................Order 5. PHAEOSPHAERIA.

Order 1. **Radioflagellata** Kent Man. Inf. 1: 225 (1880).
> Subdivision or subclass *Heliozoa* (Haeckel), and orders *Aphrothoraca* (Hertwig), *Chlamydophora* (Archer), *Desmothoraca* (Hertwig and Lesser), and *Chalarothoraca* (Hertwig and Lesser) Büschli in Bronn Kl. u. Ord. Thierreichs 1: 261, 320, *et seq.* (1881, 1882).
> Suborder *Protoplasta Filosa* Leidy in Rept. U.S. Geol. Survey Territories 12: 189 (1879).
> Class *Proteomyxa* Lankester (1885).
> Subclass *Proteomyxiae* and orders *Acystosporidia, Azoosporidia,* and *Zoosporidia;* subclass *Heliozoariae* and orders *Aphrothoracida, Chlamydophorida, Chalarothoracida,* and *Desmothoracida* Delage and Hérouard Traité Zool. 1: 66-72, 156-168 (1896).
> Order *Heliozoa* Doflein Protozoen 13 (1901).
> Orders *Vampyrellidea* and *Chlamydomyxidea* Poche in Arch. Prot. 30: 182, 193 (1913).

The proper Heliozoa together with the Proteomyxa: organisms of the character of the class, lacking central capsules, that is, firm membranes about the inner part of the protoplasts. Mostly fresh water organisms of spherical symmetry, commonly without skeletons. The type, being the only genus assigned to the order by Kent, is *Actinomonas.*

1. Pseudopodia unspecialized; amoeboid organisms with or without flagellate stages.

2. Without shells.......................... Family 1. PSEUDOSPOREA.
2. With shells; without known flagellate
stages.
   3. Shells chitinous, without siliceous
     scales..............................Family 2. LAGYNIDA.
   3. Shells bearing circular siliceous
     scales..............................Family 3. EUGLYPHIDA.
1. Pseudopodia slender, with apical knobs.......... Family 4. VAMPYRELLACEA.
1. Pseudopodia of the character of typical axo-
podia, without apical knobs; the cells or their
main bodies usually regularly spherical.
   2. Bearing flagella as well as axopodia in
     the vegetative condition...................Family 5. ACTINOMONADIDA.
   2. Without flagella in the vegetative
     condition.
      3. Cells without a lifeless outer coat....... Family 6. ACTINOPHRYIDA.
      3. Cells having a gelatinous outer coat
        without siliceous spicules..............Family 7. HETEROPHRYIDA.
      3. Cells having a gelatinous outer coat
        with siliceous spicules................ Family 8. ACANTHOCYSTIDA.
      3. Cells with a hard shell punctured
        by pores...........................Family 9. CLATHRULINIDA.

Family 1. **Pseudosporea** [Pseudosporeae] Berlese in Saccardo Sylloge 7: 460 (1888). *Monadineae Zoosporeae* Cienkowski in Arch. mikr. Anat. 1: 213 (1865). Family *Pseudosporeen* Zopf Pilzthiere 115 (1885). Orders *Azoosporidea* for the most part and *Zoosporidea* Delage and Hérouard (1896). *Azoosporidae* for the most part and *Zoosporidae* Doflein Protozoen 40, 41 (1901). Family *Pseudosporidae* Poche in Arch. Prot. 30: 197 (1913). Amoeboid organisms without shells or skeletons, the pseudopodia tapering from a broad base to a filamentous termination. Flagellate stages (with one flagellum or two unequal flagella) occur in *Protomonas, Pseudospora,* and *Diplophysalis.* In other genera, as *Arachnula* and *Chlamydomyxa,* no flagellate stages are known.

Family 2. **Lagynida** Schultze 1854. Order *Gromida* Claparède and Lachmann 1859. Family *Gromida* Carpenter 1862. Family *Gromiidae* Brady (1881). Families *Monostomina* and *Amphistomina* Lankester (1885). Amoeboid organisms having chitinous shells without siliceous scales with a broad orifice through which project pseudopodia of the character of filopodia. *Gromia, Lagynis,* etc.

Family 3. **Euglyphida** [Euglyphidae] Wallich 1874. Amoeboid organisms with a chitinous shell beset with circular siliceous scales, the filopodia projecting through a broad orifice. *Euglypha, Cyphoderia, Campuscus, Trinema,* etc.

Family 4. **Vampyrellacea** [Vampyrellaceae] Zopf Pilzthiere 99 (1885). *Monadineae Tetraplasteae* Cienkowski op. cit. 218. Family *Vampyrelleae* Berlese in Saccardo Sylloge 7: 454 (1888). Family *Vampyrellidae* Poche in Arch. Prot. 30: 182 (1913). Cells subglobular, slowly creeping, with slender pseudopodia, numerous, densely packed and stiffly radiating on mature individuals, bearing terminal knobs. *Vampyrella,* the cells colored faintly pink by some metabolic by-product, is not unfamiliar as a predator on freshwater algae cultured under unfavorable conditions.

Family 5. **Actinomonadida** [Actinomonadidae] Kent Man. Inf. 1: 226 (1880). Family *Ciliophryidae* Poche in Arch Prot. 30: 187 (1913) Family *Helioflagellidae*

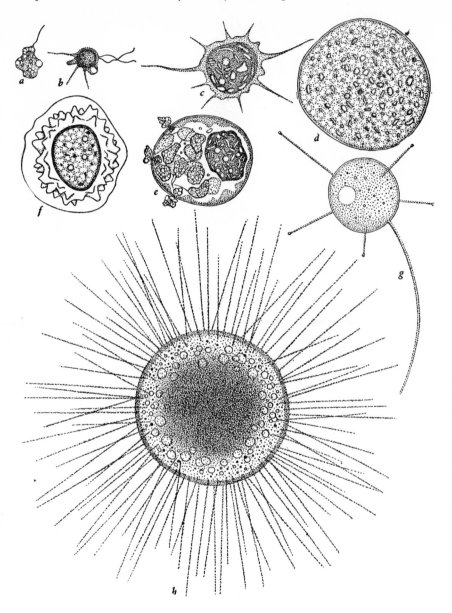

Fig. 37.—Radioflagellata: **a-f**, *Diplophysalis stagnalis* after Karling (1930);
a, b, young cells with one or two flagella; **c,** active amoeboid form; **d,** walled cell;
e, same releasing flagellate cells; **f,** resting cell.   **g,** Young cell of *Vampyrella* x 1,000.
h, *Actinosphaerium Eichhornii* x 1,000.

Doflein. Organisms bearing at the same time flagella and typical axopodia. *Dimorpha,* free-swimming, with two unequal flagella. *Actinomonas, Pteridomonas, Ciliophrys,* with one flagellum, either free-swimming or attached by a protoplasmic stalk.

Family 6. **Actinophryida** [Actinophryidae] Claus 1874. *Askeleta* Hertwig and Lesser in Arch. mikr. Anat. 10 Suppl. 164. (1874). *Aphrothoraca seu Actinophryidae* Hertwig Org. Radiolar. 142 (1879). Order *Aphrothoraca* Bütschli (1881). Suborder *Aphrothoraca* Minchin (1912). Family *Camptonematidae* Poche in Arch. Prot. 30: 187 (1913). Cells typically spherical, with typical axopodia, having no flagella nor shells nor skeletons. *Actinophrys Sol* Ehrenberg and *Actinosphaerium Eichhornii* (Ehrenberg) Stein are common in fresh water among algae, living as predators largely on diatoms; *Actinophrys* is uninucleate, the cells to 50μ in diameter; *Actonospaerium* is multinucleate, the cells to 1000μ in diameter. *Camptonema* is marine.

Mitosis in *Actinophrys* as described by Schaudinn (1896) occurs within an intact nuclear membrane which undergoes constriction; the dividing nucleus lies within a spindle-like body of cytoplasm. Schaudinn and Bĕlǎr (1923) observed conjugation. Pairs of gametes, which are usually sister cells but may sometimes be random pairs, lie within a cyst wall of secreted material. The nucleus of each gamete undergoes meiosis; at the end of each meiotic division, one of the daughter nuclei is digested; thus each gamete comes to possess a single haploid nucleus. Syngamy and karyogamy follow in due course and the zygote becomes walled. An old account of the cytology of *Actinosphaerium* by Hertwig is defaced by descriptions of the origin of nuclei from fragments of nuclei (chromidia), and of nuclear fusions at two separate stages of development.

Family 7. **Heterophryida** [Heterophryidae] Poche in Arch. Prot. 30: 189 (1913). *Heliozoa Chlamydophora* Archer in Quart. Jour. Micr. Sci. n.s. 16: 348 (1876). Order *Chlamydophora* Bütschli (1882). Suborder *Chlamydophora* Minchin (1912). Family *Lithocollidae* Poche l.c. The cells or their main bodies spherical with axopodia projecting through a gelatinous envelope. *Heterophrys* and *Astrodisculus* are simply globular cells. *Elaeorhanis* and *Lithocolla* are similar but with grains of sand or diatom shells imbedded in the envelope. *Actinolophus* is stalked. *Sphaerastrum* becomes colonial by incomplete division of the cells.

Family 8. **Acanthocystida** [Acanthocystidae] Claus 1874. *Chalarothoraca* Hertwig and Lesser in Arch. mikr. Anat. 10 Suppl. 193 (1874). *Chalarothoraca seu Acanthocystidae* Hertwig Org. Radiolar. 142. (1879). Order *Chalarothoraca* Bütschli in Bronn Kl. u. Ord. Thierreichs 1: 325 (1882). Suborder *Chalarothoraca* Minchin (1912). Resembling the preceding family, but the gelatinous envelope containing hard bodies, supposedly usually siliceous, of definite form. In *Raphidophrys,* these bodies are curved needles; in *Pinacocystis,* small plates; in *Acanthocystis* and *Pinaciophora,* disks bearing a central spine which is in some species forked. The cell of *Wagnerella* (a marine form, on rocks in bays) consists of a globular head with spines and axopodia, borne on a protoplasmic stalk attached by a foot; the nucleus lies in the foot.

In these forms the axial filaments of the pseudopodia radiate from a central granule located outside the nucleus (in *Wagnerella,* in the head). Schaudinn (1896) reported nuclear division in *Acanthocystis* as being either amitotic or mitotic: the report of amitosis is of course not to be taken seriously. In the mitotic process, the central granule acts as a centrosome; the chromosomes are numerous and minute; the nuclear membrane disappears during the middle stages. Nuclear division may be followed by division of the cell into two, or may be repeated and followed by production of buds. The buds may lose their pseudopodia and develop paired flagella. It is suspected that the flagellate cells may be gametes. The central granule is said to originate by extrusion from the nucleus of a bud.

Zuelzer (1909) found in *Wagnerella* two types of individuals, slender and stout, supposedly respectively haploid and diploid. In either type the nuclei may become numerous (and it is said that they sometimes develop from chromidia). The nuclei may migrate to the head and be released in buds, or they may become distributed throughout the protoplast, which then breaks up into biflagellate cells. It is supposed that these may be gametes, but a fusion of the heads of individuals of the slender type was observed.

Family 9. **Clathrulinida** [Clathrulinidae] Claus 1874. *Desmothoraca* Hertwig and Lesser op. cit. 225. *Desmothoraca seu Clathrulinidae* Hertwig Org. Radiolar. 142 (1879). Order *Desmothoraca* Bütschli in Bronn Kl. u. Ord. Thierreichs 1: 328 (1882). Family *Choanocystidae* Poche in Arch. Prot. 30: 192 (1913). Protoplasts lying within globular shells, apparently of chitin, usually stalked, punctured by numerous pores through which the axopodia project. In reproduction, the protoplast may divide into two, one of which escapes from the shell and secretes a new one; or it may divide into many which become unequally biflagellate. *Clathrulina, Hedriocystis, Choanocystis.*

Order 2. **Radiolaria** (J. Müller) Haeckel Radiolarien 243 (1862).
   Rhizopoda radiaria seu RADIOLARIA J. Müller in Abh. Akad. Wiss. Berlin 1858: 16 (1859).
   Orders *Thalassicollen, Sphaerozoen,* and *Peripyleen* Hertwig Org. Radiolar. 133 (1879).
   Orders *Peripylaria, Collodaria, Symbelaria,* and *Syncollaria* Haeckel in Jenaische Zeitschr. 15: 447, 469, 471, 472 (1882).
   Legion *Spumellaria* or *Peripylea,* with orders *Collodaria* and *Sphaerellaria* and seven suborders, Haeckel in Rept. Voy. Challenger Zool. 18: 5, 9 (1887).
   Legion *Spumellaria,* sublegions *Collodaria* and *Sphaerellaria,* and six orders, Haeckel Radiolarien 2: 87 (1887).
   Order *Peripylida* Delage and Hérouard Traité Zool. 1: 176 (1896).
   Suborder *Peripylaria* Minchin Protozoa 225 (1912).
   Order *Sphaeridea* Poche in Arch. Prot. 30: 206 (1913).
   Order *Peripylea* Calkins Biol. Prot. 343 (1926).
   Suborder *Peripylea* Kudo Handb. Protozool. 259 (1931).
   Suborder *Peripylina* Hall Protozoology 218 (1953).

This order and the three which follow, being the Radiolaria as conventionally construed, are unicellular marine organisms with axopodia, having within the protoplast a layer of organic material, variously punctured and of various types of symmetry, which separates the inner protoplasm from the outer. The central capsule consists of this layer (the central capsule membrane) and its contents, including the one or more nuclei of the cell. Imbedded in the protoplasm there is usually a skeleton, usually siliceous, various in structure and sometimes highly complicated. The outer cytoplasm is commonly inhabited by symbiotic cryptomonads in the resting condition (yellow cells, zooxanthellae), and sometimes contains masses of dark material, apparently debris extruded from the central capsule. The type of Radiolaria is evidently *Thalassicolla;* this genus was the first one described from living material, and was listed first by J. Müller in the original publication of the name.

The order which includes *Thalassicolla,* and to which the name Radiolaria is here restricted, is distinguished by uniformly distributed small punctures in the central capsule membrane and by the absence of skeletal spicules extending across the central capsule or meeting in its middle. Except during reproduction, each central capsule contains a single nucleus, but the cells of many examples are coenocytic, containing several or many central capsules.

Brandt (1885, 1905) observed reproduction particularly as it occurs by the production of swimming cells by some of the coenocytic forms. The nucleus divides to produce very many, and the intracapsular cytoplasm divides to produce uninucleate flagellate cells. In *Collosphaera*, all the nuclei are included in these cells. The cells are of two sizes, produced by different individuals, and are supposed to be gametes. In *Sphaerozoum* and its allies, some of the nuclei degenerate instead of being included in the swimming cells, of which two sizes are produced by single individuals. It appears that the swimming cells have characteristically two unequal flagella, though many are found to have only one, and some produce a third appendage by which they can attach themselves.

Haeckel listed thirty-two families in his legion Spumellaria. Other authors recognize about a dozen, including the following.

Family **Thalassicollida** Haeckel (1882). *Thalassicollen* J. Müller (1859). Family *Collida* Haeckel (1862); there is no corresponding generic name. Order *Collida* Haeckel (1887). Globular forms with a single central capsule, skeleton none or of numerous small spicules. *Thalassicolla, Physematium, Lampoxanthium,* etc.

Family **Sphaerozoida** Haeckel (1882). Family *Collozoida* Haeckel op. cit. Family *Sphaeroidina* Haeckel (1862); there is no corresponding generic name. Coenocytic, each cell with several nuclei in separate central capsules; skeleton none or of numerous small spicules. *Sphaerozoum,* the cells globular, to 1 mm. in diameter; *Raphidozoum,* the cells elongate.

Family **Collosphaerida** Haeckel (1862). Coenocytic, the spherical cell to 1 mm. in diameter, with several central capsules, each with an individual lattice-like skeleton. *Collosphaera.*

Family **Haliommatina** Ehrenberg 1847. Families *Ethmosphaerida, Ommatida,* and *Cladococcida* Haeckel (1862). Family *Sphaerida* Haeckel (1882). Order *Sphaeroidea,* with six families, Haeckel (1887). Globular, with small numbers of radiating main spicules, the main spicules bearing tangential branches which form a globular network of definite pattern, or, often, two or more concentric networks. *Haliomma, Actinomma, Hexacontium, Cladococcus,* and many other genera.

Further families are of the character of the Haliommatina, but with the spherical symmetry modified by abbreviation or elongation of one or more axes:

Family **Spongurida** Haeckel (1862). Order *Prunoidea,* with seven families, Haeckel (1887). Having one axis elongate. *Spongurus, Pipetta,* etc.

Family **Lithocyclidina** Ehrenberg 1847. Family *Discida* Haeckel (1862). Order *Discoidea,* with six families, Haeckel (1887). Having one axis shorter than the others. *Lithocyclia, Staurocyclia, Heliodiscus,* etc.

Family **Larcarida** Haeckel (1887). Order *Larcoidea,* with this and seven other families, Haeckel (1887). The skeleton with three unequal axes, or spiral. *Cenolarcus,* etc.

Order 3. **Acantharia** Haeckel in Jenaische Zeitschr. 15: 465 (1882).
  Order *Actipyleen* Hertwig Org. Radiolar. 133 (1879).
  Legion *Acantharia* or *Actipylea,* orders *Acanthometra* and *Acanthophracta,* and
    seven suborders, Haeckel in Rept. Voy. Challenger Zool. vol. 18 (1887).

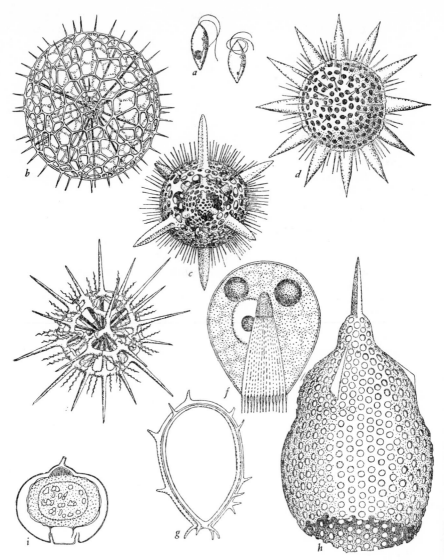

Fig. 38.—Radiolaria: **a,** motile cells of *Collosphaera Huxleyi* after Brandt (1885). **b,** Skeleton of *Haliomma capillaris* after Haeckel (1862).   **c,** Skeleton of *Actinomma Asteracanthion* after Haeckel (1862).   **d,** Skeleton of *Heliodiscus Phacodiscus* after Haeckel (1887).   Acantharia: **e,** Skeleton of *Dorataspis costata* after Haeckel (1887).   Monopylaria: **f,** Central capsule of *Tridictyopus elegans* after R. Hertwig (1879).   g, Skeleton of *Lithocircus productus* after R. Hertwig, *op. cit.* **h,** Skeleton of *Eucyrtidium carinatum* after Haeckel (1862).   Phaeosphaeria: **i,** Typical central capsule after R. Hertwig, *op. cit.*

Legion *Actipyea* or *Acantharia,* sublegions *Acanthometra* and *Acanthophracta,* and orders *Actinellida, Acanthonida, Sphaerophracta,* and *Prunophracta* Haeckel Radiolarien II Teil (1887).
Order *Actipylida* Delage and Hérouard Traité Zool. 1: 204 (1896).
Suborder *Acantharia* Minchin Protozoa 256 (1912).
Order *Acanthometrida* Poche in Arch. Prot. 30: 212 (1913).
Order *Actipylea* Calkins Biol. Prot. 345 (1926).
Suborder *Actipylea* Kudo Handb. Protozool. 216 (1931).
Suborder *Actipylina* Hall Protozoology 216 (1953).

In this group the central capsule membrane has many punctures arranged in clusters. The skeleton includes radiating spicules; in some examples these extend through the cell from side to side, passing through the central capsule; in the majority, their proximal ends meet in the center of the central capsule. In the latter forms, the number of radiating spicules is twenty, and they are arranged according to a pattern discovered by Johannes Müller and called Müller's law; they form five parallel whorls of four. Usually they bear tangential branches of definite, often highly elaborate, patterns: these may form a globular frame, or two or more concentric frames.

Haeckel, Hertwig, and Brandt found the spicules not to consist of silica. They are soluble in acids and alkalis, and were reported to be destroyed by heat. They were supposed, accordingly, to consist of some organic substance; Haeckel named it acanthin. Schewiakoff found them resistant to heat, and Bütschli (1906) analyzed them and found them to consist of strontium sulfate. This surprising fact has recently been confirmed by Odum (1951).

The cytoplasm at each point where a spicule passes through the surface is attached to the spicule by a whorl of minute fibers called myophrisks. The myophrisks are believed to be contractile, and to have the function of changing the volume, and hence the density, of the cells, enabling them to sink or float.

Young cells contain a single nucleus, eccentric in the central capsule; older ones have several to many nuclei.

Haeckel listed twenty families in his legion Acantharia. Other authors recognize about a half dozen, including the following.

Family **Litholophida** Haeckel (1882). Family *Astrolophida* Haeckel (1887). Spicules numerous, radiating, not arranged according to Müller's law. *Litholophus, Astrolophus, Actinelius,* etc.

Family **Chiastolida** Haeckel (1887). Spicules ten to twenty, extending clear through the body. *Chiastolus, Acanthochiasma.*

Family **Acanthometrida** Haeckel (1862). *Acanthometren* J. Müller (1859). Family *Acanthonida* Haeckel (1882). With twenty spicules arranged according to Müller's law; they may be branched, but do not form a continuous network. In most examples, as *Acanthometron, Xiphacantha,* etc., they are equal; in others, as *Amphilonche,* two of the spicules of the equatorial whorl are much longer than the others.

Family **Sphaerocapsida** Haeckel (1882). Family *Dorataspida* Haeckel l.c. Order *Sphaerophracta* Haeckel (1887). Like the foregoing, but the branches of the radiating spicules forming a globular network, or two or more concentric networks. *Dorataspis, Sphaerocapsa, Lychnaspis.*

Family **Diploconida** Haeckel (1862). Order *Prunophracta* Haeckel (1887). Again like the foregoing, but with the eight spicules of the two polar whorls either extended or abbreviated. *Diploconus, Hexaconus.*

Order 4. **Monopylaria** Haeckel in Jenaische Zeitschr. 15: 422 (1882).
Order *Monopyleen* Hertwig Org. Radiolar. 133 (1879).
Legion *Nassellaria* with orders *Plectellaria* and *Cyrtellaria,* and six suborders, Haeckel in Rept. Voy. Challenger Zool. vol. 18 (1887).
Legion *Nassellaria* with sublegions *Plectellaria* and *Cyrtellaria* and orders *Nassoidea, Plectoidea, Stephoidea,* and *Cyrtoidea,* Haeckel Radiolarien II Teil (1887).
Order *Monopylida* Delage and Hérouard Traité Zool. 1: 215 (1896).
Suborder *Nassellaria* Doflein.
Suborder *Monopylaria* Minchin Protozoa 256 (1912).
Order *Monopylea* Poche in Arch. Prot. 30: 218 (1913).
Suborder *Monopylea* Kudo Handb. Protozool. 261 (1931).
Suborder *Monopylina* Hall Protozoology 218 (1953).

This order is distinguished by a central membrane with one opening, or with a single circular field of pores. From this opening or field as a base, there extends into the central capsule a large conical body (apparently a bundle of protoplasmic fibers) called the porocone. The skeleton varies from none to highly elaborate; it does not in any form consist of separate spicules. Its symmetry is dorsiventral, not radial. These skeletons are well known as microfossils.

In this group, under the name of legion Nassellaria, Haeckel placed twenty-six families. Other authors recognize about a half dozen.

Family **Nassellida** Haeckel (1887). Skeleton none. *Cystidium.*

Family **Plectonida** Haeckel (1887). Family *Plectida* Haeckel (1882), not based on a generic name. Skeleton consisting of three arms radiating from a point opposite the mouth of the central capsule; sometimes with a fourth forming a caltrop. *Triplagia.*

Family **Stephanida** Haeckel (1887). Family *Stephida* Haeckel (1882), not based on a generic name. Skeleton including a ring in the sagittal plane, often with a basal tripod and with branches and crossbars. *Lithocircus, Zygostephanus.* This family is well represented by microfossils as far back as the Cambrian.

Family **Eucyrtidina** Ehrenberg 1847. Family *Polycystina* Ehrenberg in Abh. Akad. Wiss. Berlin (1838): 128 (1839), not based on a generic name. Families *Halicryptina* and *Lithochytridina* Ehrenberg 1847. Family *Cyrtida* Haeckel (1862). Order *Cyrtoidea,* with twelve families, Haeckel (1887). Skeleton a more or less basket-shaped network; the root *cyrt-* in many of the names is Greek κύρτη, a fishing basket. *Lithocampe, Cryptocalpis, Eucyrtidium, Theoconus, Dictyoconus,* and many other genera. This group is common as Mesozoic and Cenozoic microfossils, occurring mixed with diatoms and silicoflagellates.

Family **Spyridina** Ehrenberg 1847. Family *Spyrida* Haeckel (1882). Order *Spyroidea,* with four families, Haeckel (1887). The skeleton divided by sagittal grooves into two lobes.

Family **Cannobotryida** Haeckel (1887). Family *Botrida* Haeckel (1882), not based on a generic name. Order *Botryoidea,* with three families, Haeckel (1887). The skeleton divided by three or more longitudinal grooves into as many lobes.

Order 5. **Phaeosphaeria** Haeckel in Sitzber. Jenaische Gess. Med. Naturw. 1879: 156 (1879).
Phaeodariae, with orders *Phaeocystia, Phaeogromia,* Phaeosphaeria, and *Phaeoconchia* Haeckel op. cit.

Order *Tripyleen* Hertwig Org. Radiolar, 133 (1879).
Order *Phaeodaria* Haeckel in Jenaische Zeitschr. 15: 470 (1882).
Legion *Phaeodaria* and orders *Phaeocystina, Phaeosphaeria, Phaeogromia,* and
  Phaeoconchia Haeckel in Rept. Voy. Challenger Zool. vol. 18 (1887).
Order *Tripylea* Doflein.
Suborder *Tripylaria* Minchin Protozoa 256 (1912).
Suborder *Tripylea* Kudo Handb. Protozool. 263 (1931).
Suborder *Tripylina* Hall Protozoology 218 (1953).

In this order, the central capsule is of isobilateral symmetry, having a rather small main opening (astropyle) at one end and two smaller openings (parapyles) at the other. The openings are located on projections of the central capsule membrane; inside of each, the protoplasm is so differentiated as to appear to be a conical bundle of fibers with the apex at the opening (in contrast to the preceding order, in which the base of the conical structure is at the opening). A mass of variously colored bodies, supposedly excreted from the central capsule, lies in the extracapsular cytoplasm about the astroplyle. The skeletons consist in part of organic matter and are not well preserved as fossils.

Borgert (1896, 1900) described nuclear division in *Aulacantha*. A very large number of chromosomes, a matter of several hundred, form a plate which splits into two; the two plates move apart in a body of differentiated cytoplasm, but no definite spindle, and no centrosomes, were seen. The margins of the plates draw apart faster than the middles, with the effect that the plates become saucer-shaped, then bowl-shaped, and finally globular, after which nuclear membranes form about them. While the nucleus divides, the central capsule membrane becomes constricted by longitudinal grooves so placed that each daughter central capsule membrane receives one parapyle and an astropyle formed from half of the original astropyle. The rudiments of additional parapyles are first seen as granules in the intracapsular cytoplasm. Each granule grows slightly and becomes "hat-shaped," and migrates so as to come into contact with the central capsule membrane at the point appropriate for the development of its second parapyle.

Later, Borgert (1909) described a process in which the nucleus divides repeatedly, producing many. The divisions are mitotic, with small numbers of chromosomes, perhaps twenty; the eventual products become the nuclei of gametes. There are reports, in part illustrated with photographs, of similar processes in family Thalassicollida (Häcker, 1907; Huth, 1913). According to Hollande (in Grassé, 1953) the small nuclei are those of a parasitic dinoflagellate, *Solenodinium*. Le Calvez (1935) found *Coelodendrum* to produce zoospores with a pair of unequal simple flagella. They resemble cells of *Cryptomonas* or of *Bodo*.

Haeckel's legion Phaeodaria was of fifteen families. These have been maintained by the generality of authors.

A. Skeleton none or of distinct spicules; cells usually nearly spherical.

Family **Aulacanthida** [Aulacanthidae] Haeckel (1879). *Aulacantha.*

Family **Astracanthida** [Astracanthidae] Häcker. Spicules more or less thorny at the distal ends. *Aulactinium.*

  B. Skeleton spherical or of two concentric spheres, with no main opening.

Family **Aulosphaerida** Haeckel (1862). *Aulosphaera.*

Family **Cannosphaerida** [Cannosphaeridae] Haeckel (1879). *Cromodromys.*

Family **Sagosphaerida** Haeckel (1887).

  C. Skeleton with a distinct main opening, either nearly spherical, radially symmetrical, or distinctly dorsiventral.

Family **Castanellida** [Castanellidae] Haeckel (1879). Skeleton nearly globular with numerous pores. *Castanidium.*

Family **Circoporida** [Circoporidae] Haeckel (1879). Like the foregoing, but with the pores gathered about the bases of radiating spines. *Circoporus.*

Family **Tuscarorida** Haeckel (1887). The main pore on an extended neck, the skeleton accordingly flask-shaped. *Tuscarora. Tuscarilla.*

D. Shell strongly dorsiventral.

Family **Challengerida** [Challengeridae] J. Murray. Shell finely pitted. *Challengeron.*

Family **Medusettida** Haeckel (1887). Shell smooth or with small spines. *Medusetta.*

E. Shell bilaterally divided into two parts.

Family **Concharida** [Concharidae] Haeckel (1879).

Family **Coelodendrida** Haeckel (1862). The shell ·bearing extensive branched appendages. *Coelodendrum.*

Class 5. **SARKODINA** (Hertwig and Lesser) Bütschli

SARKODINA Hertwig and Lesser in Arch. mikr. Anat. 10 Suppl. 43 (1874).

Class SARKODINA Bütschli in Bronn Kl. u. Ord. Thierreichs 1, I Teil: *Inhalt* (1882).

Class, subclass, etc., *Rhizopoda* Auctt.; class, subclass, etc. *Sarcodina* Auctt.

Amoeboid organisms without flagellate stages, the pseudopodia of the character of lobopodia; without skeletons, without or with shells of various materials.

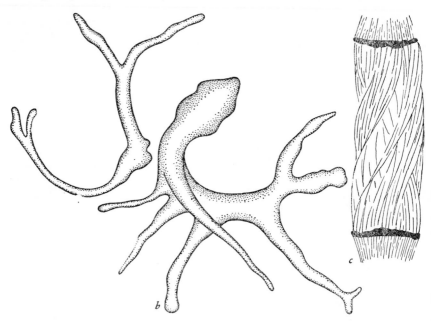

FIG. 39—*Chaos Protheus*: **a, b**, cells x 25, after the original figures of *Pelomyxa carolinensis* by Wilson (1900); **c**, mitotic figure x 2,000 after Short (1945).

Hertwig and Lesser took note that the name Rhizopoda was originally applied to organisms such as *Miliola*, which have rhizopodia; they proposed the name Sarkodina for all amoeboid organisms, with Rhizopoda as a subordinate group. Among examples of Sarkodina which are not Rhizopoda, they listed first *Difflugia*, which may accordingly be considered the standard genus.

The Sarkodina as here presented are not assumed to be a natural group. Their common characters are probably the outcome of degeneration, by which organisms of diverse evolutionary origins have lost their distinctions.

This assemblage is obviously and superficially divisible into two by the absence or presence of shells. The resulting groups are construed as orders.

Order 1. **Nuda** Schultze 1854.
  Family AMOEBAEA Ehrenberg Infusionthierchen 125 (1838).
  Order *Lobosa* Carpenter 1861.
  Order *Gymnamoebae* Haeckel Gen. Morph. 2: xxiv (1866).
  Order *Amoebina* Kent Man. Inf. 1: 27 (1880).
  Suborder *Amoebaea* Bütschli in Bronn. Kl. u. Ord. Thierreichs 1: 176 (1880).
  Order *Gymnamoebida* Delage and Hérouard (1896).
  Order *Chaidea* Poche in Arch. Prot. 30: 170 (1913).
  Subclass *Amoebaea* and order *Amoebida* Calkins Biol. Prot. 335, 337 (1926).
  Order *Amoebaea* Kudo Handb. Protozool. 204 (1931).
Sarkodina without shells. The type is the common amoeba, *Amiba diffluens*.
1. Protoplasts not tending to form pseudoplas-
  modial communities.
    2. Free-living..............................Family 1. AMOEBAEA.
                                              Family 2. MAYORELLIDA.
                                              Family 3. THECAMOEBIDA.
                                              Family 4. HYALODISCIDA.
    2. Entozoic...............................Family 5. ENDAMOEBIDA.
1. Protoplasts assembling and acting in unison
  in pseudoplasmodial communities.
    2. Parasitic in plants......................Family 6. LABYRINTHULIDA.
    2. Predatory on bacteria, in air on moist
      surfaces; mostly producing complicated
      fructifications...........................Family 7. GUTTULINACEA.
Family 1. **Amoebaea** Ehrenberg Infusionsthierchen 125 (1838). Family *Amoebidae* Bronn 1859. Family *Chaidae* Poche in Arch. Prot. 30: 171 (1913). Family *Chaosidae* Chatton in Grassé Traité Zool. 1, fasc. 2: 58 (1953). The ordinary free-living amoebas. Schaeffer (1926) limited the family to forms which produce numerous indefinite granular pseudopodia. There has been much confusion as to the identity of the species. There are apparently two species of common large amoebas:
  1. *Chaos Protheus* L. Syst. Nat. ed. 12: 1326 (1767) (*Volvox Chaos* L. Syst. Nat. ed. 10: 821. 1758. *Vibrio Protheus* O. F. Müller Verm. Terr. et Fluv. 1: 45. 1773. *Pelomyxa carolinensis* Wilson in American Nat. 34: 535. 1900. *Chaos chaos* Stiles). Schaeffer identified *Pelomyxa carolinensis* as the original *Chaos Protheus* L. It is exceptionally large, being macroscopically visible, and is multinucleate. Surely, sound nomenclature will apply to this species the name which Linnaeus gave it.
  2. *Amiba* [*Amoeba*] *diffluens* (O. F. Müller) Ehrenberg Infusionsthierchen 127 (1838) (*Proteus diffluens* O. F. Müller Animac. Infus. 9. 1786; there is an older genus

Proteus; Amiba divergens Bory Dict. Class. Hist. Nat. 1: 261. 1822; *Amoeba Proteus* Leidy). It appears that Müller intended to rename the *Chaos Protheus* of Linnaeus; that in 1773 he actually did so; but that in 1786 he applied another new name to a different organism. Ehrenberg's amended spelling *Amoeba,* although in general use, is not valid as that of a generic name; as Schaeffer suggests, the word may be used as a common noun. *Amiba diffluens* is uninucleate; large, but not visible to the naked eye.

In nuclear division in the common amoebas the nuclear membrane disappears. There are many chromosomes in a blunt-ended spindle. Short (1945) noted a peculiar twisting of the spindle of *Chaos Protheus.*

Schaeffer included in the present family three further genera, *Trichamoeba* Fromentel, *Polychaos* Schaeffer, and *Metachaos* Schaeffer. Here, in ignorance of its relationships, another well-known genus is assigned to this family.

Pelomyxa, typified by *P. palustris* Greeff (1874), resembles *Chaos Protheus* in being exceptionally large, macroscopically visible, and multinucleate. It is definitely different from *Chaos Protheus* in manner of movement (King and Jahn, 1948) and in chemical characters (Andressen and Holter, 1949).

Minute amoebas moving by means of a single pseudopodium are called *Vahlkampfia.* They are believed to have swimming stages with paired equal flagella. If so, they do not belong to the present group, but perhaps to the plant kingdom.

Family 2. **Mayorellida** [Mayorellidae] Schaeffer in Publ. Carnegie Inst. 345: 47 (1926). Producing numerous brief conical pseudopodia, but moving by a single large clear one. *Mayorella, Pontifex,* and several other genera proposed by Schaeffer; *Dactylosphaerium* Hertwig and Lesser; *Dinamoeba* Leidy? The last may be the non-flagellate stage of *Chaetoproteus* Stein.

Family 3. **Thecamoebida** [Thecamoebidae] Schaeffer op. cit. 83. Amoebas with a tough pellicle simulating a shell, moving by the outflow of clear protoplasm at the anterior margin. *Thecamoeba* Fromentel. *Rugipes* Schaeffer.

Family 4. **Hyalodiscida** [Hyalodiscidae] Poche in Arch. Prot. 30: 182 (1913). Family *Cochliopodiidae* de Saedeleer in Mém. Mus. Roy. Hist. Nat. Belgique 60: 5 (1934). Similar to the foregoing but without the tough pellicle. Commonly dome-shaped, with a row of small pseudopodia projecting from the margin. *Hyalodiscus* and *Cochliopodium* of Hertwig and Lesser, together with certain genera of Schaeffer.

Family 5. **Endamoebida** [Endamoebidae] Calkins. Entozoic amoebas.

Endamoeba Leidy is found in cockroaches and termites. The nucleus contains no karyosome, but many separate granules; in mitosis, definite chromosomes are formed (twelve in *E. disparita*), but there is apparently no centrosome; at least, no intradesmose is seen (Kirby, 1927).

Entamoeba Casagrandi and Barbagello, named at nearly the same time as the foregoing and regrettably similarly, is widely distributed in invertebrate and vertebrate hosts. *E. dysenteriae* (Councilman and Lafleur) Craig (*Endamoeba histolytica* Schaudinn) is a serious pathogen to man, the cause of amoebic dysentery. *E. coli* and *E. gingivalis* are believed to be harmless commensals. The fully mitotic character of nuclear division in these organisms was established by Kofoid and Swezy (1921, 1922, 1925). The nucleus contains a small karyosome and an intranuclear centrosome. Mitosis begins with division of the centrosome into two, which remain connected, as they draw apart, by a stainable strand, the intradesmose (the term is of Kofoid and Swezy, 1921). The karyosome breaks up into chromosomes, six in the species mentioned. Spindle fibers connecting these to the centrosomes have been seen; Child (1926) found that the two halves of the spindle swing apart as the centrosomes move apart like the legs of a compass being extended. There is no doubt that *Endamoeba* and *Entamoeba* are generically distinct.

Endolimax, Iodamoeba, and *Councilmania* occur chiefly in vertebrates and include species commensal in man. A refined technique is required to discern the characters by which they are distinguished from *Entamoeba. Karyamoebina* Kofoid and Swezy (1924, 1925,) another commensal in man, resembles *Vahlkampfia* in details of the mitotic process, and probably does not belong to the present group. *Hydramoeba,* usually listed in the present family, is not an entozoic organism, but a predator on *Hydra.*

Family 6. **Labyrinthulida** [Labyrinthulidae] Haeckel ex Doflein Protozoen 47 (1901). There is a single genus *Labyrinthula* Cienkowski, and probably only one species, *L. macrocystis,* parasitic in green and brown algae and in the marine seed plant *Zostera.* The uninucleate cells are spindle-shaped. These cells send out from one or both ends fine filaments which writhe in the water. The filaments from different cells coil together and produce "tracks" along which the cells glide. The tracks form a network on which the cells may be scattered or gathered into clusters; or the cells may abandon their tracks and generate new ones. The nature of the tracks is not clear. Possibly they are pseudopodia, on which the cells move by absorbing them at one end while generating them at the other. Young (1943) found *Labyrinthula* remarkably indifferent to variations in temperature, reaction, and salinity.

Family 8. **Guttulinacea** [Guttulinaceae] Berlese in Saccardo Sylloge 7: 325 (1888) Tribe *Dictyosteliaceae* Rostafinski Vers. 4 (1873). *Sorophoreen* with families *Guttulineen* and *Dictyosteliaceen* Zopf Pilzthiere 131-134 (1885). Families *Guttulineae* and *Dictyosteliaceae* Berlese op. cit. 451. *Sappiniaceae* Olive in Proc. American Acad. 37: 334 (1901). Families *Sappiniidae, Guttulinidae,* and *Dictyostelidae* Doflein 1909. Family *Acrasidae* Poche in Arch. Prot. 30: 177 (1913). Suborder *Acrasina* Hall Protozoology 228 (1953). Amoeboid cells predatory on bacteria and other scraps of organic matter, in air on moist surfaces, commonly on dung. The cells are capable of assembling and moving and going into a resting stage in unison. These organisms have generally been included among the Mycetozoa; the resemblance is superficial.

More recently than Olive, Raper (1940) and Bonner (1944) have surveyed the group and studied the behavior. Three families have been maintained, but one appears sufficient to accommodate the seven genera and approximately twenty species.

Cells of *Sappinia* are binucleate. They do not necessarily assemble in clusters; a single cell may secrete a stalk, by which it is raised into the air, where it rounds up and becomes dry. Alternatively, small numbers of cells may assemble and secrete a common stalk. The dry cells are "pseudospores": they are capable of resuming activity without casting off a wall. Hartmann and Nägler (1908) described a peculiar sexual process in *Sappinia diploidea.*

Guttulina and *Guttulinopsis* produce larger clusters of resting cells than *Sappinia* does; in *Guttulina* the resting cells are said to be walled spores.

Acrasis produces fruits, solitary or clustered, of the form of uniseriate rows of spores terminal on stalks consisting of rows of dead cells.

Distyostelium produces fruits consisting of a column of dead cells bearing a globular cluster of spores; *Polysphondylium* and *Coenenia* produce slightly more elaborate fruits of the same general nature. In *Dictyostelium,* Raper and Bonner saw that the amoeboid active cells, having devoured the available food, gather into a disk-shaped mass which may exceed a millimeter in diameter. Wilson (1953) found syngamy,

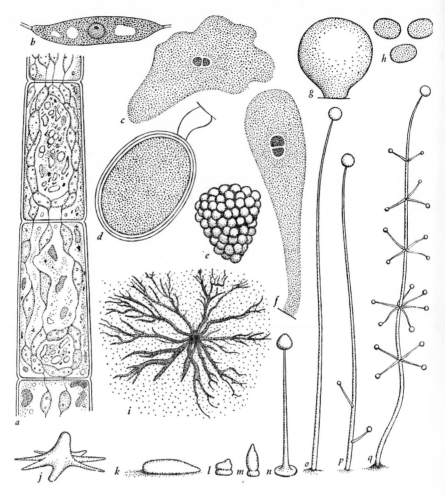

Fig. 40.—**a**, *Labyrinthula* as a parasite in cells of *Ectocarpus Mitchelliae* x 1,000 after Karling (1944). **b**, Cell of *Labyrinthula* x 2,000 after Young (1943). **c, d**, *Sappinia pedata*, active cell and cyst, x 1,000 after Dangeard (1896). **e, f**, *Sappinia pedata*, cluster of pseudospores x 100 and single pseudospore x 1,000 after Olive (1902). **g, h**, *Guttulina sessilis*, cluster of pseudospores x 100 and individual pseudospores x 1,000 after Olive (1902). **i-n**, *Dictyostelium discoideum* x 10 after Bonner (1944): **i**, the pseudoplasmodium; **j**, it heaps itself up; **k**, falls toward the light and creeps; **l, m**, again heaps itself up and becomes a fruit, **n**. **o, p**, Fruits of *Dictyostelium mucoroides;* **q**, of *Polysphondylium violaceum;* x 10, after Bonner, *op. cit.*

karyogamy, and meiosis to occur at this point; the chromosome number (n) is 7. The disk changes into a column which bends, and then falls, toward the light, and afterward creeps some distance in the same direction. When this has happened, the foremost cells, being those which were originally in the middle of the disk-shaped mass, pile up again to form a sterile stalk perhaps one millimeter tall; the cells behind them crawl up the stalk to form the globular mass of spores; the hindmost, being those which were last to arrive at the disk-shaped mass, remain behind to form a flange about the base of the stalk.

Order 2. **Lampramoebae** Haeckel Gen. Morph. 2: xxiv (1866).
Order *Testacea* Schultze 1854, non L. (1758).
Order *Thecamoebae* Haeckel.
Order *Conchulina* Cash and Hopkinson British Freshw. Rhizop. 1: 37 (1905).
Suborder *Testaceolobosa* de Saedeleer in Mém. Mus. Roy. Hist. Nat. Belgique 60: 5 (1934).
Order *Testacida* Hall Protozoology 241 (1953).
Order *Testaceolobosa* Deflandre in Grassé Traité Zool. 1, fasc. 2: 97 (1953).
Amoeboid organisms without known flagellate stages, bearing shells and producing lobopodia. Various organisms producing rhizopodia or filopodia, traditionally associated with these, have here been placed among Rhizopoda or Heliozoa, as suggested by de Saedeleer (1934) and Grassé (1953). Deflandre (in Grassé, op. cit.) distinguishes several families beside the following:
  1. Shell without secreted scales of silica.
      2. Shell of uniform secreted material..............Family 1. ARCELLINA.
      2. Shell with imbedded grains of sand.............Family 2. DIFFLUGIIDA.
  1. Shell with secreted scales of silica..................Family 3. NEBELIDA.

Family 1. **Arcellina** Ehrenberg Infusionsthierchen 129 (1838). Family *Arcellidae* Schultze 1876. *Arcella,* etc.

Family 2. **Difflugiida** [Difflugiidae] Taránek 1881. *Difflugia,* etc.

Family 3. **Nebelida** [Nebelidae] Schouteden 1906. *Nebela,* the shell beset with circular siliceous scales; *Quadrula,* the scales square; etc.

Chapter XI

PHYLUM FUNGILLI

Phylum 7. FUNGILLI Haeckel

Order *Gregarinae* Haeckel Gen. Morph. 2: xxv (1866).
Class SPOROZOA Leuckart Parasiten der Menschen 1, part 1: 241 (1879).
Phylum FUNGILLI Haeckel Syst. Phylog. 1: 90 (1894).
Class *Sporozoaria* Delage and Hérouard Traité Zool. 1: 254 (1896).
Subphylum *Sporozoa* Calkins Biol. Prot. 249 (1926).

Essentially unicellular organisms (the cells sometimes becoming multinucleate or multiple, but remaining undifferentiated except in connection with reproduction); commonly with a writhing motion; reproduction usually involving complicated sexual processes and the production of walled cysts (spores); flagella absent except sometimes on the sperms; parasitic in animals.

The class Sporozoa as originally published by Leuckart included the following groups: (a) the gregarines, first described by Dufour (1826) as worms parasitic in beetles: the generic name *Gregarina* Dufour (1828) refers to their occurrence in crowds; (b, c) coccidians and psorosperms, different sorts of parasites discovered in fishes by J. Müller and Retzius (1842); and, doubtfully, (d) Miescher's tubes (*Mieschersche Schläuche*), being certain abnormal growths in muscles. The cause of the pébrine disease of silkworms, which Nägeli (in Caspary, 1857) had named *Nosema Bombycis,* belongs to this group but was not originally included, presumably because Nägeli had considered it to be a schizomycete.

It has subsequently become known that almost every species of the animal kingdom is parasitized by one or more species of Fungilli. Not all of these parasites, but many, are serious pathogens. Thus the Fungilli are a very important group and very numerous. The number of species duly registered by name and description is apparently some two or three thousand; this is surely a small fraction of the number which exist.

The transmission of disease by biting arthropods was first demonstrated when Theobald Smith (1893) showed that the Texas fever of cattle, caused by *Babesia bigemina,* is transmitted by ticks.

All who have classified the Sporozoa or Fungilli have recognized two prime subordinate groups, the first including the gregarines and coccidians, the second including the organisms which were formerly called psorosperms (Myxosporidia or Neosporidia). In addition to the main bodies of these groups, there are certain organisms which have resisted definite placement and have been assigned sometimes to one of the main groups, sometimes to the other, and sometimes to additional main groups. In the present work the two main groups are treated as classes and the groups of uncertain relationship are included in the first. Clearly, this class is to bear the name of Sporozoa Leuckart. Schaudinn's famous paper on parasites in the owl (1903) is apparently authority for the widely entertained opinion that this class is artificial, representing at least two lines of descent. In fact, the class appears natural with the possible exception of some of the poorly known groups. The second class is marked by positive specialized characters and is clearly natural; it is not clearly certain that the second class is related to the first, and it is accordingly not certain that the phylum is natural. The classes are distinguished as follows:

1. Producing resting cells protected by cell walls and not containing polar capsules; or not producing resting cells................................Class 1. SPOROZOA.
1. Producing resting cells whose walls consist (at least usually) of modified cells, and which contain "polar capsules" enclosing coiled threads.....................................Class 2. NEOSPORIDIA.

Class 1. SPOROZOA Leuckart

Class SPOROZOA Leuckart Parasiten der Menschen 1, Abt. 1: 241 (1879).
Subclass *Gregarinida* Bütschli in Bronn Kl. u. Ord. Thierreichs 1, Abt. 1: *Inhalt* (1882).
Class *Sporozoaria* and subclass *Rhabdogeniae* Delage and Hérouard Traité Zool. 1: 254, 255 (1896).
Legion *Cytosporidia* Labbé in Thierreich 5: 3 (1899).
Subclass *Telosporidia* Schaudinn in Zool. Jahrb. Anat. 13: 281 (1900).
Class *Telosporidia* Calkins Biol. Prot. 421 (1926).
Class *Telosporidea* with subclasses *Gregarinidia, Coccidia,* and *Haemosporidia;* and class *Acnidosporidea* Hall Protozoology 270, 271, 290, 301, 323 (1953).
Sous-embranchement des Sporozoaires, with classes *Gregarinomorpha, Coccidiomorpha,* and *Sarcosporidia* Grassé Traité Zool. 1, fasc. 2: 545 *et seq.* (1953).
Fungilli which produce resting cells protected by cell walls and not containing polar capsules; or else do not produce resting cells.

The nature of the organisms included in this class may be made clear by an example, **Goussia Schubergi** (Schaudinn) comb. nov. (*Coccidium Schubergi* Schaudinn, 1900).

Goussia is parasitic in centipedes. Infection is by certain spindle-shaped cells which have a certain power of movement, and which make their way to the interior of cells of the epithelium of the gut of the host. Each parasitic cell grows and becomes globular; it becomes multinucleate; when the host cell dies and breaks up, the parasitic cell divides into many spindle-shaped cells which infect other cells of the gut epithelium.

Alternatively, a sexual process takes place. Some of the parasites emerge into the gut and do not divide but function as eggs. Others produce numerous cells which are more slender than the usual infective cells. These become flagellate, each producing two unequal flagella, and function as sperms.

The zygote becomes walled. Its nucleus divides twice. Each of the four resulting nuclei becomes the nucleus of a walled cyst. The cysts are apparently formed by a process of free cell formation: not all of the cytoplasm of the zygote is included in them. In each cyst, two of the spindle-shaped infective cells are produced, again apparently by free cell formation, excluding a part of the cytoplasm. The zygotes, with their included cysts and infective cells, pass out with the feces of the host. If a centipede eats one of them with its food, the infective cells are released to perform their function.

No feature of the life cycle described is peculiar to the Sporozoa as contracted with other nucleate organisms. Nevertheless, largely by the authority of Schaudinn, specialists in Sporozoa use an extensive system of special terms. A familiarity with these is necessary to anyone reading about Sporozoa. They include the following:

Sporozoite, the original infective cell.

Trophozoite, the vegetative individual.

Nucleogony, the multiplication of nuclei.

Plasmotomy, multiplication of cells.

Meront or schizont, the individual in process of dividing to produce further infective cells.

Schizogony or agamogony, the process of dividing to produce infective cells.

Merozoite or agamete, the infective cell produced from a trophozoite.

Gamogony, the production of gametes.

Macrogamete and microgamete mean, of course, egg and sperm; macrogametocyte and microgametocyte mean the cells which produce them.

Sporoblast or sporont, the zygote or other cell inside of which walled cysts are produced.

Sporogony, the sexual cycle which produces walled cysts.

Sporulation, the production of walled cysts by asexual processes.

Spore, the walled cyst.

Trophozoite (or schizont) and sporont are regarded as the alternating main stages in the life cycle of Sporozoa. The point at which meiosis occurs is uncertain. In the

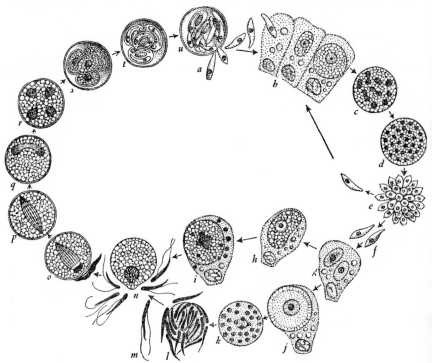

Fig. 41.—Life cycle of *Goussia Schubergi* after Schaudinn (1900): **a**, sporozoites; **b-d**, developing trophozoites; **e**, schizogony; **f**, merozoites; **g**, young gametocytes; **h, i**, development of egg; **j-m**, development of sperms; **n**, fertilization; **o**, zygote (sporoblast); **p-t**, development of spores; **u**, germination of spores.

monocystid gregarines, Muslow (1911) and Calkins and Bowling (1926) described a reduction of the chromosome number immediately before gametogenesis, quite as in typical animals. They described reduction as accomplished by a single process of nuclear division; to current cytological theory, this is an impossibility. Dobell and Jameson (1915), Jameson (1920), and Dobell (1925), dealing with organisms of the same group and also with the coccodian *Aggregata,* found meiosis to occur immediately after karyogamy. They conclude that all nuclei except those of zygotes are haploid, as among most of the lower plants.

The coccidian group, to which *Goussia* belongs, is here treated as primitive among Sporozoa because the sperms of this group are flagellate. The detailed structure of the flagella is unknown; they appear to resemble those of *Bodo* and *Cryptobia.* This fact conveys the best available hint as to what may have been the evolutionary origin of the Sporozoa. The majority of Sporozoa, having gametes which are alike or scarcely differentiated, appear to be derived from forms with markedly differentiated gametes.

The Sporozoa are classified primarily by whether or not the trophozoites are intracellular; by the occurrence or non-occurrence of asexual reproduction; and by the production or non-production of spores in the sense in which the term is used in dealing with this group, that is, of walled cysts.

1. Sexual reproduction, so far as it is known, involving oocytes which produce single large eggs and spermatocytes which produce from few to many sperms; the organisms multiplying also asexually.
    2. The gametocytes not attached in pairs.
        3. Producing walled spores. . . . . Order 1. OLIGOSPOREA.
        3. Not producing walled spores.
            4. Intracellular in erythrocytes. . . . . . . Order 3. GYMNOSPORIDIIDA.
            4. Producing macroscopic bodies in muscle. .Order 4. DOLICHOCYSTIDA.
    2. The gametocytes pairing before gametogenesis; sperms few; with or without walled spores. . . . . .Order 2. POLYSPOREA.
1. Gametes slightly differentiated or undifferentiated, produced by the gametocytes in more or less equal, usually large, numbers.
    2. The organisms multiplying also asexually.
        3. Spores producing several sporozoites. . . . Order 5. SCHIZOGREGARINIDA.
        3. Each spore producing one sporozoite. . . . Order 8. HAPLOSPORIDIIDEA.
    2. The organisms not multiplying asexually.
        3. Cells not elongate and divided into two parts. . . . . .Order 6. MONOCYSTIDEA.
        3. Cells elongate and divided into two parts. . . . . Order 7. POLYCYSTIDEA.

Order 1. **Oligosporea** Lankester in Enc. Brit. ed. 9, 19: 855 (1885).
    Tribe *Monosporées* and groups *Disporées* and *Tetrasporées* Schneider in Arch. Zool. Exp. Gen. 9: 387 (1881).

Coccididae, with tribes *Monosporea* and Oligosporea, Bütschli in Bronn Kl. u.
 Ord. Thierreichs 1: 574, 575 (1882).
Order *Monosporea* Lankester op. cit. 854.
Suborder *Coccididae* Delage and Hérouard Traité Zool. 1: 278 (1896).
Order *Coccidiidia* Labbé in Thierreich 5: 51 (1899).
Order *Coccidiomorpha* Doflein Protozoen 95 (1901).
Order, suborder, or tribe *Eimeridea* Léger in Arch. Prot. 22: 80 (1911).
Order *Eimeriidea,* suborders *Selenococcidinea* and *Eimeriinea,* and tribe *Eimer-
 ioidae,* Poche in Arch. Prot. 30: 237, 238 (1913).
Subclass *Coccidiomorpha* and order *Coccidia* Calkins Biol. Prot. 435, 436
 (1926).
Suborder *Eimeridea* Reichenow in Doflein Lehrb. Prot. ed. 5, 3: 921 (1929).
Order *Eimeriida* Hall Protozoology 297 (1953).

Sporozoa living mostly within epithelial cells of their hosts, multiplying asexually,
the gametocytes not pairing before gametogenesis, the macrogametocytes producing
single eggs and the microgametocytes numerous flagellate sperms, the zygotes usually
producing definite walled spores.

The organisms of the present order and the following are called coccidians.
Schneider (1881) classified them by the number of spores produced in each sporoblast
(i.e., zygote), either one, two, four, or many. Bütschli and Lankester gave due form
to Schneider's system. As between their names Monosporea and Oligosporea, the one
which included the typical example *Eimeria* is here chosen in preference to the one
which had page priority. Léger classified these organisms primarily by the number of
sporozoites per zygote, and distinguished eight families. Here, with the authority,
for example, of Reichenow (1929) and Kudo (1931), these are reassembled as one
family to which are appended three others including markedly exceptional or poorly
known forms.

Family 1. **Eimerida** [Eimeridae] Minchin 1903. The typical coccidians. In
Eimeria Schneider (*Coccidium* Leuckart) the zygote produces four firmly-walled
spores each with two sporozoites. The spores are symmetrically ellipsoid and release
the sporozoites through a terminal pore. Species of this genus parasitize many vertebrate hosts, rabbits, sheep, goats, swine, dogs, cats, chickens, turkeys, frogs, and
fishes. Some of the other genera differ from this as follows: *Jarrina,* attacking birds,
is distinguished by spores bearing the pore at the end of a brief neck. *Goussia,* in centipedes, has spores whose walls split lengthwise into two valves. The zygote of *Isospora,* in mammals, including man, produces two spores each with four sporozoites;
that of *Caryospora,* in snakes, produces one spore with eight sporozoites. *Barrouxia,*
in various invertebrates, produces from each zygote numerous bivalved spores each
containing one sporozoite.

Family 2. **Dobelliida** [Dobelliidae] Ikeda. The single known species, *Dobellia binucleata,* occurs in a siphunculid worm. It exhibits an exception to the characters
of the order: the male and female gametocytes become attached to each other; the
male gametocyte, however, produces many sperms, as in the generality of the order.

Family 3. **Aggregatida** [Aggregatidae] Labbé in Thierreich 5: 6 (1899). This family is distinguished by heteroecism. In *Aggregata Eberthi,* vegetative growth and
multiplication take place in crabs. When these are eaten by squids, the cells either
develop into single eggs or else divide to produce many sperms. The zygote produces
about twenty bivalved spores which pass out with the feces and infect crabs. The
number of sporozoites per spore is variable. There are several other species of *Ag-*

gregata. Various other genera, *Merocystis, Hyaloklossia, Myriospora, Caryotropha,* etc., attacking mussels, polychaet worms, and other marine invertebrates, are assigned to this family although their life cycles are not fully known.

Family 4. **Selenococcidiida** [Selenococcidiidae] Poche in Arch. Prot. 30: 238 (1913) includes the single species *Selenococcidium intermedium* Léger and Dubosq (1910) in the lobster. The vegetative cell is long and slender, and asexual reproduction is regularly by transverse division into eight.

Order 2. **Polysporea** Lankester in Enc. Brit. ed. 9, 19: 855 (1885).
 Tribe POLYSPOREA Bütschli in Bronn Kl. u. Ord. Thierreichs 1: 576 (1882).
 Suborder *Haemosporidae* Delage and Hérouard Traité Zool. 1: 284 (1896).
 Order *Haemosporidiida* Labbé in Thierreich 5: 73 (1899).
 Order, suborder, or tribe *Adeleidea* Léger in Arch. Prot. 22: 81 (1911).
 Tribe *Adeleoidae* Poche in Arch. Prot. 30: 239 (1913).
 Order *Adeleida* with suborders *Adeleina* and *Haemogregarinina* Hall Protozoology 296 (1953).

It is characteristic of this order that pairs of reproductive cells, essentially merozoites, which are to become gametocytes, become attached to each other. The macrogametocyte becomes converted into a single egg; the microgametocyte produces, at least usually, four sperms.

Family 1. **Adeleida** [Adeleidae] Mesnil in Bull. Inst. Pasteur 1: 480 (1903). Chiefly in invertebrates, either in the gut epithelium or in the kidneys, testes, or other organs. Zygote usually producing definite spores, these numerous (commonly twenty or more), thin-walled, without definite dehiscence mechanism, with two or four sporozoites. *Adelea* and *Adelina* chiefly in centipedes; *Klossia* and *Orcheobius* in snails; *Klossiella* in the kidney of the mouse; *Legerella* in various arthropods, the zygote not producing spores but numerous sporozoites.

Family 2. **Haemogregarinida** [Haemogregarinidae] Lühe in Mense Handb. Tropenkrankheiten 3: 205 (1906). Heteroecious, with vegetative multiplication in the tissues of a vertebrate host. The infection spreads to the erythrocytes of the host, and blood-sucking invertebrates are infected by these. Sexual reproduction occurs in the invertebrate host. Production of spores is suppressed; the zygote produces numerous sporozoites. *Haemogregarina* Danilewski (1885; *Drepanidium* Lankester 1882, non Ehrenberg 1861) in turtles, frogs, fishes, transmitted by leeches; *Hepatozoon* in rodents, *Karyolysus* in lizards, transmitted by mites.

Order 3. **Gymnosporidiida** Labbé in Thierreich 5: 77 (1899).
 Suborder *Gymnosporidae* Delage and Hérouard Traité Zool. 1: 284 (1896).
 Suborder *Haemosporidia* Doflein Protozoen 121 (1901).
 Order *Haemosporidia* Calkins Biol. Prot. 441 (1926).
 Subclass *Haemosporidia* with orders *Plasmodiida* and *Babesiida* Hall Protozoology 301, 302, 306 (1953).

In this order the vegetative cells occur in vertebrates and infect the erythrocytes. Sexual reproduction, so far as it has been discovered, occurs in blood-sucking arthropods. The gametocytes do not become associated in pairs; the male gametocytes produce numerous spirochaet-like sperms by a process of budding. In the zygote, the nucleus undergoes a series of divisions, after which numerous naked uninucleate sporozoites are budded off from the surface. There are no walled spores.

The name Haemosporidia, commonly applied to this order, appears to belong by priority to the preceding.

Schaudinn (1903) was disposed to connect this order with the trypanosomes, while connecting the coccidians with *Bodo* and *Cryptobia*. This view has been entertained by Lühe (in Mense, 1906), Woodcock (1909), and Léger (1910). In spite of authority thus good, it appears far-fetched. The Gymnosporidiida are of the same general nature as the Aggregatida, Adeleida, and Haemogregarinida.

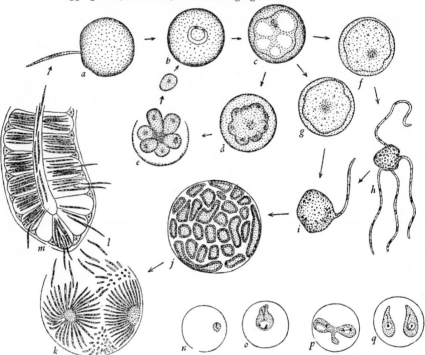

FIG. 42.—**a-m**, Life cycle of *Plasmodium* compiled from various sources: **a**, infection of an erythrocyte by a sporozoite; **b-e**, trophozoites, plasmotomy, and merozoites; **f**, spermatocyte; **g**, oocyte; **h**, production of sperms; **i**, fertilization; **j**, **k**, production of sporozoites in cells of the gut epithelium of the mosquito; **l**, sporozoites; **m**, sporozoites entering the salivary gland of the mosquito. **n-q**, Stages of division of cells of *Babesia bigemina* in erythrocytes of cattle x 2,000 after Dennis (1930).

The Gymnosporidiida are organized, somewhat arbitrarily, as three families.

Family 1. **Halteridiida** [Halteridiidae]Hartmann and Jollos 1910. Family *Leucocytozoidae* Hartmann and Jollos. Family *Haemoproteidae* Doflein. *Haemoproteus* Kruse (*Halteridium* Labbé) occurs in reptiles and birds. Vegetative growth and reproduction occur in tissue cells. Some of the merozoites infect erythrocytes, and are believed to become gametocytes, and to develop no further unless swallowed by some blood-sucking arthropod. In the best known example, *H. Columbae* of pigeons

(Argao, 1908), the alternate host is a fly. In the gut of the fly, the spermatocytes produce the elongate sperms as outgrowths. The zygotes make their way into the wall of the gut of the fly, grow, and produce very numerous sporozoites. These migrate to the salivary gland, from which they are injected into pigeons.

Leucocytozoon attacks birds; its cells become fairly large in certain blood cells which become colorless and spindle-shaped.

Family 2. **Plasmodida** [Plasmodidae] Mesnil in Bull. Inst. Pasteur 1: 480 (1903). The malaria organisms, differing from *Haemoproteus* in that they multiply in the erythrocytes of their hosts. With a few obscure exceptions, the species are construed as a single genus *Plasmodium*. Three species attack man; they have perhaps done mankind more injury than any comparable group of living creatures. Several comparatively poorly known species attack apes and monkeys. The alternate hosts of all species are mosquitoes of the genus *Anopheles*.

The vegetative individuals complete their growth within erythrocytes of their hosts in more or less definite periods of time, and undergo multiple division; the erythrocytes then break up and release the merozoites. The chill and fever of malaria are associated with the destruction of erythrocytes. In the ordinary form of malaria, called tertian malaria, development requires forty-eight hours, and the chill and fever occur every other day. Another form, called malignant tertian or tropical malaria, exhibits the same rhythm; it is distinguished by details of the appearance of the infected erythrocytes. In the third form of malaria in man, called quartan, development requires 72 hours, and the chill and fever occur every third day.

The course of development in the mosquito is quite like that of *Haemoproteus Columbae* in the fly. Some of the parasites inside the erythrocytes are gametocytes; each female gametocyte in an erythrocyte swallowed by a mosquito develops into a single egg, while each male gametocyte buds off several spirochaet-like sperms. The fertilized eggs are able to move. They break into the epithelium of the gut of the mosquito, grow into large globes, and become multinucleate; their protoplasts divide into numerous masses of protoplasm each of which buds off large numbers of sporozoites. The sporozoites are released into the body cavity of the mosquito, migrate to the salivary gland, and are injected into whatever animal the mosquito may bite.

The scientific names usually applied to the three species which cause human malaria are not valid by priority. Extensive synonymy is given by Sabrosky and Usinger, in their application to the International Commission on Zoological Nomenclature for action arbitrarily maintaining the current names (1944), and in the report by Hemming (1950) of the action of the Commission.

Certain structures in the erythrocytes of malaria patients were first recognized as parasites by Laveran, 1880, who, in 1881, named them *Oscillaria malariae*. The organism is believed to have been that of malignant tertian or tropical malaria. The word plasmodium, properly designating a certain type of body, was applied by Marchiafava and Celli 1885, in the combination *Plasmodium malariae,* believed also originally to have designated the agent of malignant tertian malaria. Feletti and Grassi, 1889, introduced the generic name *Haemamoeba,* with two species, *H. vivax,* the agent of tertian malaria, and *H. malariae,* that of quartan malaria; it is believed that the latter epithet was applied under the misapprehension that this was the organism which Marchiafava and Celli had named. It appears that Lühe, 1900, is responsible for the currently used names:

Plasmodium vivax, the organism of tertian malaria;

P. malariae, that of quartan malaria;

P. falciparum, that of malignant tertian.

In order that a great mass of literature may be read without confusion, it is expedient that these names be arbitrarily maintained. The International Commission of Zoological Nomenclature has duly taken action to this effect.

Family 3. **Babesiida** [Babesiidae] Poche in Arch. Prot. 30: 241 (1913). Family *Theileridae* du Toit in Arch. Prot. 39: 94 (1918). Minute intracellular parasites transmitted by arthropods; sexual reproduction unknown. *Theileria* Bettencourt et al. causes a fever of cattle in Africa; the parasites multiply in the tissue cells and spread to the erythrocytes, by which ticks are infected. *Babesia* Stercovici (*Piroplasma* Patton) is similar, but the parasites multiply in the erythrocytes. *B. bigemina* causes the Texas fever of cattle.

The minute nucleus of *Babesia bigemina* is largely filled by a single granule, a karyosome. This is connected by a rhizoplast to an extranuclear granule which has been identified as a blepharoplast, although no flagellum is present. In nuclear division, as described by Dennis (1930), the blepharoplast divides; the rhizoplast splits; the nucleus widens, the karyosome becoming a rod; karyosome, nucleus, and cell undergo constriction. No chromosomes are seen.

If *Bartonella bacilliformis,* the agent of the disease variously known as *verruga peruana,* Oroya fever, or Carrion's disease, is not a bacterium, perhaps it may be placed in or near this family.

Order 4. **Dolichocystida** Delage and Hérouard Traité Zool. 1: 289 (1896).
   Sarcosporidia Balbiani 1882.
   Class *Sarcosporidia* Bütschli in Bronn Kl. u. Ord. Thierreichs 1, Abt. 1: *Inhalt* (1882).
   Subclass *Sarcocystidea* Lankester in Enc. Brit. ed. 9, 19: 855 (1885).
   Order *Sarcosporidia* Doflein Protozoen 214 (1901).
   Order *Sarcocystidea* Poche in Arch. Prot. 30: 245 (1913).
   Subclass *Sarcosporidia* Calkins Biol. Prot. 461 (1926).
The characters are those of the single family and genus:
Family **Sarcocystida** [Sarcocystidae] Poche in Arch. Prot. 30: 245 (1913).

Sarcocystis Lankester produces the *Mieschersche Schläuche,* macroscopically visible bodies, globular, fusiform, or filiform, of dimensions up to several millimeters, in muscles of animals. The several supposed species, from mice, sheep, swine, deer, etc., are not morphologically distinguishable. Miescher observed these things in mice, in which they are called *Sarcocystis Muris;* material from swine is called *S. Miescheriana.*

The visible body is a mass of cells, the whole walled by modified muscle of the host. The mass originates as a single cell which divides repeatedly; the ultimate division products are crescent-shaped uninucleate reproductive cells. Erdmann (1910) observed the infection of epithelial cells of the gut of mice. Each infective cell grew and divided into several, which made their way, or were carried, to the muscles, where they gave rise to the *Mieschersche Schläuche.* Crawley (1914, 1916), on the other hand, found the infective cells to be gametocytes. In cells of the gut epithelium of the host, they may be converted as wholes into eggs, or else may give rise to numerous elongate sperms. These conflicting observations could be explained by an alternation of sexual and asexual generations, but the point is not established.

Order 5. **Schizogregarinida** Calkins Biol. Prot. 433 (1926).
Amoebosporidies Schneider.
Amoebosporidia Labbé in Thierreich 5: 120 (1899).
Suborder *Amoebosporidia* Doflein Protozoen 171 (1901).
Suborder *Schizocystinea* Poche in Arch. Prot. 30: 233 (1913).
Suborder *Schizogregarinaria* Reichenow in Doflein Lehrb. Prot. ed. 5, 3: 872 (1929).
Orders *Archegregarina* and *Neogregarina* Grassé Traité Zool. 1, fasc. 2: 622, 665 (1953).

The Sporozoa previously considered, particularly those of the first two orders, are called coccidians; those of the present order and the two which follow are called gregarines. The latter are characterized (not without exceptions) by inter- instead of intra-cellular active stages, and by the production of numerous gametes, alike or not strongly differentiated, from paired scarcely differentiated gametocytes. The present order includes the gregarines which exhibit asexual reproduction. They are a rather miscellaneous assemblage.

Family 1. **Schizocystida** [Schizocystidae] Léger and Duboscq in Arch. Prot. 12: 102 (1908). Family *Monoschizae* Weiser in Jour. Protozool. 2: 10 (1955), including the two following families. In marine worms and other invertebrates. The sporozoites enlarge in the host and become multinucleate individuals which reproduce freely by producing uninucleate buds. Some of these buds continue the infection directly; others become attached in pairs, each pair secreting a common cyst wall. Each of the individuals in the cyst become multinucleate and buds off numerous uninucleate gametes. The zygotes become walled spores which are cast out with the feces of the host, to infect others which ingest them. Each produces eight sporozoites. *Schizocystis, Siedleckia*.

Family 2. **Seleniida** [Seleniidae] Brasil in Arch. Prot. 8: 394 (1907). In marine worms. Vegetative individuals notably long and slender; spores spiny, with four sporozoites. *Selenidium, Meroselenidium*.

Family 3. **Merogregarinida** [Merogregarinidae] Fantham 1908. Family *Caulleryellidae* Keilin. *Merogregarina, Caulleryella, Tipulocystis*.

Family 4. **Spirocystida** [Spirocystidae] Calkins Biol. Prot. 435 (1925). Family *Spirocystidées* Léger and Duboscq in Arch. Prot. 35: 210 (1915). In earthworms. Spores containing a solitary sporozoite which escapes through a pore. *Spirocystis*.

Family 5. **Ophryocystida** [Ophryocystidae] Léger and Duboscq in Arch. Prot. 12: 102 (1908). Family *Amoebosporidiidae* Brasil (1907), not based on a generic name. Family *Dischizae* Weiser in Jour. Protozool 2: 10 (1955). In *Ophryocystis* Schneider (Léger, 1907), the vegetative individuals, attached to the walls of the Malpighian tubules of beetles, grow and become multinucleate and send out branches whose ends develop into additional individuals. Eventually, different individuals become attached in pairs. Each of these individuals buds off a single uninucleate gamete. The remaining protoplasm of the gametocytes forms a protective sheath around the zygote, which becomes a single spore with eight sporozoites.

Order 6. **Monocystidea** Bütschli in Bronn Kl. u. Ord. Thierreichs 1: 574 (1882).
Order *Haplocyta* Lankester in Enc. Brit. ed. 9, 19: 853 (1885).
Suborder *Acephalina* Labbé in Thierreich 5: 37 (1899).
Organisms of the character of gregarines, not multiplying asexually, the vegetative individuals not elongate and divided into serial parts.

The genus which is best known is *Monocystis* Stein, including several species which are common in earthworms. The cells grow within epithelial cells of the seminal funnels; they and their nuclei reach considerable sizes without dividing. At maturity, they escape into the seminal vescicles, where they form pairs, each pair secreting a common cyst wall. The pairing and encystment were observed, more definitely of the related genus *Zygocystis* than of *Monocystis,* by Stein (1848). The nuclei of the paired cells divide. Several observers, as Brasil (1905) and Mulsow (1911); also, as to related genera, Jameson (1920) and Noble (1938); have observed peculiarities in the first nuclear division. The peculiarities amount to this, that the large nucleus breaks up and, for the most part, undergoes dissolution, leaving a small number of definite chromosomes to undergo normal mitosis in a spindle. Repeated subsequent divisions are of normal character. The numerous nuclei thus produced become those of gametes which are budded off from the surfaces of the gametocytes. This was first observed by Wolters (1891). The gametes from the respective paired cells are presumably always of different mating types, and are usually visibly differentiated, larger and smaller. Each zygote becomes a spindle-shaped walled spore; the enucleate remainder of the gametocytes provides nourishment during their development. Each spore produces eight sporozoites.

The number of known species of Monocystidea is of the order of 150. The majority occur in annelid worms; others attack flatworms, echinoderms, insects, tunicates, and other invertebrates. Bhatia (1930) distinguished twelve families which are here merely listed.

A. The two ends of the spore alike.

Family 1. **Monocystida** [Monocystidae] (Bütschli) Poche in Arch. Prot. 30: 236 (1913). Family *Monocystiden* Stein in Arch. Anat. Phys. 1848: 187 (1848). MONO-CYSTIDAE Bütschli (1882). *Monocystis,* etc.

Family 2. **Rhynchocystida** [Rhynchocystidae] Bhatia in Parasitology 22: 158 (1930). *Rhynchocystis.*

Family 3. **Stomatophorida** [Stomatophoridae] Bhatia op. cit. 159. *Stomatophora, Choanocystis,* etc.

Family 4. **Zygocystida** [Zygocystidae] Bhatia op. cit. 160. *Zygocystis, Pleurocystis.*

Family 5. **Akinetocystida** [Akinetocystidae] Bhatia op. cit. 160. *Akinetocystis.*

Family 6. **Syncystida** [Syncystidae] Bhatia op. cit. 161. *Syncystis.*

Family 7. **Diplocystida** [Diplocystidae] Bhatia op. cit. 161. *Diplocystis, Lankesteria.*

Family 8. **Schaudinellida** [Schaudinellidae] Poche in Arch. Prot. 30: 236 (1913). *Schaudinella.*

B. The ends of the spores differentiated.

Family 9. **Doliocystida** [Doliocystidae] Labbé in Thierreich 5: 33 (1899). Family *Lecudinidae* Kamm. *Lecudina* Mingazzini (*Doliocystis* Léger).

Family 10. **Urosporida** [Urosporidae] Woodcock 1906. Family *Choanosporidae* Dogiel. *Gonospora; Lithocystis; Urospora,* the spores with long tails; *Ceratospora; Pterospora,* the spores with longitudinal flanges.

Family 11. **Ganymedida** [Ganymedidae] J. S. Huxley in Quart. Jour. Micr. Sci. n.s. 55: 169 (1910). *Ganymedes.*

Family 12. **Allantocystidae** [Allantocystidae] Bhatia op. cit. 163. *Allantocystis.*

Order 7. **Polycystidea** Bütschli in Bronn Kl. u. Ord. Thierreichs 1: 578 (1882). Order *Gregarinae* Haeckel Gen. Morph. 2: xxv (1866), the mere plural of a generic name.

Subclass *Gregarinida* Bütschli op. cit. *Inhalt* (1882).
Order *Septata* Lankester in Enc. Brit. ed. 9, 19: 853 (1885).
Order *Brachycystida,* suborder *Gregarinidae,* and tribe *Cephalina* or *Polycystina* Delage and Hérouard Traité Zool. 1: 255, 256, 269 (1896).
Order *Gregarinida* Labbé in Thierreich 5: 4 (1899).
Suborder *Eugregarinaria* Doflein Protozoen 160 (1901).
Order *Gregarinoidea* Minchin (1912).
Suborder *Gregarininea* and tribe *Gregarinoidae* Poche in Arch. Prot. 30: 234 (1913).
Subclass *Gregarinida,* order *Eugregarinida,* and suborder *Cephalina* Calkins Biol. Prot. 422, 428 (1926).

The typical gregarines, the vegetative cells elongate and divided by more or less definite constrictions into two (or, occasionally, more than two) parts; not reproducing asexually.

Typical gregarines occur chiefly in insects. The vegetative cell consists of an anterior portion (protomerite) serving for attachment and a posterior portion (deutomerite), containing the nucleus, lying in the gut cavity of the host. Both parts have a thick outer layer, commonly differentiated upon the protomerite into a more or less elaborate knob, the epimerite. Longitudinal fibrils, presumably contractile, are present. The cells writhe actively.

The individuals are commonly found in pairs, one member attached to the epithelium of the gut, the other to the posterior end of the first. This arrangement is produced by active self-placement on the part of the second member. When both are mature, they take common action to produce a globular cyst. The protoplasts remain distinct until both have become multinucleate, after which they produce numerous gametes. In some forms, as *Nina,* studied by Goodrich (1938), all of the gametes migrate from one cell, recognizably male, into the other, the female cell; the male cell is left empty and is compressed or crushed by the growth of the zygotes in the female cell. The zygotes are spores, usually fusiform, and usually producing sporozoites by eights. In *Gregarina* and *Gamocystis,* an inner layer of the cyst wall is so modelled as to form tubes (sporoducts) running from the surface to the interior. When the spores are ripe, the sporoducts become extroverted and the spores are extruded through them in uniseriate rows. In connection with this behavior, the spores have flat ends like barrels.

Family 1. **Stenophorida** [Stenophoridae] Crawley 1903. Protomerite a mere knob. *Stenophora.*

Family 2. **Gregarinida** [Gregarinidae] Greene 1859. Family *Gregarinarien* Stein in Arch. Anat. Phys. 1948: 187 (1848). Gregarines which are without epimerites and are not notably elongate. There are about a dozen genera. Cysts without sporoducts: *Hirmocystis, Hyalospora, Cnemidospora.* Cysts with sporoducts, the spores barrel-shaped: *Gregarina, Gamocystis.* The earliest observations of Sporozoa were by Dufour (1826), who, studying the anatomy of insects, found them in the gut of beetles. He took them for worms and illustrated an individual with an epimerite, which he took for a sucker. Later (1828) he applied names, *Gregarina conica* to the form first seen, *G. ovata* to a form without an epimerite found in the *forficule,* i.e., in an orthopteran. The former does not belong to the genus *Gregarina* as subsequently construed; it appears to be a member of the family Actinocephalida. *Gregarina ovata* should be regarded as the type of *Gregarina,* but the genus has usually been interpreted by *G. cuneata,* which Stein observed in cockroaches.

Family 3. **Didymophyida** [Didymophyidae] Wasilewski 1896. Family *Didymophyiden* Stein (1848). Like the foregoing, but the cells extremely elongate. *Didymophyes.*

Family 4. **Acanthosporida** [Acanthosporidae] Labbé in Thierreich 5: 27 (1899). The spores with polar or equatorial bristles. *Acanthospora.*

Family 5. **Stylocephalida** [Stylocephalidae] Ellis 1912. Family *Stylorhynchidae* Labbé op. cit. 30, based on a generic name which is a later homonym. Epimerite elongate with a small terminal knob. *Stylocephalus.*

Family 6. **Actinocephalida** [Actinocephalidae] Wasilewski 1896. Epimerite with thorns. Numerous genera, *Sciadophora, Acanthorhynchus, Actinocephalus, Hoplorhynchus, Pileocephalus,* etc.

Family 7. **Menosporida** [Menosporidae] Labbé op. cit. 29. Epimerite with a long stalk, distally branched and bearing appendages. *Menospora.*

Family 8. **Dactylophorida** [Dactylophoridae] Wasilewski 1896. Epimerite distally broadened, clinging to the host epithelium by means of numerous filiform processes. *Dactylophorus, Nina (Pterocephalus),* etc.

Family 9. **Porosporida** [Porosporidae] Labbé op. cit. 7. Heteroecious: in *Porospora,* the gregarinoid stage occurs in crabs and the production of spores occurs in mussels. The spores contain a single sporozoite and open through a pore.

Order 8. **Haplosporidiidea** Poche in Arch. Prot. 30: 178 (1913).

Order *Aplosporidies* Caullery and Mesnil 1899.

Order *Haplosporidies* Caullery and Mesnil in Arch. Zool. Exp. Gen. sér. 4, 4: 104 (1905).

Order *Haplosporidia* Auctt., the mere plural of a generic name.

Subclass *Haplosporidia* Hall Protozoology 326 (1953).

Unicellular intracellular parasites, the cells becoming multinucleate and multiplying by fragmentation, producing walled spores which germinate by releasing the protoplasts as single sporozoites.

The vegetative body is of the type properly called a plasmodium. The nuclei and the process of division, described by Granata (1914) are characteristic. The resting nucleus contains an "axial rod" as well as a nucleolus-like body. In mitosis the axial rod becomes converted into an intranuclear spindle. Individual chromosomes have not been seen; the chromatin gathers in a mass about the middle of the spindle (the figures are curiously diatom-like). The mass of chromatin, the nucleolus-like body, and the entire nucleus, divide by constriction; the ends of the spindle persist as the axial rods of the daughter nuclei. Eventually, the plasmodium secretes a thin wall and the protoplast divides into uninucleate naked cells. Granata found that these cells are gametes, and that conjugation takes place among gametes produced by the same plasmodium. The zygotes become walled spores which germinate by casting off a circular operculum and releasing the contents. If the life cycle is correctly understood, we may suppose that these organisms are degenerate gregarines.

In the present state of knowledge, it will be as well to treat the typical haplosporidians as a single family:

Family **Haplosporidiida** [Haplosporidiidae] Caullery and Mesnil in Arch. Zool. Exp. Gen. sér. 4, 4: 106 (1905). Families *Bartramiidae* and *Coelosporidiidae* Caullery and Mesnil op. cit. 107. Characters of the order. *Haplosporidium* (spores with appendages at both ends) and *Urosporidium* (spores with a single appendage) attack chiefly annelid worms. *Bartramia* attacks rotifers; *Ichthyosporidium* is a serious parasite of fishes; *Coelosporidium* attacks cockroaches.

The following family, of uncertain position, may tentatively be associated with the Haplosporidiidea:

Family **Metchnikovellida** [Metchnikovellidae] Caullery and Mesnil in Compt. Rend. Soc. Biol. 77: 527 (1914), Ann. Inst. Pasteur 33: 214 (1919). Secondary parasites, intracellular in gregarines; cells naked at first, with very minute nuclei, which become numerous, later converted into walled cysts of characteristic form, the protoplasts undergoing division into uninucleate infective cells. *Metchnikovella, Amphiamblys, Amphiacantha.*

Class 2. NEOSPORIDIA (Schaudinn) Calkins

Myxosporidia Bütschli in Zool. Jahresber. 1880: 162 (1881).
Subclass *Myxosporidia* Bütschli in Bronn Kl. u. Ord. Thierreichs 1, Abt. 1: *Inhalt* (1882).
Subclass *Amoebogeniae* Delage and Hérouard Traité Zool. 1: 291 (1896).
Subclass NEOSPORIDIA Schaudinn in Zool. Jahrb. Anat. 13: 281 (1900).
Order *Cnidosporidia* Doflein Protozoen 177 (1901).
Class *Cnidosporidia* Poche in Arch. Prot. 30: 224 (1913).
Class NEOSPORIDIA and subclass *Cnidosporidia* Calkins Biol. Prot. 445, 448 (1926).
Subphylum *Cnidosporidia* Grassé Traité Zool. 1, fasc. 1: 129 (1952).
Class *Cnidosporidea* Hall Protozoology 311 (1953).
Fungilli whose resting cells contain polar capsules; are walled, at least usually, by a layer of modified cells; and, in most examples, release a single infective cell.

As a general rule, the vegetative bodies of Neosporidia are plasmodia, i.e., naked multinucleate bodies, usually freely capable of asexual reproduction by internal or external budding. An entire small plasmodium may become converted into one or two spores, or the spores may be cut out internally and produced continually. The spores, at least in the two better-known orders, are structures formed from several cells; they are not homologous with the spores of the proper Sporozoa. In most examples, only one of the cells involved in the formation of a spore is fertile, and only one infective protoplast is released on germination. Of the sterile cells, one or more become converted into the structures called polar capsules. These resemble the nematocysts of coelenterates: they contain a coiled hollow thread capable of swift extroversion. Extroversion occurs during germination. Its significance is unknown. The presence of polar capsules marks the class as a natural group.

Three orders are recognized:

1. Spores covered by two valves formed from accessory cells...................................Order 1. PHAENOCYSTES.
1. Spores covered by three valves formed from accessory cells.................................Order 2. ACTINOMYXIDA.
1. Spores very minute, with a continuous membrane.......................................Order 3. CRYPTOCYSTES.

Order 1. **Phaenocytes** Gurley in Bull. U. S. Fish Comm. 11: 410 (1893).
Order *Nematocystida* Delage and Hérouard Traité Zool. 1: 291 (1896).
Order *Phaenocystida* Labbé in Thierreich 5: 85 (1899).

Order *Cnidosporidia* Doflein Protozoen 177 (1901).

Order *Myxosporidia* Calkins Biol. Prot. 449 (1926).

Most species of this order parasitize fishes, living either in internal cavities or in the tissue cells; fewer than a dozen species are known from miscellaneous other animals, amphibia, reptiles, insects, and worms. Most of these parasites are not extremely injurious.

The infective protoplast which issues from a spore is, at least usually, binucleate. The nuclei fuse and the fusion nucleus divides repeatedly as the plasmodium grows.

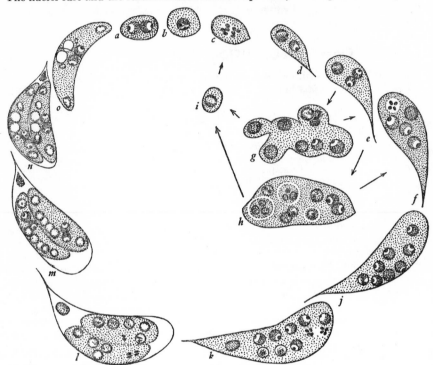

Fɪɢ. 43.—Diagram of the life cycle of *Myxoceros Blennius* after E. Noble (1941).

In the examples which are believed to be more primitive, the plasmodia are freely capable of budding, and the mature plasmodia are rather small and are converted as wholes into single or paired spores. In the remaining examples, the plasmodia do not multiply by budding, but produce spores continually.

Noble (1941) described the mitotic process. There is a rather large intranuclear centrosome, which divides, the daughter centrosomes moving to opposite sides of the nuclear cavity. Four chromosomes appear; this is apparently constant throughout the order. The nuclear membrane and the centrosomes disappear. No spindle has been seen. The chromosomes divide, and the daughter chromosomes move apart and melt into two masses. The masses swell, a nuclear membrane appears about each, and a centrosome appears inside of each.

The spore-forming structure (sporoblast) is a protoplast with several nuclei; it is either a whole small plasmodium, or half of one, or a protoplast cut out endogenously within a plasmodium. Two of the nuclei are set apart in cells which become converted into the valves of the spore. Two or four are set apart in cells which become converted into polar capsules. Two, of which it is established that they have two chromosomes each, are the nuclei of the infective protoplast.

In a review of the literature as to life cycles, Noble (1944) remarks as follows. "A survey of the literature reveals that there is little agreement on the details of nuclear changes in the Myxosporidia. Some authors maintain that the cycle is mainly haploid, others have described a diploid cycle. Some reports indicate that there are two reduction divisions and two zygotes in one cycle. When only one zygote is reported the reduction division in one case occurs just before fertilization, in another case it occurs just after fertilization. Some authors have maintained that there is no sexual process." Noble's own conclusions include the following. The organisms are diploid at most stages. The meiotic divisions are among those by which the sporoblast becomes multinucleate. The two haploid nuclei of the spore, which unite after germination, are derived from a single diploid nucleus. Authors who have described fusions of protoplasts, or transfers of nuclei from one protoplast to another, have had no evidence beyond an understandable unwillingness to accept fusions of sister nuclei.

Nearly two hundred species of the present order are listed in the monograph of Kudo (1920), who established three suborders.

A. Valves conical, spores biconic (suborder EURYSPOREA Kudo).

Family 1. **Myxoceratida** nom. nov. Family *Ceratomyxidae* Doflein Protozoen 182 (1901), based on a generic name which is a later homonym. Characters of the suborder. **Myxoceros** nom. nov. (*Ceratomyxa* Thélohan 1892, non *Ceratiomyxa* Schröter 1889; if ever names are homonymous without being absolutely identical, these are.) Some thirty-five species; the type is **M. sphaerulosa** (Thélohan) comb. nov.; Noble studied mitosis in **M. Blennius** (Noble) comb. nov. *Leptotheca, Myxoproteus, Wardia, Mitraspora.*

B. Valves hemispherical, spores spherical (suborder SPHAEROSPOREA Kudo).

Family 2. **Chloromyxida** [Chloromyxidae] Gurley in Bull. U. S. Fish Comm. 11: 418 (1893). *Chloromyxées* Thélohan in Bull. Soc. Philomath. Paris sér. 8, 4: 176 (1892). *Chloromyxea* Braun in Centralbl. Bakt. 14: 739 (1893). With four polar capsules. *Chloromyxum.*

Family 3. **Sphaerosporida** [Sphaerosporidae] Davis 1917. With two polar capsules. *Sphaerospora, Sinuolinea.*

C. Valves saucer- or boat-shaped, spores disk-shaped or fusiform (suborder PLATYSPOREA Kudo).

Family 4. **Myxidiida** [Myxidiidae] Gurley op. cit. 420. *Myxidiées* Thélohan op. cit. 175. *Myxidiea* Braun l.c. *Myxidium, Sphaeromyxa, Zschokkella.*

Family 5. **Coccomyxida** [Coccomyxidae] Léger and Hesse 1907. *Coccomyxa.*

Family 6. **Myxosomatida** [Myxosomatidae] Poche in Arch. Prot. 30: 230 (1913). *Myxosoma, Lentospora.*

Family 7. **Myxobolida** [Myxobolidae] Gurley op. cit. 413. *Myxobolées* Thélohan op. cit. 176. *Myxobolea* Braun l.c. *Myxobolus, Henneguya, Hoferellus.*

Order 2. **Actinomyxida** Stolc 1911.

This order includes about a dozen parasites in annelid worms. A plasmodial stage and asexual reproduction are believed not to occur; the infective protoplast grows into an individual whose one or two nuclei remain undivided until the commencement of the ill-understood process by which the complicated spores, with three valves and three polar capsules, are produced.

Family 1. **Tetractinomyxida** [Tetractinomyxidae] Poche in Arch. Prot. 30: 231 (1913). Family *Haploactinomyxidae* Granata in Arch. Prot. 50: 205 (1925). Spores subglobular, with a single binucleate sporozoite. *Tetractinomyxon.*

Family 2. **Synactinomyxida** [Synactinomyxidae] Poche l.c. Family *Euactinomyxidae* Granata l.c. Family *Triactinomyxidae* Kudo Handb. Protozool. 314 (1931). Spores producing eight or more sporozoites. *Sphaeractinomyxon* and *Neactinomyxon,* the spores subglobular. *Synactinomyxon,* with two of the valves protruding as considerable horns, the whole horse-shoe shaped. *Triactinomyxon* and *Hexactinomyxon,* all three valves drawn out into long horns, the whole caltrop- or anchor-shaped.

Order 3. **Cryptocystes** Gurley in Bull. U. S. Fish Comm. 11: 409 (1893).
Microsporidies Balbiani 1882.
Order *Microsporidiida* Labbé in Thierreich 5: 104 (1899).
The parasites of this order attack chiefly arthropods and fishes. They multiply asexually and produce serious epizootics. The spores are very minute, and the details of the processes by which they are formed are unknown. A polar capsule is present in each spore (those of *Telomyxa* have two polar capsules). The polar capsules are not visible in living material, but are revealed by treatment with alkali.

In Kudo's monograph of this order (1924), more than 150 species are treated. They form four families.

Family 1. **Glugeida** [Glugeidae] Gurley op. cit. 409. *Glugeidées* Thélohan op. cit. 174. *Glugeidea* Braun l.c. Family *Nosematidae* Labbé in Thierreich 5: 104 (1899). Family *Plistophoridae* Doflein Protozoen 205 (1901). Spores oval, ovoid, or pyriform. *Nosema Bombycis* Nägeli causes the pébrine disease of silkworms; another species of *Nosema* causes an epizootic of honeybees. *Glugea* attacks several species of fishes. *Gurleya, Thelohania, Duboscquia, Plistophora,* etc.

Family 2. **Coccosporida** [Coccosporidae] Kudo Handb. Protozool. 323 (1931). Family *Cocconemidae* Léger and Hesse 1922, based on a generic name which is a later homonym. Spores globular. *Coccospora Slavinae* (Léger and Hesse) Kudo, in the oligochaet worm *Slavina.*

Family 3. **Mrazekiida** [Mrazekiidae] Léger and Hesse 1922. Spores elongate, exceedingly minute, resembling bacteria. *Mrazekia, Octospora, Spironema, Toxonema.*

Family 4. **Telomyxida** [Telomyxidae] Léger and Hesse 1910. *Telomyxa glugeiformis,* in the fat body of the larva of *Ephemera vulgata,* producing ellipsoid spores with a polar capsule at each end.

Chapter XII
PHYLUM CILIOPHORA

Phylum 8. CILIOPHORA (Doflein) nomen phylare novum

Class *Infusoires* Lamarck Phil. Zool. 1: 127 (1809).

Class INFUSORIA Lamarck Anim. sans Verteb. 1: 392 (1815).

Class *Protozoa* Goldfuss in Isis 1818: 1008 (1818).

Class *Polygastrica* Ehrenberg Infusionsthierchen p.* (1838).

Hauptgruppe Protozoa, class INFUSORIA, and order STOMATODA Siebold in Siebold and Stannius Lehrb. vergl. Anat. 1: 3, 10 (1848).

Subkingdom *Archezoa* Perty Kennt. kl. Lebensf. 22 (1852), not phylum *Archezoa* Haeckel (1894).

Order *Ciliata* Perty op. cit. 137.

Subphylum *Infusoria* Haeckel Gen. Morph. 2: lxxviii (1866).

Phylum *Infusoria* Haeckel Syst. Phylog. 1: 90 (1894).

Subphylum CILIOPHORA Doflein Protozoen 227 (1901).

Dependent organisms, mostly predatory, unicellular but mostly of complicated structure; swimming by means of cilia at least at some stage of life; mostly with nuclei of two types in each cell. *Vorticella,* the only genus named by Linnaeus, is to be considered the type.

These organisms are the typical examples of the accepted groups Infusoria and Protozoa. The name Infusoria, referring to creatures which appear in infusions, is said to have been introduced by Ledermüller, 1763, or Wrisberg, 1764. As a scientific name it has status from its application to a class by Lamarck (1815). The name Protozoa, applied to a class in its original publication by Goldfuss, is a later synonym of Infusoria. In treating the group as a phylum, one finds it necessary to apply a new name, and takes up as such the name which Doflein applied to it as a subphylum.

The essential point in the definition of the phylum is the word *cilia.* Cilia are cell-organs of the same nature as flagella, differing in being smaller in proportion to the cell which bears them, more numerous, and distributed generally on the surface. In Loeffler's classic investigation (1889), they were found to bear solitary terminal appendages; by subsequent terminology, they are acroneme. Doflein appears to have been mistaken in emphasizing the difference between flagella and cilia; there is no fundamental difference. A verbal distinction, nevertheless, is expedient: the application of the term cilium is to be restricted to two things, (a) the swimming organelles of the Ciliophora, and (b) moving fibrils protruding abundantly from certain epithelial cells of animals. Botanical usage, which treats cilium and flagellum as synonyms, is unsound. The structures which in botany have been called cilia are definitely flagella.

The cells of Ciliophora reach moderately large sizes; those of the classroom example *Paramaecium* attain a length of 0.25 mm. and are perceptible to the naked eye. The cells of some of the Ciliophora are the most highly complicated of all individual cells. In addition to the cilia, the cell organs which require discussion are the pellicle, neuromotor fibrils, trichocysts, structures involved in nutrition, contractile vacuoles, and nuclei.

The cell has a firm ectoplasm or pellicle which gives it a definite form. The cilia spring from basal granules imbedded in the pellicle. In simpler examples, the cilia are essentially uniform and uniformly distributed on the surface. Other examples are without separate cilia upon part or all of the surface, but bear a variety of structures which consist of coalescent cilia. Membranelles are triangular appendages consisting of brief rows of cilia; undulating membranes represent long rows; cirri represent tufts. The organisms of class Tentaculifera bear cilia only in the juvenile condition. At maturity they bear extensible tubular structures called tentacles, by means of which they capture free-swimming ciliates and absorb their contents.

The basal granules of the cilia are linked together by a system of fibrils; the cilia and fibrils make up the neuromotor apparatus. This term was coined by Sharp, in his study of *Diplodinium* (1914). The neuromotor fibrils form a highly elaborate network, not connected with the nucleus, as in flagellates, but to a central structure, apparently regulative, called the motorium. The motorium of *Diplodinium* is a massive body near the anterior end; that of the tintinnids is fairly large in proportion to the cells (Campbell, 1926, 1927); that of *Paramaecium,* presumably a comparatively primitive organism, is a minute body lying near the dorsal side of the cytopharynx (Lund, 1933).

Imbedded in the pellicle, in addition to the neuromotor fibrils, there are certain minute ellipsoid bodies called trichocysts. These, when the cell is irritated, discharge their contents in the form of elongate rods or threads. Their mechanism and effect are not understood.

In most Ciliophora, each cell has a mouth and gullet; or better, since these structures are not homologous with those of animals, a cytostome and cytopharynx. The cytopharynx is a more or less funnel-shaped impression in the cell. It is bounded laterally by ciliate pellicle; its outer opening is the cytostome; it is closed at the inner end by a layer of cell membrane directly over fluid cytoplasm. Prey, chiefly bacteria and small algae, encountered by the organism as it swims, is swept into the cytopharynx by the action of the cilia. When a certain mass of prey has accumulated, the cell membrane at the inner end of the cytopharynx becomes impressed and undergoes constriction, enclosing the prey in a food vacuole. The material in the food vacuole undergoes digestion; while this is taking place, movement of the cytoplasm carries the vacuole along a definite circuitous course through the interior of the cell. After some time, the vacuole arrives at a certain point on the pellicle, the anus or cytoproct, where it discharges its contents and disappears by bursting through the pellicle.

In freshwater species, each cell contains one or more contractile vacuoles which appear at definite points and disappear periodically by discharge of their contents to the exterior. Associated with the proper contractile vacuoles, there may be systems of "canals" which are in fact additional contractile vacuoles. These structures have been much studied; there are notable accounts by Day (1930) and Mac Lennan (1933). When a vacuole has disappeared by discharge, it reappears as one or more minute vacuoles in the same area: minute bodies of gelled protoplasm turn into sol, and then become lifeless liquid. The discharge of a "canal" into the proper contractile vacuole occurs by dissolution of the bounding membranes of gelled protoplasm where the two are in contact, followed by contraction of the membrane of the canal. The proper contractile vacuole discharges by essentially the same mechanism. Its membrane meets and becomes fused with the bounding membrane of the cell, generally at the end of a brief channel through the pellicle; the combined membrane breaks, and the membrane of the vacuole contracts.

Earlier biologists supposed that the contractile vacuole is an excretory mechanism. More probably, its function is purely hydrostatic, to rid the cell of the water which is constantly entering by osmosis. Marine Ciliophora have no contractile vacuoles.

In many members of the family Opalinoea each cell has many similar nuclei. These divide, from time to time, by mitosis. Cell division takes place independently of nuclear division, by transverse constriction, when a certain size has been reached.

In the generality of Ciliophora, each cell has one or more nuclei of each of two types, macronuclei, which are conspicuous, and micronuclei, discerned with difficulty.

Cell division occurs by transverse constriction and is necessarily associated with nuclear division. The macronucleus becomes elongate and divides by constriction without any formation of chromosomes; in other words, amitotically. The micronucleus also becomes elongate and divides by constriction. Early observers supposed this process also to be amitotic. Actually, there appear within the intact nuclear membrane a spindle and a definite number of chromosomes. Reichenow (editing Doflein, 1927) compiled the following diploid counts:

Stentor coeruleus	28
Didinium nasutum	16
Chilodon uncinatus	4
Carchesium polypinum	16

Turner (1930) found 8 in *Euplotes Patella.* Thus the chromosome numbers of Ciliophora appear usually to be small powers of 2.

The chromosomes duly undergo division, the daughter chromosomes going to different ends of the nuclear cavity. The nucleus becomes greatly elongate and its membrane presses in from the sides and cuts it in two. Turner observed in the axis of the spindle of *Euplotes* a rather small endosome which becomes elongate and undergoes constriction while the chromosomes are forming.

Opalina has a sexual process in which the multinucleate cells divide into many uninucleate gametes. These are sexually differentiated, larger and smaller; they duly unite in pairs and the zygotes grow and become ordinary multinucleate individuals.

In the generality of Ciliophora, early observers discovered a sexual process in which the cells, apparently undifferentiated, join in pairs but maintain their individuality. The uniting cells become attached to each other in definite positions: in *Paramaecium,* by their ventral or mouth-bearing surfaces; in *Euplotes,* by the left halves of their broad ventral surfaces; in the ophryoscolecids and various other groups, by their anterior ends. They remain attached, while continuing to swim, for several hours, during which an exchange of nuclei takes place, and then resume their separate life. Calkins (1926) was disposed, contrary to historical usage, to confine application of the term conjugation to this exceptional form of syngamy.

The nuclear details of conjugation were described by Maupas (1889) and Richard Hertwig (1889), whose observations have repeatedly been confirmed. When a pair have joined, their macronuclei divide several times; the ultimate fragments are digested and disappear. The micronuclei also divide, concurrently in both conjugants, a fixed number of times, in *Paramaecium* three, in *Euplotes* four. These divisions include a meiotic process. Most of the haploid nuclei produced are digested; as a general rule, only one survives to undergo the final division, which is mitotic, producing in each conjugant two genetically identical haploid nuclei. By this time a cytoplasmic connection has been established between the conjugants. In *Paramaecium,* the spindles of the mitotic final nuclear divisions extend through this connection, so that when mitosis is complete each protoplast contains two haploid nuclei of different origin. In other ciliates the same result is attained, apparently, by the migration of one nucleus of each pair. Karyogamy takes place in each conjugant. The cytoplasmic connection is broken and the conjugants separate from each other. During several subsequent hours, the zygote nucleus undergoes a characteristic number of divisions, three in *Paramaecium*. Among the nuclei produced, one usually enlarges and becomes a macronucleus; others, of the number characteristic of the form, survive as micronuclei; the remainder are digested.

In *Vorticella* and its allies, syngamy consists of the complete fusion of a smaller swimming individual with a larger one attached by a stalk. The nuclear processes are believed to be essentially as in other ciliates. The reproduction of the Tentaculifera has not been much studied, but here also the nuclear changes are as in the generality of ciliates (Noble, 1932).

The possibility of conjugation is limited by the occurrence of mating types. Certain early observations had suggested the existence of these; the definite discovery was by Sonneborn, in *Paramaecium Aurelia* (1937). Results of further study are available in a symposium edited by Jennings (1940) and in a review by Kimball (1943). To current knowledge, then:

Paramaecium caudatum includes four mating types divided into two groups; types I and II conjugate with each other, and types III and IV with each other, but the two groups are mutually sterile.

Paramaecium Aurelia includes eight mutually sterile groups, each of two mutually fertile mating types.

Paramaecium Bursaria includes three mutually sterile groups. The first group is of four types, each self-sterile but able to conjugate with any other; the second group is of eight such types, and the third again of four.

Paramaecium multimicronucleatum is without mating types; any race can conjugate with any other

Euplotes Patella includes six mating types all in one group; each can conjugate with any other.

The heredity of mating types is not understood. It is not a matter of simple Mendelian heredity. In *Paramaecium Bursaria* group I, the progeny of a cell of a given mating type may include after conjugation either two or all four of the mating types. The mating type of a line becomes fixed in connection with the first or second cell division after conjugation, at the time that macronuclei are being differentiated; it is accordingly believed that something in the macronuclei fixes the mating types.

So far as mating types are present, pure lines of ciliates cannot conjugate. Early attempts to maintain pure cultures failed by death after intervals of some months. These observations led to speculations that the vitality of protoplasm is limited, and that sexual reproduction restores it. Woodruff, however, proved it possible to maintain *Paramaecium Aurelia* indefinitely without conjugation: he reported (1926) a culture so maintained for sixteen years, an estimated eleven thousand generations.

The cultures are not thus persistent without nuclear change. At intervals, the macronuclei break up and dissolve, and are replaced by new ones formed by division of the micronuclei. Woodruff and Erdmann (1914) applied to this process of replacement of nuclei the term endomixis. It is not possible that this process is the genetic equivalent of karyogamy. It is, presumably, the physiological equivalent of conjugation in its feature of providing new macronuclei.

Diller (1936) observed in *P. Aurelia* a different manner of replacement of nuclei, by autogamy. In this process, the nuclei of a solitary cell go through the preliminaries of conjugation; two haploid nuclei, sister products of one act of mitosis, unite to form a zygote nucleus; and this divides in the usual manner to produce micronuclei and macronuclei. Wichtermann (1939, 1940) observed that two cells, joined as in conjugation, may simultaneously undergo autogamy instead of exchanging nuclei.

In the normal conjugation of ciliates, the gamete nuclei produced in each cell, being sister products of mitosis, are genetically identical; and the zygote nuclei produced after interchange are also genetically identical with each other. Autogamy is believed to produce diploid nuclei which are completely homozygous. Thus the sexual processes of the ciliates tend strongly to limit the variability of the progeny. This is a peculiar and surprising feature of the group.

The ciliates have attracted experimental study, beyond what has already been implied, of various functions, including nutrition, inheritance of acquired characters, and regeneration after injury.

Hall and his associates (1940-1945) have shown that *Colpidium campylum* and *Tetraphymena Geleii* (the latter is in their earlier papers called *Glaucoma piriformis*) require thiamin and probably riboflavin. Nutritional requirements, rather than such an entity as vitality, are presumably responsible for the limited life of early attempted pure cultures. As to minerals, the same scholars demonstrated the necessity of Ca and Fe: others have demonstrated the necessity of K, Mg, and P.

It has been observed of certain cultures in which the rate of division has been increased by exposure to high temperature that they would continue to divide abnormally rapidly when returned to normal temperatures. The peculiarity disappeared in individuals which conjugated. By refrigeration or by application of chemicals, there have been produced "monsters," individuals of abnormal structure, which have reproduced themselves through many generations, and have proved capable eventually of giving rise to normal individuals. Jollos (1913) designated as *Dauermodifikationen,* that is, enduring changes, modifications of the type described. They are actually acquired characters which can be inherited within limits. It is evident that they are determined by macronuclei or by cytoplasm, and that they are not in conflict with the principle that the truly enduring heredity of nucleate organisms lies in nuclei which divide mitotically.

Balamuth (1940) reviewed the literature of experimental mutilation of Protozoa and gave a bibliography of 173 titles. Most of the experiments have been performed on ciliates. The conclusions from them include these, that regeneration of parts artificially cut away takes place with different degrees of facility in different groups, and that it is effected, if at all, by the same mechanism by which the parts are produced after division or excystment. The less elaborate ciliates, as *Opalina* and *Paramaecium,* are usually killed by mutilation, since this allows the fluid inner cytoplasm to escape. In *Stentor,* injury to the crown of membranelles results in the appearance of a new crown of membranelles on the side of the body, followed by its migration to the injured area. In *Stylonychia* and *Euplotes,* destruction of one cirrus is followed by the appearance, in a certain area of the surface, of the primordia of a complete set of cirri; the original cirri are absorbed, and the new ones migrate along the surface to their proper stations. The regulation of regeneration is explained, as are various other phenomena, in a review by Weisz (1954).

Micronuclei are necessary for unlimited life and for sexual reproduction, but not for regeneration and a long period of life. Schwartz kept a culture of *Stentor* alive without micronuclei for more than a year. Macronuclear material is necessary for regeneration, but any fragment of a macronucleus is sufficient. This is a very significant observation. It means that all the factors controlling the vegetative structure and behavior of a cell can be spread out and intermingled in all parts of a body of considerable size; it furnishes an analogy to the state of affairs which may be supposed to exist in bacteria.

The Ciliophora are treated as two classes, Infusoria and Tentaculifera. Hartog (1909) estimated the number of known species of the former as about five hundred. This number would have included practically all of the fresh-water species known up to the present. Entozoic and marine species were known, but hundreds of species of these ecological groups have subsequently been discovered. Including some two hundred species of Tentaculifera, the phylum Ciliophora appears to be of about twelve hundred known species.

Class 1. INFUSORIA Lamarck

Class *Ciliata* Haeckel Gen. Morph. 2: lxxviii (1866).
Class *Ciliatea* Hall Protozoology 333 (1953).
Further synonymy essentially as of the name of the phylum.
Ciliophora lacking tentacles, bearing cilia or modified cilia in the mature condition.

Stein (1867) provided four orders of Infusoria. These orders are surely natural. Subsequent authors have proposed many modifications of Stein's system, and many of these are surely sound; but among groups proposed as additional orders, only the opalinids are positively entitled to this status.

1. Nuclei all alike, commonly numerous.............. Order 1. OPALINALEA.
1. Nuclei differentiated into macronuclei and
   micronuclei.
    2. Without a spiral band of membranelles
       or cilia about the cytostome.................. Order 2. HOLOTRICHA.
    2. With a spiral band of membranelles or
       cilia about the cytostome.
           3. The spiral sinistrorse.
               4. Not of the character of the fol-
                 lowing order...................... Order 3. HETEROTRICHA.
               4. Flattened, cirri and most cilia
                 confined to the ventral surface........ Order 4. HYPOTRICHA.
           3. The spiral dextrorse..................... Order 5. STOMATODA.

Order 1. **Opalinalea** *nom. nov.*
Suborder *Opalininea* Poche in Arch. Prot. 30: 250 (1913).
Protociliata Metcalf in Anat. Record 14: 89 (1918) and Jour. Washington Acad. Sci. 8: 431 (1918).
Subclass *Protociliata* Kudo Handb. Protozool. 335 (1931).
Order *Opalinida* Hall Protozoology 113 (1953), preoccupied by family *Opalinidae* Claus.

Nuclei not differentiated into two types; cilia abundant, undifferentiated; sexual reproduction by the complete union of differentiated minute uninucleate gametes. Commensal in the gut of amphibia and fishes.

The group has been treated monographically by Metcalf (1923). A single family is usually recognized.

Family **Opalinoea** Pritchard 1842. Family *Opalinaea* Siebold in Siebold and Stannius Lehrb. vergl. Anat 1: 10 (1848). Family *Opalinina* Stein Org. Inf. 2: 169 (1867). Family *Opalinidae* Claus 1874. Family *Protoopalinidae* Metcalf 1940. There are about 150 known species of four approximately equally numerous genera: *Protoopalina* Metcalf, cylindrical, with one or two nuclei which are always found in a stage of mitosis; *Zelleriella* Metcalf, similar, the cells flattened; *Cepedia* Metcalf, cylindrical, with many nuclei; *Opalina* Purkinje and Valentin, flattened and multinucleate.

Order 2. **Holotricha** Stein Org. Inf. 2: 169 (1867).
Orders *Gymnostomata* and *Trichostomata,* and suborder *Aspirotricha* Bütschli in Bronn Kl. u. Ord. Thierreichs 1: 1674 (1889).
Suborder *Hymenostomata* Hickson 1903.
Orders *Gymnostomataceae* and *Aspirotrichaceae* Hartog in Cambridge Nat. Hist. 1: 137 (1909).
Order *Holotricha* with suborders *Anoplophryinea, Gymnostomata,* and *Hymenostomata* Poche in Arch. Prot. 30: 250-255 (1913).
Order *Holotrichida* Calkins Biol. Prot. 376 (1926).
Infusoria with differentiated macronuclei and micronuclei, with simple cilia distributed generally over the surface of the body, not having membranelles in a spiral band about the cytostome.
This group is the mass of the more primitive typical Infusoria, of numerous families, not all of which are to be listed here. Arrangements of the families in other groups than the three here maintained have been proposed and are presumably more nearly natural.
 a. Cytostome anterior. Suborder GYMNOSTOMATA (Bütschli) Poche. Suborder *Gymnostomina* Hall.
Family **Enchelia** Ehrenberg Infusionsthierchen 298 (1838). Family *Enchelina* Stein Org. Inf. 2: 169 (1867). Family *Enchelyidae* Kent. Families *Holophryidae* and *Cyclodinidae* Schouteden. Family *Didiniidae* Poche. Comparatively unspecialized forms, radially symmetrical or nearly so. *Enchelis* O. F. Müller; *Holophrya, Chaenia, Prorodon; Ichthiophthirius,* becoming parasitic in the skins of fishes; *Lacrymaria,* the cytostome at the end of an extensible proboscis; *Didinium,* barrel-shaped, with the cilia confined to two belts, having an extensible proboscis by means of which it seizes other Infusoria and through which it swallows them.
Family **Colepina** Ehrenberg op. cit. 316 includes the single genus *Coleps.* The cells look like hand grenades of World War I: they are approximately barrel-shaped (the axis more or less curved), the pellicle forming hardened quadrangular plates between which the cilia project. The anterior cytostome can be opened widely to ingest other Infusoria.
 b. Cytostome lateral. Suborder ASPIROTRICHA Bütschli.
Family **Parameciina** Perty (1852). Family *Paramoecidae* Grobben. *Paramaecium* [Hill] O. F. Müller Verm. Terr. Fluv. 1: 54 (1773). The name is variously spelled; the spelling here used is Müller's in what is believed to be the first publication under binomial nomenclature.
Family **Colpodaea** Ehrenberg Infusionsthierchen 345 (1838). Family *Colpodidae* Claus 1879. Family *Ophryoglenidae* Kent 1882. Small forms, oval, bean-shaped, or flattened. *Ophryoglena, Glaucoma, Colpoda, Tetrahymena,* and many others.

Family **Cyclidina** Ehrenberg op. cit. 244. Family *Pleuronemidae* Kent. Family *Pleuronemina* Bütschli (1889). Similar, with a conspicuous undulating membrane along one side. *Cyclidium* and many other genera.

Family **Urocentrina** Claparède and Lachmann Etudes Inf. 1: 134 (1858). Family *Urocentridae* Schouteden. *Urocentrum,* the single genus, top-shaped, with cilia confined to two belts and a tail-like tuft, constantly whirling in the water.

Family **Trachelina** Ehrenberg op. cit. 319. Family *Tracheliidae* Kent. Having an anterior proboscis, the mouth at the base of this. *Trachelius, Dileptus, Lionotus, Loxodes,* etc.

Family **Chlamydodontida** [Chlamydodontidae] Claus 1874. Family *Chlamydodonta* Stein, the mere plural of a generic name. Family *Chilodontida* Bütschli. Family *Nassulidae* Schouteden. Flattened. The cytopharynx surrounded by longitudinal rods, apparently of hardened protein, collectively forming a conical basket, enclosed except when the cytostome is open for ingestion. *Chilodon, Chlamydodon, Nassula.*

c. Cytostome lacking; parasitic, mostly in invertebrates. Suborder ANOPLOPHRYINEA Poche; suborder *Astomina* Hall.

Family **Anoplophryida** [Anoplophryidae] and seven other families, all named by Cépède, 1910.

Order 3. **Heterotricha** Stein Org. Inf. 2: 169 (1867).

Suborder *Spirotricha,* sections *Heterotricha* and *Oligotricha* Bütschli in Bronn Kl. u. Ord. Thierreichs 1: 1674 (1889).

Section *Chonotricha* Wallengren in Acta Univ. Lund 31, part 2, no. 7: 48 (1895).

Order *Oligotricha* Doflein Protozoen 240 (1901).

Orders *Heterotrichaceae* and *Oligotrichaceae* Hartog in Cambridge Nat. Hist. 1: 137 (1909).

Orders *Heterotrichida* and *Oligotrichida* Calkins Biol. Prot. 386, 388 (1926).

Suborder *Entodiniomorpha* Reichenow in Doflein Lehrb. Prot. ed. 5, 3: 1195 (1929); Order *Chonotricha* Reichenow op. cit. 1211; suborder *Ctenostomata* Kahl ex Reichenow op. cit. 1024.

Orders *Spirotrichida* and *Chonotrichida* Hall Protozoology 380, 411 (1953).

Infusoria having a sinistrorse spiral band of cilia about the cytostome, these cilia united (except in family Spirochonina) in triangular-attenuate membranelles; not having the body flattened and the cilia or cirri confined to the ventral surface.

The peristomal apparatus of this order is an evidently derived character, so peculiar as to appear to have evolved only once: in short, the order appears natural. There are numerous subordinate groups. Several of these, of many species or of exceptional character, have been segregated as additional orders; it is by an arbitrary decision that they are here treated as suborders.

a. Comparatively unspecialized examples. Suborder SPIROTRICHA Bütschli. Suborders *Heterotrichina* and *Oligotrichina* Hall.

Family **Plagiotomina** Bütschli op. cit. 1719 (1889). Family *Plagiotomidae* Poche (1913). Peristomal area narrow and elongate, extending from the anterior end to a cytostome located near the middle of one side. *Blepharisma. Spirostomum.*

Family **Bursarina** Stein Org. Inf. 2: 169, 295 (1867). Family *Bursariidae* Kent. Cytostome seated in a deep pit in one side of the body. *Bursaria. Balantidium,* parasitic in the gut of Amphibia and mammals; *B. coli,* a serious pathogen in man.

Family **Stentorina** Stein op. cit. 169, 217. Family *Stentoridae* Claus. Peristomal area anterior, more or less transverse. *Stentor,* sessile and obconic, familiar. *Follicu-*

lina, the posterior end seated in a chitinous lorica, the peristomal area broadly expanded as two wings.

Family **Halterina** Claparède and Lachmann Etudes Inf. 1: 367 (1858). Family *Halteriidae* Claus. *Halteria,* subglobular, with a single whorl of long cilia; familiar in infusions, recognizable by the motion of the cells, alternately revolving slowly and snapping violently from place to place.

b. Loricate, free-swimming. Suborder TINTINNOINEA Kofoid and Campbell. Suborder *Tintinnina* Hall.

Family **Tintinnodea** Claparède and Lachmann Etudes Inf. 1: (1858). Family *Tintinnidae* Claus. Peristomal membranelles elongate and ciliate, the cylindrical or conical body attached in and retractile into the lorica; characteristically with two macronuclei and two micronuclei. Mostly marine. Kofoid and Campbell, who monographed the group (1929), found it possible to distinguish the natural and subordinate groups entirely by the structure of the lorica. They divided the former single family into twelve and recognized more than three hundred species.

c. Laterally flattened, with a tough membrane and few cilia and membranelles. Suborder CTENOSTOMATA Kahl. Suborder *Ctenostomina* Hall.

Family **Ctenostomida** [Ctenostomidae] Lauterborn in Zeit. wiss. Zool. 90: 665 (1908). Kahl (1932) monographed the group and found twenty-five species, which he arranged in six genera and three families.

d. Cylindrical, entozoic, with no ciliation except the membranelles. Suborder ENTODINIOMORPHA Reichenow. Suborder *Entodiniomorphina* Hall.

Family **Ophryoscolecina** Stein Org. Inf. 2: 168 (1867). Family *Ophryoscolecidae* Claus. Becker (1932) reviewed previous studies of this group, examples of which were first mentioned by Gruby and Delafond, 1843. He noted 71 species, of the genera *Entodinium, Diplodinium, Ophryoscolex, Epidinium,* etc. (the genera were first named by Stein) in the domestic ox; and 52 (*Didesmis, Paraisotricha, Spirodinium, Cycloposthium,* etc.) in the horse. Dogiel (1927) monographed the family, but it is certain that large numbers of species remain to be discovered in wild animals, oxen and others.

The barrel-shaped cells are about 0.1-0.25 mm. long. The cytostome is anterior, surrounded by the usual spiral band of membranelles; this may be broken up into several partial files, and there may be belts or clusters of membranelles on other parts of the body. The posterior end is drawn out into processes, one, few, or many, obscure or prominent, horn-like or fringe-like. Internally, beside contractile vacuoles and a neuromotor apparatus including a large motorium, there are characteristic skeletal plates. These consist of minute cylindrical bodies imbedded in an amorphous matrix, the whole staining with iodine and consisting supposedly of some polysaccharide carbohydrate.

Animals are infected by eating food contaminated with the saliva of others. The ciliates may be present in the rumen in numbers from one thousand to three million per cc. It has been supposed that they are symbiotic, benefitting their hosts by carrying on useful syntheses, or perhaps merely by controlling numbers of bacteria in the rumen. There is no good evidence for these ideas: the probability is, that they are harmless commensals.

e. Cylindrical or obconic, sessile, cilia of the peristomal band separate, body otherwise naked. Suborder **Chonotricha** (Wallengren) *subordo novus.*

Family **Spirochonina** Stein Org. Inf. 2: 168 (1867). Family *Spirochonidae* Grob-

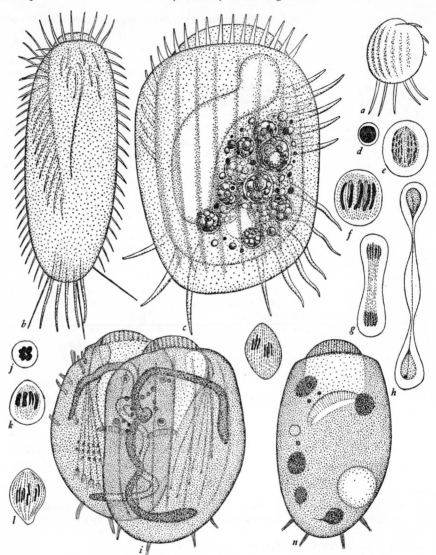

Fig. 44.—Infusoria, order Hypotricha: **a,** *Aspidisca* x 800. **b,** *Stylonychia* x 400. **c,** *Euplotes* x 400. **d-n,** *Euplotes Patella* after Turner (1930); **d-h,** stages of mitosis x 2,000, **i,** conjugating cells x 400; **j, k,** polar and equatorial views of the heterotypic division in a conjugant x 2,000; **l,** early anaphase of the homeotypic division x 2,000; **m,** first division of the zygote nucleus x 2,000; **n,** a cell after conjugation x 400, the macronucleus breaking up, the zygote nucleus divided into four, of which one is to become a macronucleus, one a micronucleus, and two are to undergo dissolution.

ben. *Spirochona* and a few other genera, attached to aquatic animals, fresh-water or marine, best known from the crustacean *Gammarus*.

Order 4. **Hypotricha** Stein Org. Inf. 2: 168 (1867).
  Section *Hypotricha* Bütschli in Bronn Kl. u. Ord. Thierreichs 1: 1674 (1889).
  Order *Hypotrichaceae* Hartog in Cambridge Nat. Hist. 1: 137 (1909).
  Order *Hypotrichida* Calkins Biol. Prot. 389 (1926).
  Suborder *Hypotricha* Kudo Man. Protozool. ed. 3: 668 (1946).
  Suborder *Hypotrichina* Hall Protozoology 381 (1953).

Flattened Infusoria bearing a band of membranelles crossing the upper surface near the anterior end from right to left and continued rearward on the lower surface beside the cytostome, along which lie also undulating membranes; mostly bearing cirri, which are confined to the lower surface, as are most free cilia, if these are present.

This group is evidently natural, and evidently a specialized offshoot from the preceding order. It might reasonably be treated as a subordinate group of the preceding order; Bütschli, Kudo, and Hall have done so. There are comparatively few species. Several are familiar in infusions and have been much studied.

Family 1. **Peritromina** Stein Org. Inf. 2: 168 (1867). Family *Peritromidae* Kent. Cilia abundant on the lower surface, cirri none. *Peritromus*.

Family 2. **Urostylida** [Urostylidae] Calkins Biol. Prot. 390 (1926). As above, but with frontal and sometimes also anal cirri. Numerous genera, *Urostyla, Uroleptus, Epiclintes, Stilotricha; Kerona* O. F. Müller, an ectoparasite on the animal *Hydra*.

Family 3. **Oxytrichina** Ehrenberg Infusionsthierchen 362 (1838). Family *Oxytrichidae* Kent. Family *Pleurotrichidae* Bütschli. Cirri present; cilia in one or two marginal rows, few or absent on the ventral surface. *Oxtricha, Stylonychia, Pleurotricha, Euplotes*, etc.

Order 5. **Stomatoda** Siebold in Siebold and Stannius Lehrb. vergl. Anat. 1: 10 (1848).
  Order *Ciliata* Perty Kennt. kl. Lebensf. 137 (1852).
  Order *Peritricha* Stein Org. Inf. 2: 168 (1867).
  Section *Peritricha* Bütschli in Bronn Kl. u. Ord. Thierreichs 1: 1674 (1889).
  Order *Peritrichaceae* Hartog in Cambridge Nat. Hist. 1: 138 (1909).
  Order *Peritrichida* Calkins Biol. Prot. 395 (1926).

Infusoria having a dextrorse spiral band of membranelles about the cytostome, which can in most examples be concealed and protected by contraction of the body; free-swimming only in the immature condition, at maturity attached and without separate cilia; syngamy occurring by the complete union of a smaller swimming individual with a larger attached one. *Vorticella* is the apparent type of the old ordinal names Stomatoda and Ciliata, which are accordingly held to belong to this order.

Family **Vorticellina** Ehrenberg Infusionsthierchen 259 (1838). Family *Vaginifera* Perty (1852). Family *Vorticellidae* Fromentel 1874. *Vorticella* L., a familiar microscopic organism in material from ponds and ditches, consists of solitary bell-shaped cells on contractile stalks. *Carchesium* and *Zoothamnium* are similar organisms in colonies. *Ophrydium, Epistylis*, etc., consist of similar colonies of non-contractile cells. *Cothurnia* and *Vaginicola* are solitary stalkless cells having conical loricae into which they can withdraw themselves.

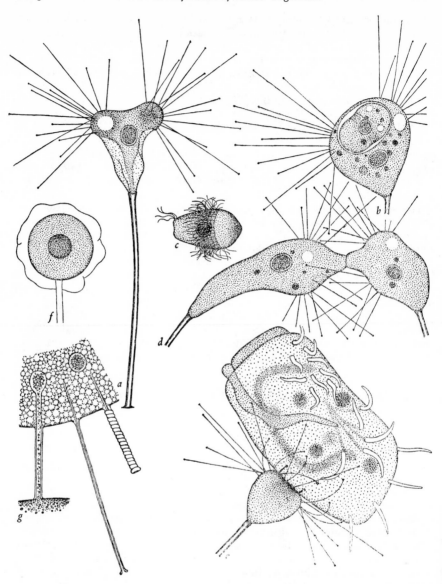

Fig. 45—*Tokophyra Lemnarum* after A. Noble (1932): **a**, representative individual; **b**, budding; **c**, swimming bud; **d**, conjugation; **e**, feeding on a cell of *Euplotes*;
f, cyst; **g**, tentacles, feeding, expanded, and contracted.   **g** x 2,000, all others x 400.

Family **Urceolarina** Perty (1852). Family *Trichodinidae* Claus. Family *Urceolaridae* Kudo. *Urceolaria, Trichodina,* etc., disk- or barrel-shaped cells attached on or in aquatic animals by means of a whorl of hard hooks.

Class 2. TENTACULIFERA (Huxley) Kent

Order *Iufusoires suceurs* and group *Acinetina* Claparède and Lachmann Etudes Inf. 1: 377, 381 (1858).
Class *Acinetae* Haeckel Gen. Morph. 2: lxxix (1866), the mere plural of a generic name.
TENTACULIFERA Huxley Man. Anat. Invert. 100 (1877).
Class TENTACULIFERA with orders SUCTORIA and *Acinetaria* Kent Man. Inf. 1: 36 (1880).
Class *Acinetaria* and order SUCTORIA Lankester in Enc. Brit. ed. 9, 19: 865 (1885).
Subclass *Suctoria* Bütschli in Bronn Kl. u. Ord. Thierreichs 1: 1842 (1889).
Class *Acinetoidea* Poche in Arch. Prot. 30: 263 (1913).
Class *Suctorea* Hall Protozoology 413 (1953).
Organisms swimming by means of cilia while immature, at maturity lacking cilia and usually attached, provided with tentacles by which they capture and paralyze their prey and absorb food. *Acineta* is the type genus.

These organisms are rather unfamiliar. They occur both in fresh water and in salt, and prey chiefly upon Infusoria. There are differentiated macronuclei and micronuclei; in branching or colonial individuals, a single macronucleus may extend to all parts. Asexual reproduction is by budding, often endogenous. Conjugation occurs either between attached individuals or between an attached individual and a swimming bud. The fact that one individual may bend past another to conjugate with a third indicates the presence of mating types. Conjugating individuals exhibit pregamic and postgamic nuclear divisions quite as among Infusoria (Noble, 1932). The group is undoubtedly derived from Infusoria; whether from something of the nature of *Didinium, Vorticella,* or *Spirochona* remains uncertain.

Collin (1912) accounted for about 170 species and recognized eight families. One of these families has subsequently been transferred to order Holotricha. The remainder may be construed as a single order:

Order **Suctoria** Kent (1880). Lankester chose this as between two ordinal names which Kent published at the same time.
a. Individuals subglobular, usually stalked, their tentacles essentially uniform.
Family 1. **Podophryina** Bütschli in Bronn Kl. u. Ord. Thierreichs 1: 1926 (1889). Family *Podophryidae,* Rousseau and Schouteden 1907. Buds produced exogenously. *Podophrya, Sphaerophrya,* naked; *Urnula,* loricate.
Family 2. **Acinetida** [Acinetidae] Claus 1874. *Acinetina* Claparède and Lachmann (1858). Family *Acinetina* Bütschli (1889). Bodies with a thin pellicle, with or without loricae; budding endogenous. *Acineta, Tokophrya,* etc.
Family 3. **Discophryida** [Discophryidae] Collin in Arch. Zool. Exp. Gen. 51: 364 (1912). Body with a firm pellicle, budding endogenous. *Discophrya,* etc.
b. Individuals branching or colonial.
Family 4. **Dendrosomida** [Dendrosomidae] Kent Man. Inf. 2: 215 (1882). Family *Dendrosomina* Bütschli (1889). Family *Dendrosomatidae* Poche (1913). *Dendrosoma,* etc.

Family 5. **Ophryodendrida** [Ophryodendridae] Kent l.c. Family *Ophryodendrina* Bütschli (1889). *Ophryodendron,* etc.

Family 6. **Dendrocometida** [Dendrocometidae] Kent l.c. Family *Dendrocometina* Bütschli (1889). *Dendrocometes, Stylocometes.*

c. With differentiated tentacles for piercing and sucking.

Family 7. **Ephelotida** [Ephelotidae] Kent. l.c. Family *Ephelotina* Sand 1899. Marine, individuals subglobular, stalked. *Ephelota,* naked; *Podocyathus,* loricate.

With this peculiar and highly evolved group, the here-proposed classification of organisms which lack the distinctive characters both of plants and of animals is concluded.

LIST OF NOMENCLATURAL NOVELTIES

	Page
Family Kurthiacea fam. nov.	21
Family Pasteurellacea nom. nov.	22
Family Chromatiacea nomen familiare novum	31
Family Rhodobacillacea nom. nov.	31
Family Chlorobiacea nom. nov.	31
Order Sphaerotilalea nom nov.	33
Lagenocystis, nom. nov., and *L. radicicola,* comb. nov.	82
Family Dinamoebidina nom nov.	101
Phylum Opisthokonta phylum novum	110
Chilomastix hominis comb. nov.	165
Pentatrichomonas obliqua comb. nov.	167
Goussia Schubergi comb. nov.	207
Family Myxoceratida and *Myxoceros,* nomina nova; *M. sphaerulosa* and *M. Blennius,* combinationes novae	221
Phylum Ciliophora nomen phylare novum	223
Order Opalinalea nom. nov.	228
Suborder Chonotricha subordo novus	231

BIBLIOGRAPHY

Adanson, Michel. Familles des plantes. Paris. 1763.

Agardh, Carolus A. Synopsis algarum Scandinaviae, adjecta dispositione universali algarum. Lund. 1817.

—————. Systema algarum. Lund. 1824.

Agardh, Jacob Georg. Species genera et ordines algarum . . . 3 vols. Lund. 1848-1901.

Agassiz, Louis. Contributions to the natural history of the United States of America. 4 vols. Boston. 1857-1862.

Ajello, Libero. A cytological and nutritional study of Polychytrium aggregatum. I. Cytology. American Jour. Bot. 35: 1-11. 1948.

Alexeieff, A. Notes sur quelques protistes copricoles. Arch. Prot. 50: 27-49. 1924.

Allen, Ruth F. A cytological study of heterothallism in Puccinia graminis. Jour. Agr. Res. 40: 585-614. 1930.

Allman, George. Notes on Noctiluca. Quart. Jour. Micr. Sci. n. s. 12: 326-332. 1872.

Andresen, N., and H. Holter. The genera of amoebae. Science 110: 114-115. 1949.

Angst, Laura. The gametophyte of Soranthera ulvoidea. Publ. Puget Sound Biol. Sta. 5: 159-163. 1926.

—————. The holdfast of Soranthera ulvoidea. Publ. Puget Sound Biol. Sta. 5: 267-275. 1927.

Anigstein, Ludwik. Untersuchungen über die Morphologie und Biologie der Rickettsia Melophagi Noller. Arch. Prot. 57: 209-246. 1927.

Archer, William. Résumé of recent contributions to our knowledge of "Freshwater Rhizopoda." Part I. Heliozoa. Quart. Jour. Micr. Sci. n. s. 16: 283-309, 347-376. 1876.

Areschoug, John Erh. Enumeratio phycearum in maribus Scandinaviae crescentium. Act. Reg. Soc. Sci. Upsala ser. 2, 13: 223-382. 1847; 14: 385-454. 1850.

Argao, Henrique de Beaurepaire. Über den Entwicklungsgang und die Überträgung von Haemoproteus columbae. Arch. Prot. 12: 154-167. 1908.

Ascherson, Paul. Flora der Provinz Brandenburg der Altmark und des Herzogthums Magdeburg. Berlin. 1864.

Atkinson, George F. Phylogeny and relationships in the Ascomycetes. Ann. Missouri Bot. Gard. 2: 315-376. 1915.

Baker, Woolford B. Studies in the life history of Euglena. I. Euglena agilis, Carter. Biol. Bull. 51: 321-362. 1926.

Balamuth, William. Regeneration in Protozoa: a problem in morphogenesis. Quart. Rev. Biol. 15: 290-337. 1940.

Barkley, Fred A. Keys to the phyla of organisms. Missoula, Montana. 1939.

—————. Un Esbozo de clasificatión de los organismos. Rev. Fac. Nal. Agron. Medellín 10: 83-103. 1949.

Barrett, J. T. Development and sexuality in some species of Olpidiopsis, (Cornu) Fischer. Ann. Bot. 26: 209-238. 1912.

Bartling, Fr. Th. Ordines naturales plantarum . . . Göttingen. 1830.

de Bary, Anton. Ueber die Myxomyceten. Bot. Zeit. 16: 357-358, 361-364, 365-369. 1858.

—————. Bericht über die Fortschritte der Algenkunde in den Jahren 1855, 1856, und 1857. Bot. Zeit. 16, Beilagen 55-99. 1858.

—————. Die Mycetozoen. Zeit. wiss. Zool. 10: 88-175. 1859.

—————. and M. Woronin. Beitrag zur Kenntniss der Chytridieen. Berichte Verhandl. naturf. Gess. Freiburg 3, Heft 2: 22-55. 1864.

————. Review of Baranetsky, J. Beitrag zur Kenntniss des selbstständigen Lebens der Flechtengonidien. Bot. Zeit. 26: 196-198. 1868.

————. Vergleichende Morphologie und Biologie der Pilze Mycetozoen und Bacterien. Leipzig. 1884.

Beadle, G. W., and E. L. Tatum. Neurospora. II. Methods of producing and detecting mutations concerned with nutritional requirements. American Jour. Bot. 32: 678-686. 1945.

Becker, Elery Ronald. The present status of problems relating to the ciliates of ruminants and Equidae. Quart. Rev. Biol. 7: 282-297. 1932.

Beijerinck, Martinus W. Die Bacterien der Papilionaceen-Knollchen. Bot. Zeit. 46: 725-735, 741-750, 757-771, 781-790, 797-804. 1888.

————. Ueber oligonitrophile Mikroben. Centralbl. Bakt. Abt. 2, 7: 561-582. 1901.

Bělǎr, Karl. Bau und Vermehrung von Prowazekia josephi n. sp. Arch. Prot. 35: 103-118. 1914.

————. Protozoenstudien I. Arch. Prot. 36: 13-51. 1915.

————. Protozoenstudien II. Arch. Prot. 36: 241-302. 1916.

————. Die Kernteilung von Prowazekia. Arch. Prot. 41: 308-320. 1920.

————. Protozoenstudien III. Arch. Prot. 43: 431-462. 1921.

————. Untersuchungen an Actinophrys sol Ehrenberg. I. Die Morphologie des Formwechsels. Arch. Prot. 46: 1-96. 1923.

Berdan, Helen. Revision of the genus Ancylistes. Mycologia 30: 396-415. 1938.

Bergey, David H., and others. Bergey's manual of determinative bacteriology. Baltimore. 1923. Second edition, 1925; third edition, 1930; fourth edition, 1934; fifth edition, 1938-1939; sixth edition, 1948.

Bergh, R. S. Der Organismus der Cilioflagellaten. Morph. Jahrb. 7: 177-288. 1882.

Bernard, Francis. Rôle des flagellées dans la transparence marine en Mediteranée. Bull. Soc. Hist. Nat. Afrique du Nord 38: 74-79. 1947.

Bersa, Egon. Über das Vorkommen von kohlensaurem Kalk in einer Gruppe von Schwefelbakterien. Sitzber. Akad. Wiss. Wien Math.-nat. Kl., Abt. I, 129: 231-259. 1920.

Berthold, G. Die Geschlechtliche Fortpflanzung der eigentlichen Phaeosporeen. Mittheil. zool. Stat. Neapel 2: 401-413. 1881.

Bessey, Charles E. The structure and classification of the lower green algae. Trans. American Micr. Soc. 36: 121-136. 1905.

————. A synopsis of plant phyla. Univ. Nebraska Studies 7: 275-373. 1907.

Bessey, Ernst A. Morphology and taxonomy of the Fungi. Philadelphia and Toronto. 1950.

Betts, Edwin M., and Samuel L. Meyer. Heterothallism and segregation of sexes in Ascobolus geophilus. American Jour. Bot. 26: 617-619. 1939.

Bewley, W. F., and H. B. Hutchinson. On the changes through which the nodule organism (Ps. radicicola) passes under cultural conditions. Jour. Agr. Sci. 10: 144-162. 1920.

Bhatia, B. L. Synopsis of the genera and classification of haplocyte gregarines. Parasitology 22: 156-167. 1930.

Bisby, Guy R. Some observations on the formation of the capillitium and the development of Physarella mirabilis Peck and Stemonitis fusca Roth. American Jour. Bot. 1: 274-288. 1914.

Blakeslee, Albert Francis. Zygospore formation a sexual process. Science n. s. 19: 864-866. 1904.

——————. Sexual reproduction in the Mucorineae. Proc. American Acad. 40: 205-319. 1904.

——————. Zygospore germinations in the Mucorineae. Ann. Myc. 4: 1-28. 1906.

Blochmann, F. Die mikroskopische Tierwelt des Süsswassers. Abt. I. Protozoa. 2d ed. Hamburg. 1895.

Bodman, Sister Mary Cecilia. Morphology and cytology of Guepinia Spathularia. Mycologia 30: 635-652. 1938.

Bohlin, Knut. Zur Morphologie und Biologie einzelliger Algen. Öfvers kgl. vetensk.-Akad. Förhandl. 54: 507-529. 1897.

——————. Utkast till de gröna algernas och arkegoniaternas fylogeni. Upsala. 1901.

Bölsche, Wilhelm. Haeckel, his life and work. Translated by Joseph McCabe. Philadelphia. 1906.

Bonner, John T. A descriptive study of the development of the slime mold Dictyostelium discoideum. American Jour. Bot. 31: 175-182. 1944.

Borgert, Adolf. Über die Dictyochiden, insbesondere über Distephanus speculum; sowie Studien an Phaeodarien. Zeit. wiss. Zool. 51: 629-676. 1891.

——————. Fortpflanzungsverhältnisse bei tripyleen Radiolarien (Phaeodarien). Verhandl. deutschen zool. Gess. 1896: 192-195.

——————. Untersuchungen über die Fortpflanzung der tripyleen Radiolarien, speciell von Aulacantha scolymantha H. I. Theil. Zool. Jahrb. Anat. 14: 203-276. 1900.

——————. Untersuchungen über die Fortpflanzung der tripyleen Radiolarien, speziell von Aulacantha scolymantha H. II. Teil. Arch. Prot. 14: 134-263. 1909.

Bornet, Ed., and Gustave Thuret. Recherches sur la fécondation des floridées. Ann. Sci. Nat. Bot. sér. 5, 7: 137-166. 1867.

——————. and Cr. Flahault. Revision des nostocacées heterocystes. Ann. Sci. Nat. Bot. sér. 7, 3: 323-381. 1886; 4: 343-373. 1886; 5: 51-129. 1887; 7: 177-262. 1888.

Bory de Saint-Vincent (editor). Dictionnaire classique d'histoire naturelle. 17 vols. Paris. 1822-1831.

Brady, Henry B. Notes on some of the reticularian Rhizopoda of the "Challenger" expedition. Part III. Quart. Jour. Micr. Sci. n. s. 21: 31-71. 1881.

——————. Report on the Foraminifera dredged by H. M. S. Challenger, during the years 1873-1876. Rep. Sci. Results Voy. Challenger Zool. 9: i-xxi, 1-814. 1884.

Brandt, K. Die kolonienbildenden Radiolarien . . . Flora und Fauna des Golfes von Neapel 13: i-viii, 1-276. 1885.

——————. Zur Systematik der kolonienbildenden Radiolarien. Zool. Jahrb. Suppl. 8: 311-352. 1905.

Brasil, Louis. Recherches sur la reproduction des grégarines monocystidées. Arch. Zool. Exp. Gen. sér. 4, 3: 17-38. 1905.

——————. Nouvelles recherches sur la reproduction des grégarines monocystidées. Arch. Zool. Exp. Gen. sér. 4, 4: 69-99. 1905.

——————. Recherches sur le cycle évolutif des Seleniidae, grégarines parasites d'annélides polychètes. Arch. Prot. 8: 370-397. 1907.

Braun, Alexander. Über Chytridium, eine Gattung einzelliger Schmarotzergewächse auf Algen und Infusorien. Abh. Akad. Wiss. Berlin 1855: 21-83. 1856.

Braun, M. Review of Thélohan (1892). Centralbl. Bakt. 14: 737-739. 1893.

Brown, William H. The development of the ascocarp of Lachnea scutellata. Bot. Gaz. 52: 275-305. 1911.

—————. The development of Pyronema confluens var. igneum. American Jour. Bot. 2: 289-298. 1915.

Buchanan, R. E. Studies in the nomenclature and classification of bacteria II. The primary subdivisions of the Schizomycetes. Jour. Bact. 2: 155-164. 1917.

—————. ————— III. The families of the Eubacteriales. Jour. Bact. 2: 347-350. 1917.

—————. ————— VIII. The subgroups and genera of the Actinomycetales. Jour. Bact. 3: 403-406. 1918.

—————. General systematic bacteriology. Baltimore. 1925.

—————, Ralph St. John-Brooks, and Robert S. Breed. International bacteriological code of nomenclature. Jour. Bact. 55: 287-306. 1948.

Buller, A. H. R. Some observations on the spore discharge of the higher fungi. Proc. Internat. Congr. Pl. Sci. Ithaca 2: 1627-1628. 1929.

Burkholder, P. R. Movement in the Cyanophyceae. Quart. Rev. Biol. 9: 438-459. 1934.

Butler, E. J. On Allomyces, a new aquatic fungus. Ann. Bot. 25: 1023-1035. 1911.

Butler, Ellys T., William J. Robbins, and B. O. Dodge. Biotin and the growth of Neurospora. Science 94: 262-263. 1941.

Bütschli, O. Protozoa. Zool. Jahresber. 1880, Abt. 1: 122-173. 1881.

—————. Protozoa. in Dr. H. G. Bronn's Klassen und Ordnungen des Thier-Reichs 1: i-xviii, 1-2035. 1880-1889.

—————. Ueber den Bau der Bacterien und verwandter Organismen. Leipzig. 1890.

—————. Ueber die chemische Natur der Skeletsubstanz der Acantharia. Zool. Anz. 30: 784-789. 1906.

Calkins, Gary N. Mitosis in Noctiluca miliaris and its bearing on the nuclear relations of the Protozoa and Metazoa. Jour. Morph. 15: 711-772. 1899.

—————. The biology of the Protozoa. Philadelphia and New York. 1926.

————— and Rachel C. Bowling. Gametic meiosis in Monocystis. Biol. Bull. 51: 385-399. 1926.

le Calvez, J. Flagellispores du radiolaire Coelodendrum ramosissimum (Haeckel). Arch. Zool. Exp. Gen. 77, notes 99-103. 1935.

—————. Recherches sur les foraminifères II. Place de la méiose et sexualité. Arch. Zool. Exp. Gen. 87: 211-243. 1950.

Cameron, Frank. Potash from kelp. United States Dept. Agr. Office of the Secretary Rept. 100: 1-122. 1915.

Campbell, Arthur Shackleton. The cytology of Tintinnopsis nucula (Fol) Laackmann ... Univ. California Publ. Zool. 29: 179-236. 1926.

—————. Studies on the marine ciliate Favella (Jörgensen) ... Univ. California Publ. Zool. 29: 429-452. 1927.

Campbell, Douglas H. A university text-book of botany, New York. 1902.

de Candolle, Augustin Pyrame. Théorie élémentaire de la botanique ... Paris. 1813.

Cantino, Edward C. The vitamin nutrition of an isolate of Blastocladia Pringsheimii. American Jour. Bot. 35: 238-242. 1948.

—————. The physiology of the aquatic phycomycete, Blastocladia Pringsheimii, with emphasis on its nutrition and metabolism. American Jour. Bot. 36: 95-112. 1949.

Carter, P. W., I. M. Heilbron, and B. Lythgoe. The lipochromes and sterols of the algal classes. Proc. Roy. Soc. London B 128: 82-109. 1939.

Cash, James, and others. The British freshwater Rhizopoda and Heliozoa. 5 vols. London. 1905-1921.

Caspary, Rob. Bericht über die Verhandlungen der botanischen Sektion der 33. Versammlung deutscher Naturforscher und Aerzte, gehalten in Bonn vom 18. bis 24. September 1857. Bot. Zeit. 15: 749-776, 784-792. 1857.

Castle, Edward S. The structure of the cell walls of Aspergillus and the theory of cellulose particles. American Jour. Bot. 32: 148-151. 1945.

Caullery, Maurice, and Félix Mesnil. Recherches sur les haplosporidies. Arch. Zool. Exp. Gen. sér. 4, 4: 101-181. 1905.

————— and —————. Sur les Metchnikovellidae et autres protistes parasites des grégarines d'annélides. Compt. Rend. Soc. Biol. 77: 527-532. 1914.

————— and —————. Metchnikovellidae et autres protistes parasites des grégarines d'annélides. Ann. Inst. Pasteur 33: 209-240. 1919.

Chadefaud, M. Les protistes trichocytifères ou progastréades. Ann. Protistol. 5: 323-343. 1936.

Chatton, Édouard. Les blastodinides, ordre nouveau de dinoflagelées parasites. Compt. Rend. 143: 981-985. 1906.

—————. Les cnidocystes du péridinien Polykrikos schwartzi Bütschli. Arch. Zool. Exp. Gen. 54: 157-194. 1914.

—————. Les péridiniens parasites. Morphologie, reproduction, ethologie. Arch. Zool. Exp. Gen. 59: 1-475. 1920.

Child, Horace J. Studies on the ingestion of leucocytes, and on mitosis, in Endamoeba gingivalis. Univ. California Publ. Zool. 28: 251-284. 1926.

Christman, A. H. Sexual reproduction in the rusts. Bot. Gaz. 39: 267-275. 1905.

Cienkowski, L. Zur Entwickelungsgeschichte der Myxomyceten. Jahrb. wiss. Bot 3: 325-337. 1863.

—————. Das Plasmodium. Jahrb. wiss. Bot. 3: 400-441. 1863.

—————. Beiträge zur Kenntniss der Monaden. Arch. mikr. Anat. 1: 203-232. 1865.

Claparède, Édouard, and Johannes Lachmann, Études sur les infusoires et les rhizopodes. 2 vols. Geneva. 1858-1861.

Clements, Frederic E. The genera of Fungi. Minneapolis. 1909.

————— and Cornelius L. Shear. The genera of Fungi. New York. 1931.

Cleveland, L. R. The method by which Trichonympha campanula, a protozoön in the intestine of termites, ingests solid particles of wood for food. Biol. Bull. 48: 282-288. 1925.

—————. The ability of termites to live perhaps indefinitely on a diet of pure cellulose. Biol. Bull. 48: 289-293. 1925.

—————. The effects of oxygenation and starvation on the symbiosis between the termite, Termopsis, and its intestinal flagellates. Biol. Bull. 48: 309-326. 1925.

—————. Symbiosis among animals with special reference to termites and their intestinal flagellates. Quart. Rev. Biol. 1: 51-60. 1926.

—————. Sex produced in the protozoa of Cryptocercus by molting. Science 105: 16-17. 1947.

—————. The origin and evolution of meiosis. Science 105: 287-289. 1947.

—————. An ideal partnership. Sci. Month. 67: 173-177. 1948.

Clint, Hilda B. The life history and cytology of Sphacelaria bipinnata Sauv. Univ. Liverpool Publ. Hartley Bot. Lab. 3: 1-23. 1927.

Cohn, Ferdinand. Beiträge zur Entwickelungsgeschichte der Infusorien II. Ueber den Encystirungsprocess der Infusorien. Zeit. wiss. Zool. 4: 253-281. 1953.

—————. Conspectus familiarum Cryptogamarum secundum methodum naturalem dispositarum. Hedwigia 11: 17-20. 1872.

—————. Untersuchungen über Bacterien. Beitr. Biol. Pfl. 1, Heft 2: 127-224. 1872.

—————. Untersuchungen über Bacterien II. Beitr. Biol. Pfl. 1, Heft 3: 141-207. 1875.

Coker, William Chambers. The Saprolegniaceae with notes on other water molds. Chapel Hill, North Carolina. 1923.

Collin, Bernard. Étude monographique sur les acinétiens II. Morphologie, physiologie, systematique. Arch. Zool. Exp. Gen. 51: 1-457. 1912.

Colson, Barbara. The cytology and development of Phyllactinia corylea Lév. Ann. Bot. n.s. 2: 381-402. 1938.

Committee on Medical Research, Washington, and the Medical Research Council, London. Chemistry of penicillin. Science 102: 627-629. 1945.

Conard, Henry S. Plants of Iowa, being a fifth edition of the Grinnell Flora . . . Grinnell, Iowa. 1939.

Conn, H. J., and Gladys E. Wolfe. The flagellation of bacteria. Science 87: 283-284. 1938.

—————. Taxonomic relationships of certain non-spore-forming rods in soil. Jour. Bact. 36: 320-321. 1938.

Connell, Frank H. The morphology and life cycle of Oxymonas dimorpha sp. nov., from Neotermes simplicicornis (Banks). Univ. California Publ. Zool. 36: 51-66. 1930.

Conrad, Walter. Recherches sur les flagellates de nos eaux saumatres. Arch. Prot. 56: 167-231. 1926.

Cook, W. R. Ivimey. A monograph of the Plasmodiophorales. Arch. Prot. 80: 179-254. 1933.

Cooke, Wm. Bridge. A nomenclatorial survey of the genera of pore fungi. Lloydia 3: 81-104. 1940.

—————. Myxomycetes of Mount Shasta. Madroño 10: 55-62. 1949.

Copeland, Edwin Bingham. What is a plant? Science n. s. 65: 388-390. 1927.

Copeland, Herbert F. The kingdoms of organisms. Quart. Rev. Biol. 13: 384-420. 1938.

—————. Progress report on basic classification. American Nat. 81: 340-361. 1947.

—————. Spirodiscus Ehrenberg identified as Ophiocytium Nägeli. Science 119: 338. 1954.

—————. Observations on Prasiola mexicana . . . . Madroño 13: 138-140. 1955.

Cotner, F. B. The development of the zoospores in the Oömycetes at optimum temperatures and the cytology of their active stages. American Jour. Bot. 17: 511-546. 1930.

—————. Cytological study of the zoospores of Blastocladia. Bot. Gaz. 89: 295-309. 1930.

Couch, John N. The development of the sexual organs of Leptolegnia caudata. American Jour. Bot. 19: 584-599. 1932.

—————. Observations on cilia of aquatic Phycomycetes. Science 88: 476. 1938.

—————. A new Conidiobolus with sexual reproduction. American Jour. Bot. 26: 119-130. 1939.

—————. The structure and action of the cilia in some aquatic Phycomycetes. American Jour. Bot. 28: 704-713. 1941.

————— and Alma J. Whiffen. Observations on the genus Blastocladiella. American Jour. Bot. 29: 582-591. 1942.

Crawley, Howard. The evolution of Sarcocystis muris in the intestinal cells of the mouse (preliminary note). Proc. Acad. Nat. Sci. Philadelphia 66: 432-436. 1914.

—————. The sexual evolution of Sarcocystis muris. Proc. Acad. Nat. Sci. Philadelphia 68: 2-43. 1916.

Cupp, Easter E. Spirotrichonympha polygyra sp. nov. from Neotermes simplicicornis Banks. Univ. California Publ. Zool. 33: 351-378. 1930.

Cushman, Joseph A. Foraminifera. Their classification and economic use. Sharon, Massachussetts. 1928.

Cuvier, Georges. Sur un nouveau rapprochement à établir entre les classes qui composent le règne animal. Ann. Mus. Hist. Nat. Paris 19: 73-84. 1812.

Dangeard, P.-A., et Sappin-Trouffy. Urédinées. Le Botaniste 3: 119-125. 1893.

—————. Recherches sur la reproduction sexuelle des champignons. Le Botaniste 3: 221-281. 1893.

—————. La reproduction sexuelle des ascomycètes. Le Botaniste 4: 21-58. 1894.

—————. La truffe. Recherches sur son développement, sa structure, sa reproduction sexuelle. Le Botaniste 4: 63-87. 1894.

—————. Contributions à l'étude des acrasiées. Le Botaniste 5: 1-20. 1896.

—————. Recherches sur la structure du Polyphagus Euglenae Nowak. Le Botaniste 7: 213-261. 1900.

—————. L'origine du périthèse chez les ascomycètes. Le Botaniste 10: 1-385. 1907.

—————. Études sur le développement et la stracture des organismes inférieurs. Le Botaniste 11: 1-311. 1910.

Dangeard, Pierre. Recherches sur les Bangia et les Porphyra. Le Botaniste 18: 183-244. 1927.

Danilewski, B. Die Hämatozoen der Kaltblüter. Arch. mikr. Anat. 24: 588-598. 1885.

Davis, Bradley Moore. The fertilization of Albugo candida. Bot. Gaz. 29: 297-311. 1900.

—————. Oogenesis in Saprolegnia. Bot. Gaz. 35: 233-249, 320-349. 1903.

Day, Howard Calvin. Studies on the contractile vacuole in Spirostomum and Paramecium. Physiol. Zool. 3: 56-71. 1930.

Deflandre, Georges. Sur la structure des flagelles. Ann. Protistol. 4: 31-54. 1934.

Delage, Yves, and Edgard Hérouard. Traité de zoologie concrète. vols. 1, 2, 5, 8. Paris. 1896-1901.

DeLamater, Edward D., and Mary Elizabeth Hunter. Preliminary report of true mitosis in the vegetative cell of Bacillus megatherium. American Jour. Bot. 38: 659-662. 1951.

—————, Sidney Yaverbaum, and Lucille Schwartz. The nuclear cytology of Eremascus albus. American Jour. Bot. 40: 475-492. 1953.

Dennis, Emery Westervelt. The morphology and binary fission of Babesia bigemina of Texas cattle fever. Univ. California Publ. Zool. 33: 179-192. 1930.

Diesing, K. M. Revision der Prothelminthen. Abtheilung: Mastigophoren. Sitzber. Akad. Wiss. Wien Math.-nat. Cl. 52, Abt. I: 287-401. 1866.

Diller, William F. Nuclear reorganization processes in Paramecium aurelia, with descriptions of autogamy and 'hemixis.' Jour. Morph. 59: 11-67. 1936.

Dobell, Clifford. The structure and life history of Copromonas subtilis nov. gen. et nov. spec.: a contribution to our knowledge of flagellates. Quart. Jour. Micr. Sci. n. s. 52: 75-120. 1908.

——————. On Cristispira veneris nov. spec., and the affinities and classification of the spirochaets. Quart. Jour. Micr. Sci. n. s. 56: 507-541. 1911.

—————— and A. Pringle Jameson. The chromosome cycle in coccidia and gregarines. Proc. Roy. Soc. London B 89: 83-94. 1915.

——————. The life-history and chromosome cycle of Aggregata eberthi [Protozoa: Sporozoa: Coccidia]. Parasitology 17: 1-136. 1925.

Dodge, B. O. Nuclear phenomena associated with heterothallism and homothallism in the ascomycete Neurospora. Jour. Agr. Res. 35: 289-305, 1937.

——————. Some problems in the genetics of the Fungi. Science 90: 379-385. 1939.

Doflein, Franz. Die Protozoen als Parasiten und Krankheitserreger . . . Jena. 1901.

——————. Lehrbuch der Protozoenkunde. [The work of 1901 is construed as the first edition.] Jena. Third edition, 1911; fourth edition, 1916; fifth edition, revised by Eduard Reichenow, 1927-1929.

——————. Rhizochrysis. Zool. Anz. 47: 153-158. 1916.

——————. Beiträge zur Kenntnis von Bau und Teilung der Protozoenkerne. Zool. Anz. 49: 289-306. 1918.

——————. Untersuchungen über Chrysomonadinen. Arch. Prot. 44: 149-213. 1922.

Dogiel, V. A. Monographie der Familie Ophryoscolecidae Teil I. Arch. Prot. 59: 1-288. 1927.

Drechsler, Charles. Morphology of the genus Actinomyces. Bot. Gaz. 67: 65-83, 147-168. 1919.

——————. Some conidial Phycomycetes destructive to terricolous amoebae. Mycologia 27: 6-40. 1935.

——————. Some non-catenulate conidial Phycomycetes preying on terricolous amoebae. Mycologia 27: 176-205. 1935.

——————. New Zoopagaceae destructive to soil rhizopods. Mycologia 29: 229-249. 1937.

Drew, Kathleen M. The organization and inter-relationships of the carposporophytes of living Florideae. Phytomorphology 4: 55-69. 1954.

Duboscq, O., and O. Tuzet. L'ovogénèse, la fécondation et les premiers stades du développement des éponges calcaires. Arch. Zool. Exp. Gen. 79: 157-316. 1937.

Dufour, Léon. Recherches anatomiques sur les carabiques et sur plusieurs autres insectes coléoptères. Ann. Sci. Nat. 8: 5-54. 1826.

——————. Note sur la grégarine, nouveau genre de ver qui vit en troupeau dans les intestins de divers insectes. Ann. Sci. Nat. 13: 366-368. 1828.

Dujardin, Félix. Observations sur les rhizopodes et les infusoires. Compt. Rend. 1: 338-340. 1835.

——————. Recherches sur les organismes inférieurs. Ann. Sci. Nat. Zool. sér. 2, 4: 343-377. 1835.

——————. Histoire naturelle des zoophytes. Infusoires. Paris. 1841.

Eckstein, Gustav. Noguchi. New York and London. 1931.

Edgerton, C. W. Plus and minus strains in the genus Glomerella. American Jour. Bot. 1: 244-254. 1914.

Editorial Board of the monograph on the chemistry of penicillin. The chemical study of penicillin: a brief history. Science 105: 653-659. 1947.

Ehrenberg, Christian Gottfried. Beiträge zur Kenntniss der Organisation der Infusorien und ihrer geographischen Verbreitung, besonders in Siberien. Abh. Akad. Wiss. Berlin 1830: 1-88. 1832.

————. Die Infusionsthierchen als volkommene Organismen. Ein Blick in das tiefe organische Leben der Natur. Leipzig. 1838.

————. Über die Bildung der Kreidefelsen und des Kreidemengels durch unsichtbare Organismen. Abh. Akad. Wiss. Berlin 1838: 59-147. 1939.

Eidam, Eduard. Basidiobolus, eine neue Gattung der Entomophthoraceen. Beitr. Biol. Pfl. 4: 181-251. 1886.

Elder, Albert L. Penicillin. Sci. Month. 58: 405-409. 1944.

Elliott, Eugene W. The swarm-cells of Myxomycetes. Mycologia 41: 141-170. 1949.

Ellis, David. Iron bacteria. New York. 1916.

Ellison, Bernard R. Flagellar studies on zoospores of some members of the Mycetozoa, Plasmodiophorales, and Chytridiales. Mycologia 37: 444-459. 1945.

Emerson, Ralph. Life cycles in the Blastocladiales. Trans. British Myc. Soc. 23: 123. 1939.

————. An experimental study of the life cycles and taxonomy of Allomyces. Lloydia 4: 77-144. 1941.

———— and Edward C. Cantino. The isolation, growth, and metabolism of Blastocladia in pure culture. American Jour. Bot. 35: 157-171. 1948.

———— and Charles M. Wilson. The significance of meiosis in Allomyces. Science 110: 86-88. 1949.

Enderlein, Günther. Grundelemente der vergleichenden Morphologie und Biologie der Bakterien. Sitzber. Gess. naturf. Freunde Berlin 1916: 403-406. 1916.

————. Ein neues Bakteriensystem auf vergleichend morphologischer Grundlage. Sitzber. Gess. naturf. Freunde Berlin 1917: 309-319. 1917.

————. Bakterien-Cyklogenie . . . Berlin and Leipzig. 1925.

Endlicher, S. Genera plantarum secundum ordines naturales disposita. Vienna. 1836-1840.

Engelmann, Th. W. Die Purpurbacterien und ihre Beziehungen zum Licht. Bot. Zeit. 46: 661-669, 677-689, 693-701, 709-720. 1888.

Engler, Adolf, and K. Prantl. Die Natürlichen Pflanzenfamilien. 20 vols. Leipzig. 1889-1909.

————. Syllabus der Vorlesungen über specielle und medicinisch-pharmaceutisch Botanik . . . Berlin. 1892.

————. Übersicht über die Gliederung des "Pflanzenreich" . . . Das Pflanzenreich Heft 1: v-vii. 1900.

————. Syllabus der Pflanzenfamilien . . . [The syllabus of 1892 is construed as the first edition.] Third edition, 1903; ninth and tenth edition, with E. Gilg, 1924.

Entz, Géza. Über die mitotische Tielung von Ceratium hirundinella. Arch. Prot. 43: 415-430. 1921.

Erdmann, Rhoda. Die Entwicklung der Sarcocystis muris in der Muskulatur. Sitzber. Gess. naturf. Freunde Berlin 1910: 377-387. 1910.

Fairchild, David G. Ueber Kerntheilung und Befruchtung bei Basidiobolus ranarum Eidam. Jahrb. wiss. Bot. 30: 285-296. 1897.

Falkenberg, P. Die Befruchtung und der Generationswechsel von Cutleria. Mittheil. zool. Stat. Neapel 1: 420-447. 1879.

Farmer, J. Bretland, and J. Ll. Williams. On fertilization and segmentation of the spore in Fucus. Ann. Bot. 10: 479-487. 1896.

Faull, J. H. The morphology, biology and phylogeny of the Pucciniastreae. Proc. Internat. Congr. Pl. Sci. Ithaca 2: 1735-1745. 1929.

Ferris, Virginia Rogers. A note on the flagellation of Phytophthora infestans (Mont.) de Bary. Science 120: 71-72. 1954.

Fischer, Alfred. Ueber die Geisseln einiger Flagellaten. Jahrb. wiss. Bot. 26: 187-235. 1894.

————. Untersuchungen über Bakterien. Jahrb. wiss. Bot. 27: 1-163. 1895.

Fitzpatrick, Harry M. The cytology of Eocronartium muscicola. American Jour. Bot. 5: 397-419. 1918.

Flexner, Simon. Hideo Noguchi. A biographical sketch. Science 69: 653-660. 1929.

Fott, Bohuslav. Uber den inneren Bau von Vacuolaria viridis (Dangeard) Senn. Arch. Prot. 84: 242-250. 1935.

Fournier, Eugène (editor). Actes du Congrès International de Botanique tenu à Paris en août 1867. Paris. 1867.

França, Carlos. La flagellose des euphorbes. Arch. Prot. 34: 108-132. 1914.

Franck, James, and Walter E. Loomis. Photosynthesis in plants. Ames, Iowa. 1949.

Frank, B. Ueber die Parasiten in den Wurzelanschwellungen der Papilionaceen. Bot. Zeit. 37: 377-388, 393-400. 1879.

————. Ueber die auf Wurzelsymbiose beruhende Ernährung gewisser Bäume durch unterirdische Pilze. Ber. deutschen bot. Gess. 3: 128-145. 1885.

Fries, Elias. Systema mycologicum, sistens Fungorum ordines, genera et species, huc usque cognitas . . . 3 vols. Lund. 1821-1832.

————. Epicrisis systematis mycologici, seu synopsis Hymenomycetum. Lund. 1836-1838.

————. Hymenomycetes europaei sive epicriseos systematis mycologici editio altera. Upsala. 1874.

Fritsch, F. E. The structure and reproduction of the algae. 2 vols. Cambridge. 1935, 1945.

Frye, T. C., G. B. Rigg, and W. C. Crandall. The size of kelps on the Pacific Coast of North America. Bot. Gaz. 60: 473-482. 1915.

Galloway, J. J. A manual of Foraminifera. Bloomington, Indiana. 1933.

Gams, Helmut. Die Protochlorinae als autotrophe Vorfahren von Pilzen and Tieren ? Mikroscopie 2: 383-387. 1947.

Gardner, Nathaniel Lyon. Cytological studies in Cyanophyceae. Univ. California Publ. Bot. 2: 237-296. 1906.

Gäumann, Ernst. Vergleichende Morphologie der Pilze. Jena. 1926. English translation by C. W. Dodge. New York and London. 1928.

Gay, Frederick P. The unsolved problems of leprosy. Science 81: 283-285. 1935.

Geitler, Lothar. Der Zellbau von Glaucocystis Nostochinearum und Gloeochaete Wittrockiana und die Chromatophoren-symbiosetheorie von Mereschkowsky. Arch. Prot. 47: 1-24. 1923.

————. Die reduktionsteilung und Copulation von Cymbella lanceolata. Arch. Prot. 58: 465-507. 1927.

————. Somatische Teilung, Reduktionsteilung, Copulation und Parthenogenese bei Cocconeis placentula. Arch. Prot. 59: 506-549. 1927.

————. Porphyridium sordidum n. sp. eine neue Süsswasserbangiale. Arch. Prot. 76: 595-604. 1932.

Gicklhorn, Josef. Notiz über den durch Chromulina smaragdina nov. spec. bedingten Smaragdglanz des Wasserspiegels. Arch. Prot. 44: 219-226. 1922.

Gilbert, Frank A. Factors influencing the germination of myxomycetous spores. American Jour. Bot. 16: 280-286. 1929.

————. Spore germination in the Myxomycetes: a comparative study of spore germination by families. American Jour. Bot. 16: 421-432. 1929.

Gilbert, Henry C. Critical events in the life history of Ceratiomyxa. American Jour. Bot. 22: 52-74. 1935.

Goldfuss, August. Ueber die Classification der Zoophyten. Isis 1818: 1008-1013. 1818.

Gomont, Maurice. Monographie des oscillariées (nostochacées homocytées). Ann. Sci. Bot. sér. 7, 15: 263-368. 1892.

Goodrich, Helen Pixell. Nina: a remarkable gregarine. Quart. Jour. Micr. Sci. n. s. 81: 107-126. 1938.

van Goor, A. C. J. Die Cytologie von Noctiluca miliaris im Lichte der neueren Theorien über den Kernbau der Protisten. Arch. Prot. 39: 147-208. 1918.

Granata, Leopoldo. Richerche sul ciclo evolutivo di Haplosporidium limnodrili Granata. Arch. Prot. 35: 47-79. 1914.

————. Gli attinomissidi. Morphologia—sviluppo—sistematica. Arch. Prot. 50: 139-212. 1925.

Grassé, Pierre P. (Editor). Traité de Zoologie. Anatomie, systematique, biologie. Paris. vols. 1 — . 1952 — .

Grassi, B., and A Foà. Intorno ai protozoi dei termitidi. Atti Accad. Lincei ser. 5, Rendiconti Cl. Sci. 20, 1° Semestre: 725-741. 1911.

————. Flagellati viventi nei termitidi. Mem. Accad. Lincei Cl. Sci. ser. 5, 12: 231-394. 1917.

Greeff, Richard. Pelomyxa palustris (Pelobius), ein amöbenartiger Organismus des süssen Wassers. Arch. mikr. Anat. 10: 51-73. 1874.

Greville, Robert Kaye. Algae Britannicae, or descriptions of the marine and other inarticulated plants of the British Islands belonging to the order Algae . . . Edinburgh and London. 1830.

Gross, J. Sporenbildung bei Cristispira. Arch. Prot. 29: 279-292. 1913.

Gross, Catherine. The cytology of Vaucheria. Bull. Torrey Bot. Club 64: 1-15. 1937.

Gurley, R. R. On the classification of the Myxosporidia, a group of protozoan parasites infesting fishes. Bull. United States Fish Comm. 11: 407-420. 1893.

Haeckel, Ernst Heinrich. Die Radiolarien (Rhizopoda radiaria). Ein Monographie. 4 vols. Berlin. 1862-1888.

————. Generelle Morphologie. 2 vols. Berlin. 1866.

————. Monographie der Moneren. Jenaische Zeitschr. 4: 64-137. 1868.

————. Die Kalkschwämme. Eine Monographie . . . 3 vols. Berlin. 1872.

————. Ueber die Phaeodarien, eine neue Gruppe kieselschaliger mariner Rhizopoden. Sitzber. Jenaische Gess. Med. u. Naturw. 1879: 151-157. 1879.

————. Entwurf eines Radiolarien-Systems auf Grund von Studien der Challenger-Radiolarien. Jenaische Zeitschr. 15: 418-472. 1882.

————. Report on the Radiolaria collected by H. M. S. Challenger during the years 1873-1876. Rept. Voy. Challenger Zool. 18: i-viii, 1-1803, plates 1-140. 1887.

————. Systematische Phylogenie. 3 vols. Berlin. 1894-1896.

Häcker, V. Über Chromosomen- und Sporenbildung bei Radiolarien. Verhandl. deutschen zool. Gess. 1907: 74-84. 1907.

Hall, Richard P. Morphology and binary fission of Menoidium incurvum (Fres.) Klebs. Univ. California Publ. Zool. 20: 447-476. 1923.

——————. Binary fission in Oxyrrhis marina Dujardin. Univ. California Publ. Zool. 26: 281-324. 1925.

——————. Mitosis in Ceratium hirundinella O. F. M., with notes on nuclear phenomena in encysted forms and the question of sexual reproduction. Univ. California Publ. Zool. 28: 29-64. 1925.

—————— and William N. Powell. Morphology and binary fission of Peranema trichophorum (Ehrbg.) Stein. Biol. Bull. 54: 36-64. 1928.

—————— and Theodore Louis Jahn. On the comparative cytology of certain euglenid flagellates and the systematic position of the families Euglenidae Stein and Astasiidae Bütschli. Trans. American Micr. Soc. 48: 388-405. 1929.

——————. The method of ingestion in Peranema trichophorum and its bearing on the pharyngeal-rod ("Staborgan") problem in the Euglenida. Arch. Prot. 81: 308-317, 1933.

——————. A note on the flagellar apparatus of Peranema trichophorum and the status of the family Peranemidae Stein. Trans. American Micr. Soc. 53: 237-244. 1934.

——————. A note on behavior of chromosomes in Euglena. Trans. American Micr. Soc. 56: 288-290. 1937.

—————— and H. W. Schoenborn. Studies on the question of autotrophic nutrition in Chlorogonium euchlorum, Euglena anabaena, and Euglena deses. Arch. Prot. 90: 259-271. 1938.

——————. The trophic nature of the plant-like flagellates. Quart. Rev. Biol. 14: 1-12. 1939.

——————. The trophic nature of Euglena viridis. Arch. Zool. Exp. Gen. 80: notes 61-67. 1939.

—————— and H. W. Schoenborn. Selective effects of inorganic culture media on bacteria-free strains of Euglena. Arch. Prot. 93: 72-80. 1939.

—————— and ——————. The question of autotrophic nutrition in Euglena gracilis. Physiol. Zool. 12: 76-84. 1939.

—————— and J. B. Loefer. Effects of culture filtrates and old medium on growth of the ciliate, Colpidium campylum. Proc. Soc. Exp. Biol. Med. 43: 128-133. 1940.

——————. Vitamin deficiency as one explanation for inhibition of protozoan growth by conditioned medium. Proc. Soc. Exp. Biol. Med. 47: 306-308. 1941.

—————— and A. Schottenfeld. Maximal density and phases of death in populations of Glaucoma piriformis. Physiol. Zool. 14: 384-393. 1941.

——————. Incomplete proteins as nitrogen sources, and their relation to vitamin requirements in Colpidium campylum. Physiol. Zool. 15: 95-107. 1942.

—————— and W. B. Cosgrove. The question of the synthesis of thiamin by the ciliate, Glaucoma piriformis. Biol. Bull. 86: 31-40. 1944.

——————. Comparative effects of certain vitamins on populations of Glaucoma piriformis. Physiol. Zool. 17: 200-209. 1944.

—————— and W. B. Cosgrove. Mineral deficiency as a factor limiting growth of the ciliate Tetrahymena geleii. Physiol. Zool. 18: 425-430. 1945.

——————. Protozoology. New York. 1953.

Hanatschek, Herta. Der Phasenwechsel bei der Gattung Vaucheria. Arch. Prot. 78: 497-513. 1932.

Hansen, H. N., and William C. Snyder. The dual phenomenon and sex in Hypomyces solani f. cucurbitae. American Jour. Bot. 30: 419-422. 1943.

Hanson, Anne Marie. A morphological, developmental, and cytological study of four saprophytic chytrids. IV. Phlyctorhiza endogena gen. nov., sp. nov. American Jour. Bot. 33: 732-739. 1946.

Harper, Robert A. Beitrag zur Kenntnis der Kerntheilung und Sporenbildung im Ascus. Ber. deutschen bot. Gess. 13: (67) - (78). 1895.

——————. Die Entwicklung des Peritheciums bei Sphaerotheca Castagnei. Ber deutschen bot. Gess. 13: 475-481. 1895.

——————. Kerntheilung und freie Zellbildung im Ascus. Jahrb. wiss. Bot. 30: 249-284. 1897.

——————. Cell division in sporangia and asci. Ann. Bot. 13: 467-525. 1899.

——————. Sexual reproduction in Pyronema confluens and the morphology of the ascocarp. Ann. Bot. 14: 321-400. 1900.

——————. Cell and nuclear division in Fuligo varians. Bot. Gaz. 30: 217-251. 1900.

——————. Cleavage in Didymium melanospermum (Pers.) Macbr. American Jour. Bot. 1: 127-144. 1914.

Hartmann, Max, and K. Nägler. Copulation bei Amoeba diploidea n. sp. mit Selbstständigbleiben der Gametenkerne während des ganzen Lebenszyklus. Sitzber. Gess. naturf. Freunde Berlin 1908: 112-125. 1908.

——————. S. von Prowazek. Arch. Prot. 36: i-xii. 1915.

Hartog, Marcus. Protozoa. in Harmer, S. F., and A. E. Shipley. The Cambridge Natural History 1: 3-162. 1909.

Harvey, William Henry. A manual of the British algae . . . London. 1841.

Haskins, R. H. Cellulose as a substratum for saprophytic chytrids. American Jour. Bot. 26: 635-639. 1939.

Hatch, Winslow R. Gametogenesis in Allomyces arbuscula. Ann. Bot. 49: 623-649. 1935.

Haupt, Arthur W. Cell structure and cell division in the Cyanophyceae. Bot. Gaz. 75: 170-190. 1923.

——————. Structure and development in Zonaria Farlowii. American Jour. Bot. 19: 239-254. 1932.

Heald, F. D., and R. C. Walton. The expulsion of ascospores from the perithecia of the chestnut blight fungus, Endothia parasitica (Murr.) And. American Jour. Bot. 1: 499-521. 1914.

Hein, Illo. Studies on the mycelium of Psalliota campestris. American Jour. Bot. 17: 197-211. 1930.

——————. Studies on morphogenesis in Agaricus (Psalliota) campestris. American Jour. Bot. 17: 882-915. 1930.

Hemming, Francis. The official record of the proceedings of the International Commission on Zoological Nomenclature at their session held in Paris in July, 1948. Bull. Zool. Nomencl. 4: 1-648. 1950.

——————. Zoological nomenclature: decisions taken by the Fourteenth International Congress of Zoology, Copenhagen, August, 1953. Science 119: 131-133. 1954.

Henrici, Arthur T. Morphologic variation and the rate of growth of bacteria. Springfield and Baltimore. 1928.

—————— and Delia Johnson. Stalked bacteria, a new order of Schizomycetes. Jour. Bact. 29: 3-4. 1935.

—————— and ——————. Studies of freshwater bacteria II. Stalked bacteria, a new order of Schizomycetes. Jour. Bact. 30: 61-94. 1935.

Hertwig, Richard, and E. Lesser. Ueber Rhizopoden und denselben nahestandende Organismen. Morphologische Studien. Arch. mikr. Anat. 10, Suppl. 35-243. 1874.

————. Zur Histologie der Radiolarien. Leipzig. 1876.

————. Ueber Leptodiscus medusoides, eine neue den Noctiluceen verwandte Flagellate. Jenaische Zeitschr. 11: 307-323. 1877.

————. Der organismus der Radiolarien. Jena. 1879.

————. Ueber die Conjugation der Infusorien. Abh. Kgl. Bayerischen Akad. Wiss. Math.-phys Cl. 17: 151-233. 1889.

Higgins, Bascombe Britt. Contribution to the life history and physiology of Cylindrosporium on stone fruits. American Jour. Bot. 1: 145-173. 1914.

————. Morphology and life history of some Ascomycetes with special reference to the presence and function of spermatia II. American Jour. Bot. 16: 287-296. 1929.

Higgins, E. Marion. A cytological investigation of Stypocaulon scoparium (L.) Kütz., with especial reference to the unilocular sporangia. Ann. Bot. 45: 345-353. 1931.

Hillier, James, Stuart Mudd, and Andrew G. Smith. Internal structure and nuclei in cells of Escherichia coli as shown by improved electron microscopic techniques. Jour. Bact. 57: 319-338. 1949.

Hinshaw, H. Corwin. On the morphology and mitosis of Trichomonas buccalis (Goodey) Kofoid. Univ. California Publ. Zool. 29: 159-174. 1926.

Hirsch, Hilde E. The cytogenetics of sex in Hypomyces solani var. cucurbitae. American Jour. Bot. 36: 113-121. 1949.

Hofeneder, Heinrich. Über eine neue kolonienbildende Chrysomonadine. Arch. Prot. 29: 293-307. 1913.

Hofmeister, Wilhelm, Anton de Bary, Th. Irmisch, and Julius Sachs. Handbuch der physiologischen Botanik. 4 vols. Leipzig. 1865-1877.

Hogg, John. On the distinction of a plant and an animal, and on a fourth kingdom of nature. Edinburgh New Philos. Jour. n. s. 12: 216-225. 1860.

Hollenberg, George J. Culture studies of marine algae. I. Eisenia arborea. American Jour. Bot. 26: 34-41. 1939.

Horne, A. S. Nuclear division in the Plasmodiophorales. Ann. Bot. 44: 199-231. 1930.

Hovasse, Raymond. Contribution à l'étude de l'appareil de Golgi des flagellés libres: l'existence d'un corps parabasal chez Cercomonas longicauda Duj. Arch. Zool. Exp. Gen. 79, notes 43-46. 1937.

Howard, Frank L. The life history of Physarum polycephalum. American Jour. Bot. 18: 116-133. 1931.

————. Nuclear division in plasmodia of Physarum. Ann. Bot. 46: 461-477. 1932.

Hutchinson, Henry Brougham, and James Clayton. On the decomposition of cellulose by an aerobic organism (Spirochaeta cytophaga, n. sp.). Jour. Agr. Sci. 9: 143-173. 1919.

Huth, Walther. Zur Entwicklungsgeschichte der Thalassicollen. Arch. Prot. 30: 1-124. 1913.

Huxley, Julian S. On Ganymedes anaspidis (nov. gen., nov. sp.): a gregarine from the digestive tract of Anaspides tasmaniae (Thompson). Quart. Jour. Micr. Sci. n. s. 55: 155-175. 1910.

Huxley, Thomas Henry. Zoological notes and observations made on board H. M. S. Rattlesnake. II. Upon Thalassicolla, a new zoophyte. Ann. Mag. Nat. Hist. ser. 2, 8: 433-442. 1851.

————————. A manual of the anatomy of invertebrated animals. London. 1877.

Ischikawa, C. Vorläufige Mittheilungen über die Conjugationserscheinungen bei den Noctiluceen. Zool. Anz. 14: 12-14. 1891.

Ishikawa, Matsuharu. Cytological studies on Porphyra tenera Kjellm. I. Bot. Mag. Tokyo 35: 206-218. 1921.

Iyengar, M. O. P., and R. Subrahmanyan. On reduction division and auxospore formation in Cyclotella Meneghiniana Kütz. Jour. Indian Bot. Soc. 21: 231-237. 1942.

———————— and ————————. On reduction division and auxospore formation in Cyclotella Meneghiniana Kütz. Jour. Indian Bot. Soc. 23: 125-152. 1944.

Jahn, Eduard. Myxomycetenstudien. 3. Kernteilung und Geisselbildung bei den Schwärmern von Stemonitis flaccida Lister. Ber. deutschen bot. Gess. 22: 84-92. 1904.

————————. Myxomycetenstudien. 4. Die Keimung der Sporen. Ber. deutschen bot. Gess. 23: 489-497. 1905.

————————. Beiträge zur botanischen Protistologie. 1. Die Polyangiden. Leipzig. 1924.

Jahn, Theodore Louis. The euglenoid flagellates. Quart. Rev. Biol. 21: 246-274. 1946.

James-Clark, H. [Preliminary abstract of the following.] Proc. Boston Soc. Nat. Hist. 11: 16. 1866.

————————. On the Spongiae Ciliatae as Infusoria Flagellata; or, observations on the structure, animality, and relationship of Leucosolenia botryoides, Bowerbank. Mem. Boston Soc. Nat. Hist. 1: 305-340. 1868.

Jameson, A. Pringle. The chromosome cycle of gregarines, with special reference to Diplocystis schneideri Kunstler. Quart. Jour. Micr. Sci. n. s. 64: 207-266. 1920.

Janicki, C. Zur Kenntnis des Parabasalapparats bei parasitischen Flagellaten. Biol. Centralbl. 31: 321-330. 1911.

————————. Untersuchungen an parasitischen Flagellaten. II. Teil. Die Gattungen Devescovina, Parajoenia, Stephanonympha, Calonympha.—Über den Parabasalapparat.—Über Kernkonstitution und Kernteilung. Zeit. wiss. Zool. 112: 573-691. 1915.

Jennings, H. S. (editor), T. M. Sonneborn, A. C. Giese, L. C. Gilman, and R. F. Kimball. Mating types and their interactions in the ciliate infusoria. Biol. Symp. 1: 117-188. 1940.

Johns, Robert M., and R. K. Benjamin. Sexual reproduction in Gonapodya. Mycologia 46: 201-208. 1954.

Jollos, Victor. Experimentelle Unterschungen an Infusorien. Biol. Centralbl. 33: 222-236. 1913.

Jones, George Neville. On the number of species of plants. Sci. Month. 72: 289-294. 1951.

Jordan, Edwin O., and I. S. Falk (editors). The newer knowledge of bacteriology and immunology. Chicago. 1928.

Juel, H. O. Ueber Zellinhalt, Befruchtung und Sporenbildung bei Dipodascus. Flora 91: 47-55. 1902.

Kahl, A. Ctenostomata (Lauterborn) n. subordo. Vierte Unterordnung der Heterotricha. Arch. Prot. 77: 231-304. 1932.

Kampter, Erwin. Die Coccolithineen der Südwestküste von Istrien. Ann. Naturhist. Staatsmus. Wien 51: 54-149. 1940.

Kanda, Tiyoiti. On the gametophytes of some Japanese species of Laminariales. Sci. Papers Inst. Alg. Res. Hokkaido Imp. Univ. 1: 221-260. 1936; 2: 87-111. 1938.

Kanouse, Bessie Bernice. A monographic study of special groups of the water molds I. Blastocladiaceae. American Jour. Bot. 14: 278-306. 1927.

—————. A monographic study of special groups of the water molds II. Leptomitaceae and Pythiomorphaceae. American Jour Bot. 14: 335-357. 1927.

Karling, John S. A preliminary descriptive study of a parasitic monad in cells of American Characeae. American Jour. Bot. 17: 928-937. 1930.

—————. Studies in the Chytridiales VI. The occurrence and life history of a new species of Cladochytrium in cells of Eriocaulon septangulare. American Jour. Bot. 18: 526-557. 1931.

—————. The cytology of the Chytridiales with special reference to Cladochytrium replicatum. Mem. Torrey Bot. Club 19, no. 1: 1-92. 1937.

—————. A new fungus with anteriorly uniciliate zoospores: Hyphochytrium catenoides. American Jour. Bot. 26: 512-519. 1939.

—————. The Plasmodiophorales . . . New York. 1942.

—————. Parasitism among the chytrids. American Jour. Bot. 29: 24-35. 1942.

—————. The life history of Anisolpidium Ectocarpii gen. nov. et sp. nov., and a synopsis and classification of other fungi with anteriorly uniflagellate zoospores. American Jour. Bot. 30: 637-648. 1943.

—————. Phagomyxa algarum n. gen., n. sp., an unusual parasite with plasmodiophoralean and proteomyxean characteristics. American Jour. Bot. 31: 38-52. 1944.

—————. Brazilian anisochytrids. American Jour. Bot. 31: 391-397. 1944.

—————. Brazilian chytrids. VI. Rhopalophlyctis and Chytriomyces, two new chitinophyllic operculate genera. American Jour. Bot. 32: 362-369. 1945.

—————. Keratinophilic chytrids. I. Rhizophidium keratinophilum n. sp., a saprophyte isolated on human hair, and its parasite, Phlyctidium mycetophagum n. sp. American Jour. Bot. 33: 751-757. 1946.

—————. Keratinophilic chytrids. II. Phlyctorhiza variabilis n. sp. American Jour. Bot. 34: 27-32. 1947.

—————. An Olpidium parasite of Allomyces. American Jour. Bot. 35: 503-510. 1948.

Karsten, G. Die Auxosporenbildung der Gattungen Cocconeis, Surirella, und Cymatopleura. Flora 87: 253-283. 1900.

—————. Die sogenannte "Mikrosporen" der Planktondiatomeen und ihre weitere Entwicklung, beobachtet an Corethron Valdiviae n. sp. Ber. deutschen bot. Gess. 22: 544-554. 1904.

Keene, Mary Lucille. Cytological studies of the zygospores of Sporodinia grandis. Ann. Bot. 28: 455-470. 1914.

—————. Studies of zygospore formation in Phycomyces nitens Kunze. Trans. Wisconsin Acad. 19: 1195-1220. 1919.

Keitt, G. W., and M. H. Langford. Venturia inaequalis (Cke.) Wint. I. A groundwork for genetic studies. American Jour. Bot. 28: 805-820. 1941.

Kelley, Arthur P. Mycotrophy in plants. Waltham, Massachussetts. 1950.

Kent, William Saville. A manual of the Infusoria: including a description of all known flagellate, ciliate, and tentaculiferous Protozoa, British and foreign, and an account of the organization and affinities of the sponges. 3 vols. London. 1880-1882.

Keuten, Jacob. Die Kerntheilung von Euglena viridis Ehrenberg. Zeitfl wiss. Zool. 60: 215-235. 1895.

Keysselitz, G. Über Trypanophis grobbeni (Trypanosoma grobbeni Poche). Arch. Prot. 3: 367-375. 1904.

Kidder, George W. Streblomastix strix, morphology and mitosis. Univ. California Publ. Zool. 33: 109-124. 1929.

Kimball, Richard F. Mating types in the ciliate Infusoria. Quart. Rev. Biol. 18: 30-45. 1943.

King, Robert L., and Theodore Louis Jahn. Concerning the genera of amebas. Science 107: 293-294. 1948.

Kirby, Harold, Jr. The intestinal flagellates of the termite, Cryptotermes hermsi Kirby. Univ. Calif. Publ. Zool. 29: 103-120. 1926.

—————. Studies on some amoebae from the termite Mirotermes, with notes on some other Protozoa from the Termitidae. Quart. Jour. Micr. Sci. n. s. 71: 189-222. 1927.

—————. A species of Proboscoidella from Kalotermes (Cryptotermes) dudleyi Banks, a termite of Central America, with remarks on the oxymonad flagellates. Quart. Jour. Micr. Sci. n. s. 72: 355-386. 1928.

—————. Snyderella and Coronympha, two new genera of multinucleate flagellates from termites. Univ. California Publ. Zool. 31: 417-432. 1929.

—————. Trichomonad flagellates from termites I. Tricercomitus gen. nov., and Hexamastix Alexeieff. Univ. California Publ. Zool. 33: 393-444. 1930.

—————. Trichomonad flagellates from termites II. Eutrichomastix, and the subfamily Trichomonadinae. Univ. California Publ. Zool. 36: 171-262. 1931.

—————. Host-parasite relationships in the distribution of Protozoa in termites. Univ. California Publ. Zool. 41: 189-212. 1937.

—————. The devescovinid flagellates Caduceia theobromae França, Pseudodevescovina ramosa new species, and Macrotrichomonas pulchra Grassi. Univ. California Publ. Zool. 43: 1-40. 1939.

—————. Observations on a trichomonad from the intestine of man. Jour. Parasitol. 29: 422-423. 1943.

—————. Some observations on cytology and morphogenesis in flagellate protozoa. Jour. Morph. 75: 361-421. 1944.

—————. Flagellate and host relationships of trichomonad flagellates. Jour. Parasitol. 33: 214-228. 1947.

Klebs, Georg. Flagellatenstudien. Zeit. wiss. Zool. 55: 265-351, 353-445. 1893.

Knasyi, Georges. The cell structure and cell division of Bacillus subtilis. Jour. Bact. 19: 113-115. 1930.

—————. Elements of bacterial cytology. Ithaca, New York, 1944.

Kniep, H. Allomyces javanicus n. sp., ein anisogamer Phycomycet mit Planogameten. Ber. deutschen bot. Gess. 47: 199-212. 1929.

Knight, Margery. Studies in the Ectocarpaceae. I. The life-history and cytology of Phylaiella litoralis, Kjellm. Trans. Roy. Soc. Edinburgh 53: 343-360. 1923.

—————. Studies in the Ectocarpaceae. II. The life-history and cytology of Ectocarpus siliculosus, Dillw. Trans. Roy. Soc. Edinburgh 56: 307-332. 1929.

Kofoid, Charles Atwood. New species of dinoflagellates. Bull. Mus. Comp. Zool. Harvard 50: 163-207. 1907.

————— and Elizabeth Bohn Christianson. On binary and multiple fission in Giardia muris (Grassi). Univ. California Publ. Zool. 16: 30-54. 1915.

————— and Olive Swezy. Mitosis in Trichomonas. Proc. Nat. Acad. Sci. 1: 315-321. 1915.

———— and ————. Studies on the parasites of the termites I. On Streblomastix strix, a polymastigote flagellate with a linear plasmodial phase. Univ. California Publ. Zool. 20: 1-20. 1919.

———— and ————. Studies on the parasites of the termites II. On Trichomitus termitidis, a polymastigote flagellate with a highly developed neuromotor system. Univ. California Publ. Zool. 20: 21-40. 1919.

———— and————. Studies on the parasites of the termites III. On Trichonympha campanula sp. nov. Univ. California Publ. Zool. 20: 41-98. 1919.

———— and ————. Studies on the parasites of the termites IV. On Leidyopsis sphaerica gen. nov., sp. nov. Univ. California Publ. Zool. 20: 99-116. 1919.

————. A critical review of the nomenclature of human intestinal flagellates, Cercomonas, Chilomastix, Trichomonas, Tetratrichomonas, and Giardia. Univ. California Publ. Zool. 20: 145-168. 1920.

———— and Olive Swezy. On the free, encysted, and budding stages of Councilmania lafleuri, a parasitic amoeba of the human intestine. Univ. California Publ. Zool. 20: 169-198. 1921.

———— and ————. The free-living unarmored dinoflagellates. Mem. Univ. California 5: i-viii, 1-562. 1921.

———— and ————. Mitosis and fission in the active and encysted phases of Giardia enterica (Grassi) of man, with a discussion of the method of origin of bilateral symmetry in the polymastigote flagellates. Univ. California Publ. Zool. 20: 199-234. 1922.

———— and ————. Mitosis in Endamoeba dysenteriae in the bone marrow in arthritis deformans. Univ. California Publ. Zool. 20: 301-307. 1922.

————. The life cycle of the Protozoa. Science n. s. 57: 397-408. 1923.

———— and Olive Swezy. Karyamoeba falcata, a new amoeba from the human intestinal tract. Univ. California Publ. Zool. 26: 221-242. 1924.

———— and ————. On the number of chromosomes and type of mitosis in Endamoeba dystenteriae. Univ. California Publ. Zool. 26: 331-352. 1925.

———— and ————. Karyamoebina substituted for Karyamoeba, with a note on its occurrence in man. Univ. California Publ. Zool. 26: 435-436. 1925.

———— and ————. On Oxymonas, a flagellate with an extensile and retractile proboscis from Kalotermes from British Guiana. Univ. California Publ. Zool. 28: 285-300. 1926.

———— and Arthur Shackleton Campbell. A conspectus of the marine and freshwater ciliata belonging to the suborder Tintinnoinea, with descriptions of new species principally from the Agassiz expedition to the eastern tropical Pacific 1904-1905. Univ. California Publ. Zool. 34: 1-403. 1929.

Koidzumi, Makoto. Studies on the intestinal Protozoa found in the termites of Japan. Parasitiology 13: 235-309. 1921.

Kolk, Laura A. A comparison of the filamentous iron organisms, Clonothrix fusca Roze and Crenthrix polyspora Cohn. American Jour. Bot. 25: 11-17. 1938.

Kukzynski, Max H. Untersuchungen an Trichomonaden. Arch. Prot. 33: 119-204. 1914.

Kudo, Richard R. Studies on Myxosporidia. Illinois Biol. Monogr. 5: 239-506. 1920.

————. A biological and taxonomic study of the Microsporidia. Illinois Biol. Monogr. 9: 77-344. 1924.

—————. Observations on Lophomonas blattarum, a flagellate inhabiting the colon of the cockroach, Blatta orientalis. Arch. Prot. 53: 191-214. 1926.

—————. Handbook of protozoology. Springfield and Baltimore. 1931. Third edition. 1946.

Kühn, Alfred, and W. von Schuckmann. Über den Bau und die Teilungserschinungen von Trypanosoma brucei (Plimmer u. Bradford). Sitzber. heidelberger Akad. Wiss., Math.-nat. Kl., 1911, Abh. 11: 1-21. 1911.

—————. Über Bau, Teilung und Encystierung von Bobo edax Klebs. Arch. Prot. 35: 212-255. 1915.

Kunieda, Hiroshi. On the development of the sexual organs and embryogeny in Sargassum Horneri Ag. Jour. Coll. Agr. Tokyo Imp. Univ. 9: 383-396. 1928.

Kützing, Friedrich Traugott. Phycologia generalis oder Anatomie, Physiologie und Systemkunde der Tange. Leipzig. 1843.

—————. Phycologia germanica, d. i. Deutschland's Algen . . . Nordhausen. 1845.

Kylin, Harald. Über den Bau der Spermatozoiden der Fucaceen. Ber. deutschen bot. Gess. 34: 194-201. 1916.

—————. Über den Generationswechsel bei Laminaria digitata. Svensk Bot. Tidskr. 10: 551-561. 1916.

—————. Über die Keimung der Florideensporen. Arkiv för Bot. 14, no. 22: 1-25. 1917.

—————. Studien über die Entwicklungsgeschichte der Florideen. Kgl. Svenska Vetensk.-Akad. Handl. 63, no. 11: 1-139. 1923.

—————. Studien über Delesseriaceen. Kgl. Fysiog. Sällsk. Handl. n. f. 35, no. 6: 1-111. 1924.

—————. The marine red algae in the vicinity of the biological station at Friday Harbor, Wash. Kgl. Fysiog. Sällsk. Handl. n. f. 36, no. 9: 1-87. 1925.

—————. Entwicklungsgeschichtliche Florideenstudien. Kgl. Fysiog. Sällsk. Handl. n. f. 39, no. 4: 1-127. 1928.

—————. Über die Entwicklungsgeschichte der Florideen. Kgl. Fysiog. Sällsk. Handl. n. f. 41, no. 6: 1-104. 1930.

—————. Some physiological remarks on the relationship of the Bangiales. Bot. Notiser 1930: 417-420. 1930.

—————. Die Florideenordnung Gigartinales. Kgl. Fysiog. Sällsk. Handl. n. f. 43, no. 8: 1-88. 1932.

—————. Ueber die Entwicklungsgeschichte der Phaeophyceen. Kgl. Fysiog. Sällsk. Handl. n. f. 44, no. 7: 1-102. 1933.

—————. Zur Kenntniss der Entwicklungsgeschichte einiger Phaeophyceen. Kgl. Fysiog. Sällsk. Handl. n. f. 45, no. 9: 1-18. 1934.

—————. Über Rhodomonas, Platymonas, und Prasinocladus. Kgl. Fysiog. Sällsk. Förhandl. 5, no. 22: 1-13. 1935.

—————. Über eine marine Porphyridium-Art. Kgl. Fysiog. Sällsk. Förhandl. 7, no. 10: 1-5. 1937.

—————. Bermerkungen über die Entwicklungsgeschichte einiger Phaeophyceen. Kgl. Fysiog. Sällsk. Handl. n. f. 48, no. 1: 1-34. 1937.

—————. Zur Biochemie der Rhodophyceen. Kgl. Fysiog. Sällsk. Förhandl. 13, no. 6: 1-13. 1943.

—————. Zur Biochemie der Cyanophyceen. Kgl. Fysiog. Sällsk. Förhandl. 13, no. 7: 1-14. 1943.

Labbé, Alphonse. Sporozoa. Thierreich Lief. 5: i-xx, 1-180. 1899.

de Lamarck, J.-B.-P.-A. Philosophie Zoologique . . . 2 vols. Paris. 1809.

——————. Histoire naturelle des animaux sans vertèbres. 7 vols. Paris. 1815-1822.

Lamouroux, J. V. F. Essai sur les genres de la famille des thalassiophytes non articulées. Ann. Mus. Hist. Nat. Paris 20: 21-47, 115-139, 267-293. 1813.

Lander, Caroline A. Spore formation in Scleroderma lycoperdoides. Bot. Gaz. 95: 330-337. 1933.

Lanjouw, J., and others. International code of botanical nomanclature. Utrecht. 1952.

Lankester, E. Ray. On Drepanidium ranarum, the cell-parasite of the frog's blood and spleen (Gaule's Würmschen). Quart. Jour. Micr. Sci. n. s. 22: 53-65. 1882.

——————. Protozoa. in Encyclopedia Britannica ed. 9, 19: 830-866. 1885.

Lauterborn, Robert. Untersuchungen über Bau, Kernteilung, und Bewegung der Diatomen. Leipzig. 1896.

——————. Protozoen-Studien. V. Teil. Zur Kenntnis einiger Rhizopoden und Infusorien aus dem Gebiete des Oberrheins. Zeit. wiss. Zool. 90: 645-669. 1908.

Lebour, Marie V. The dinoflagellates of northern seas. Plymouth, England. 1925.

Lederberg, Joshua, and E. L. Tatum. Sex in bacteria: genetic studies, 1945-1952. Science 118: 169-175. 1953.

Ledingham, G. A. Studies on Polymyxa graminis, n. gen. n. sp., a plasmodiophoraceous parasite on wheat. Canadian Jour. Res. 17 C: 38-51. 1939.

van Leeuwenhoeck, Antony. Observations communicated to the publisher in a Dutch letter of the 9th of Octob. 1676. here English'd: Concerning little animals by him observed in rain, well, sea, and snow-water; as also in water wherein pepper hath lain infused. Phil. Trans. Roy. Soc. London 12: 821-831. 1677.

Léger, Louis. Les schizogrégarines des trachéates. I. Le genre Ophryocystis. Arch. Prot. 8: 159-202. 1907.

—————— and O. Duboscq. L'évolution schizogonique de Aggregata (Eucoccidium) eberthi (Labbé). Arch. Prot. 12: 44-108. 1908.

——————. Mycétozoairs endoparasites des insectes. I. Sporomyxa scauri nov. gen. nov. spec. Arch. Prot. 12: 109-130. 1908.

—————— and O. Duboscq. Selenococcidium intermedium Lég. et Dub. et la systématique des sporozoaires. Arch. Zool. Exp. Gen. sér. 5, 5: 187-238. 1910.

——————. Caryospora simplex, coccidie monosporée, et la classifiaction des coccidies. Arch. Prot. 22: 71-88. 1911.

—————— and O. Duboscq. Étude sur Spirocystis nidula Lég. et Dub. schizogrégarine du Lumbriculus variegatus Müll. Arch. Prot. 35: 199-211. 1915.

Leidy, Joseph. Fresh-water rhizopods of North America. Rept. U. S. Geol. Surv. Territories, vol. 12. 1879.

Lemmermann, E. Silicoflagellatae. Ber. deutschen bot. Gess. 19: 247-271. 1901.

Leuckart, Rudolf. Die Parasiten der Menschen und die von ihnen herrührenden Krankheiten. vol. 1. [no more were issued]. Leipzig and Heidelberg. 1879-1901.

Levine, Michael. The origin and development of lamellae in Agaricus campestris and in certain species of Coprinus. American Jour. Bot. 9: 509-533. 1922.

Lewis, Charles E. The basidium of Amanita bisporigera. Bot. Gaz. 41: 348-352. 1906.

Lewis, Ivey F., and Conway Zirkle. Cytology and systematic position of Porphyridium cruentum Naegeli. American Jour. Bot. 7: 333-340. 1920.

—————— and Hilah F. Bryan. A new protophyte from the Dry Tortugas. American Jour. Bot. 28: 343-348. 1941.

Light, Sol F. On Hoplonympha natator gen. nov. sp. nov. a non-xylophagous hyper-mastigote from the termite, Kalotermes simplicicornis Banks, characterized by biradial symmetry and a highly developed pellicle. Univ. California Publ. Zool. 29: 123-140. 1926.

——————. Kofoidia, a new flagellate, from a California termite. Univ. California Publ. Zool. 29: 467-492. 1927.

Linder, David H. Evolution of the Basidiomycetes and its relation to the terminology of the basidium. Mycologia 32: 419-447. 1940.

Linnaeus, Carolus. Systema naturae sive regna tria naturae systematice proposita per classes, ordines, genera, & species. Leyden. 1735 [Facsimile reprint. Stockholm. 1907]. Tenth edition. 2 vols. Stockholm. 1758. Twelfth edition. 3 vols. Stockholm. 1766-1768.

——————. Species plantarum . . . 2 vols. Stockholm. 1753.

——————. Genera plantarum . . . 6th ed. Stockholm. 1764.

Lister, Arthur. On the division of nuclei in the Mycetozoa. Jour. Linnaean Soc. Bot. 29: 529-542. 1893.

——————. A monograph of the Mycetozoa being a descriptive catalogue of the species in the herbarium of the British Museum. London. 1894.

Lister, J. J. Contributions to the life-history of the Foraminifera. Phil. Trans. Roy. Soc. London B 186: 401-453. 1895.

Loeffler, F. Eine neue Methode zum Färbung der Mikroorganismen, in besonderen ihrer Wimperhaare und Geisseln. Centralbl. Bakt. 6: 209-224. 1889.

Lohman, H. Die Coccolithophoridae, eine Monographie der Coccolithen bildenden Flagellaten. Arch. Prot. 1: 89-165. 1902.

Löhnis, F., and N. R. Smith. Life cycles of the bacteria. Jour. Agr. Res. 6: 675-702. 1916.

—————— and ——————. Studies upon the life cycles of the bacteria—Part II: life history of Azotobacter. Jour. Agr. Res. 23: 401-432. 1923.

Longest, Pauline Moser. Structure of the cilia in Ectocarpus Mitchelliae and Codium decorticatum. Jour. Elisha Mitchell Soc. 62: 249-252. 1946.

Lucas, George Blanchard. Genetics of Glomerella. IV. Nuclear phenomena in the ascus. American Jour. Bot. 33: 802-806. 1946.

Lund, Everett Eugene. A correlation of the silverline and neuromotor systems of Paramoecium. Univ. California Publ. Zool. 39: 35-76. 1933.

Luther, A. Ueber Chlorosaccus, eine neue Gattung der Süsswasseralgen, nebst einigen Bemerkungen zur Systematik verwandter Algen. Bihang till kgl. Svenska Vetensk.-Akad. Handl. 24, Afd. III, no. 13: 1-22. 1899. ,

Luttrell, E. S. Taxonomy of the Pyrenomycetes. Univ. Missouri Studies 24 (no. 3): 1-120. 1951.

Lyngbye, Hans Christian. Tentamen hydrographiae Danicae . . . Copenhagen. 1819.

Macbride, Thomas H. The North American slime-molds. New York and London. 1899. Second edition. 1922.

—————— and G. W. Martin. The Myxomycetes. New York. 1934.

MacLennan, Ronald F. The pulsatory cycle of the contractile vacuoles in the Ophryoscolecidae, ciliates from the stomach of cattle. Univ. California Publ. Zool. 39: 205-250. 1933.

Mackay, James Townsend. Flora Hibernica . . . Dublin. 1836.

Manton, I. The fine structure of plant cilia. Symp. Soc. Exp. Biol. 6: 306-319. 1952.

Martin, C. H. Observations on Trypanoplasma congeri. Part I.—The division of the active form. Quart. Jour. Micr. Sci. n. s. 55: 485-496. 1910.

Martin, Ella M. The morphology and cytology of Taphrina deformans. American Jour. Bot. 27: 743-751. 1940.

Martin, G. W. Morphology of Conidiobolus villosus. Bot. Gaz. 80: 311-318. 1925.

—————. The genus Protodontia. Mycologia 24: 508-511. 1932.

—————. Three new heterobasidiomycetes. Mycologia 26: 261-265. 1934.

—————. Atractobasidium, a new genus of Tremellaceae. Bull. Torrey Bot. Club 62: 339-343. 1935.

—————. A new type of heterobasidiomycete. Jour. Washington Acad. Sci. 27: 112-114. 1937.

—————. New or noteworthy Fungi from Panama and Colombia. I. Mycologia 29: 618-625. 1937.

—————. New or noteworthy Fungi from Panama and Colombia. II. Mycologia 30: 431-441. 1938.

—————. Myxomycetes from Colombia. Trans. American Micr. Soc. 57: 123-126. 1938.

—————. The morphology of the basidium. American Jour. Bot. 25: 682-685. 1938.

—————. New or noteworthy Fungi from Panama and Colombia. III. Mycologia 31: 239-249. 1939.

—————. New or noteworthy Fungi from Panama and Colombia. IV. Mycologia 31: 507-518. 1939.

—————. Notes on Iowa Fungi. VIII. Proc. Iowa Acad. Sci. 46: 89-95. 1939.

—————. The genus "Lycogalopsis." De Lilloa 4: 69-73. 1939.

—————. Notes on Iowa Fungi. IX. Proc. Iowa Acad. Sci. 49: 145-152. 1942.

—————. New or noteworthy tropical Fungi II. Lloydia 5: 158-164. 1942.

—————. The generic name Auricularia. American Midl. Nat. 30: 77-82. 1943.

—————. The numbers of Fungi. Proc. Iowa Acad. Sci. 58: 175-178. 1951.

Matthews, Velma Dare. Studies on the genus Pythium. Chapel Hill, North Carolina. 1931.

Maupas, E. La rajeunissement karyogamique ches les ciliés. Arch. Zool. Exp. Gen. sér. 2, 7: 149-517. 1889.

McClintock, Barbara. Neurospora. I. Preliminary observations of the chromosomes of Neurospora crassa. American Jour. Bot. 32: 671-678. 1945.

McKay, Hazel Hayden. The life history of Pterygophora californica Ruprecht. Univ. California Publ. Bot. 17: 111-148. 1933.

McLarty, D. A. Studies in the Woroninaceae—II. The cytology of Olpidiopsis Achlyae sp. nov. (ad int.). Bull. Torrey Bot. Club 68: 75-99. 1941.

McNab, W. R. On the classification of the vegetable kingdom. Jour. of Bot. 15: 340-344. 1877.

Mense, Carl (editor). Handbuch der Tropenkrankheiten. 3 vols. Leipzig. 1905-1906.

Mesnil, Félix. Les travaux récents sur les coccidies. Bull. Inst. Pasteur 1: 473-480. 1903.

Metcalf, Maynard M. Opalina and the origin of the Ciliata. Anat. Rec. 14: 88-89. 1918.

—————. Opalina and the origin of the ciliate Infusoria. Jour. Washington Acad. Sci. 8: 427-431. 1919.

—————. The opalinid ciliate infusorians. United States Nat. Mus. Bull. 120: i-vii, 1-484. 1923.

Minchin, Edward Alfred. Investigations on the development of trypanosomes in the tsetse-flies and other Diptera. Quart. Jour. Micr. Sci. n. s. 52: 159-260. 1908.

————————. The structure of Trypanosoma lewisi in relation to microscopical technique. Quart. Jour. Micr. Sci. n. s. 53: 755-808. 1909.

———— and H. M. Woodcock. Observations on certain blood-parasites of fishes occurring at Rovigno. Quart. Jour. Micr. Sci. n. s. 55: 113-154. 1910.

———— and ————————. Observations on the trypanosome of the little owl (Athene noctua) with remarks on the other protozoan parasites occurring in this bird. Quart. Jour. Micr. Sci. n. s. 57: 141-185. 1911.

————————. An introduction to the study of the Protozoa . . . London. 1912.

———— and J. D. Thomson. The rat-trypanosome, Trypanosoma lewisi, in relation to the rat-flea, Ceratophyllus fasciatus. Quart. Jour. Micr. Sci. n. s. 60: 463-692. 1915.

Mitchell, Herschel K., and Mary B. Houlahan. Neurospora. IV. A temperature-sensitive riboflavineless mutant. American Jour. Bot. 33: 31-35. 1946.

Miwa, Tomoo. Biochemische Studien über die Zellmembran von Braun- und Rotalgen. Japanese Jour. Bot. 11: 41-127. 1940.

Molisch, Hans. Die Purpurbakterien nach neuen Untersuchungen. Jena. 1907.

Moreau, Fernand. Recherches sur la reproduction des Mucorinées et de quelques autres Thallophytes. Le Botaniste 13: 1-136. 1913.

Moreau, Mme. F. Les phénomènes de la sexualité chez les uredinées. Le Botaniste 13: 145-284. 1914.

Mottier, David M. Das Centrosom bei Dictyota. Ber. deutschen bot. Gess. 16: 123-128. 1898.

————————. Nuclear and cell division in Dictyota dichotoma. Ann. Bot. 14: 163-192. 1900.

Müller, Johannes, and A. Retzius. Ueber parasitische Bildung. Arch. Anat. Physiol. 1842: 193-212. 1842.

————————. Über die Thalassicollen, Polycystinen und Acanthometren des Mittelmeeres. Abh. Akad. Wiss. Berlin 1858: 1-62. 1859.

Müller, Otho Fredericus. Vermium terrestrium et fluviatilum seu animalium infusorium helminthicorum et testaceorum, non marinorum, succincta historia. 2 vols. Copenhagen and Leipzig. 1773, 1774.

————————. Animacula infusoria fluviatilia et marina . . . Copenhagen. 1786.

Müller, Otto. Durchbrechung der Zellwand in ihren Beziehungen zur Ortsbewegung der Bacillariaceen. Ber. deutschen bot. Gess. 7: 169-180. 1889.

————————. Die Ortsbewegung der Bacillariaceen III. Ber. deutschen bot. Gess. 14: 54-64. 1896.

Mulsow, Karl. Über Fortpflanzungserscheinungen bei Monocystis rostrata n. sp. Arch. Prot. 22: 20-55. 1911.

Myers, Earl H. The life cycle of Patellina corrugata, a foraminifera. Science 79: 436-437. 1934.

————————. The life history of Patellina corrugata Williamson a foraminifer. Bull. Scrips Inst., Tech. ser. 3: 355-392. 1935.

————————. The life cycle of Spirillina vivipara Ehrenberg, with notes on morphogenesis, systematics and distribution of the Foraminifera. Jour. Roy. Micr. Soc. ser. 3, 56: 120-146. 1936.

————————. Biology, ecology, and morphogenesis of a pelagic foraminifer. Stanford Univ. Publ. Biol. Sci. 9: 1-40. 1943.

Myers, Margret E. The life history of the brown alga, Egregia Menziesii. Univ. California Publ. Bot. 14: 225-246. 1928.

Nabel, Kurt. Über die Membran der niederer Pilze, besonders von Rhizidiomyces bivellatus nov. spez. Arch. Mikrobiol. 10: 515-541. 1939.

Nägeli, Carl. Gattungen einzelliger Algen physiologisch und systematisch bearbeitet. Zurich. 1849.

Noble, Alden E. On Tokophrya lemnarum Stein (Suctoria) with an account of its budding and conjugation. Univ. California Publ. Zool. 37: 477-520. 1932.

Noble, Elmer R. The life cycle of Zygosoma globosum sp. nov. a gregarine parasite of Urechis caupo. Univ. California Publ. Zool. 43: 41-66. 1938.

——————. Nuclear cycles in the life history of the protozoan genus Ceratomyxa. Jour. Morph. 69: 455-479. 1941.

——————. Life cycles in the Myxosporidia. Quart. Rev. Biol. 19: 213-235. 1944.

Noguchi, Hideo, Raymond C. Shanor, Evelyn B. Tilden, and Joseph R. Tyler. Phlebotomus and Oroya fever and verruga peruana. Science 68: 493-496. 1928.

Nowakowski, Leon. Beitrag zur Kenntniss der Chytridiaceen. II. Polyphagus Euglenae, eine Chytridiacee mit geschlechtlicher Fortpflanzung. Beitr. Biol. Pfl. 2: 201-219. 1876.

Odum, Howard T. Notes on the strontium content of sea water, celestite radiolaria, and strontianite snail shells. Science 114: 211-213. 1951.

Olive, Edgar W. A preliminary enumeration of the Sorophoreae. Proc. American Acad. 37: 333-344. 1901.

——————. Monograph of the Acrasieae. Proc. Boston Soc. Nat. Hist. 30: 451-513. 1902.

——————. Mitotic division of the nuclei of the Cyanophyceae. Beih. bot. Centralbl. 18: 9-44. 1904.

——————. Cytological studies on the Entomophthoreae. Bot. Gaz. 41: 192-208, 229-261. 1906.

Olive, Lindsay S. Karyogamy and meiosis in the rust Coleosporium Vernoniae. American Jour. Bot. 36: 41-54. 1949.

Oltmanns, Friedrich. Zur Entwicklungsgeschichte der Florideen. Bot. Zeit. 56: 99-140. 1898.

——————. Morphologie und Biologie der Algen. 2 vols. Jena, 1904, 1905. Second edition. 3 vols. 1922, 1923.

Oparin, Alexander Ivanovitch. The origin of life. Translated by Sergius Morgulis. New York. 1938.

d'Orbigny, Alcide Dessalines. Tableau methodique de la classe des céphalopodes. Ann. Sci. Nat. 7: 96-169, 245-314. 1826.

——————. Foraminifères. in R. de la Sagra. Histoire . . . de l'ile de Cuba, vol. 8. Paris. 1839.

Orla-Jensen, S. Die Hauptlinien des natürlichen Bakteriensystems. Centralbl. Bakt. Abt. 2, 22: 305-346. 1909.

Overholts, L. O. Research methods in the taxonomy of the Hymenomycetes. Proc. Internat. Congr. Pl. Sci. Ithaca 2: 1688-1712. 1929.

Owen, Richard. Palaeontology, or a systematic summary of extinct animals and their geological relations. Edinburgh. 1860. Second edition. 1861.

Papenfuss, George F. Structure and taxonomy of Taenioma, including a discussion of the phylogeny of the Ceramiales. Madroño 7: 193-214. 1944.

——————. Review of the Acrochaetium-Rhodochorton complex of the red algae. Univ. California Publ. Bot. 18: 299-334. 1945.

————. Proposed names for the phyla of algae. Bull. Torrey Bot. Club 73: 217-218. 1946.

————. Extension of the brown algal order Dictyosiphonales to include the Punctariales. Bull. Torrey Bot. Club. 74: 398-402. 1947.

Park, William Hallock, Anne Wessels Williams, and Charles Krumweide. Pathogenic microorganisms . . . 8th ed. Philadelphia and New York. 1924.

Pascher, Adolf. Zur Gliederung der Heterokonten. Hedwigia 53: 6-22. 1912.

———— and others. Die Süsswasser-Flora Deutschlands, Österreichs und der Schweiz. 15 Hefte. Jena. 1913-1936.

————. Über Flagellaten und Algen. Ber. deutschen bot. Gess. 32: 136-160. 1914.

————. Über eine neue Amöbe—Dinamoeba (varians)—mit dinoflagellaten-artigen Schwärmern. Arch. Prot. 36: 118-136. 1916.

————. Fusionsplasmodien bei Flagellaten und ihre Bedeutung für die Ableitung der Rhizopoden von den Flagellaten. Arch. Prot. 37: 31-64. 1916.

————. Die braune Algenreihe Chrysophyceen. Arch. Prot. 52: 489-564. 1925.

————. Eine Chrysomonade mit gestielten und verzweigten Kolonien. Arch. Prot. 57: 319-330. 1927.

————. Eine eigenartige rhizopodiale Flagellate. Arch. Prot. 63: 227-240. 1928.

————. Zur Verwandtschaft der Monadaceae mit den Chrysomonaden: eine gehäusebewohnende, farblose Chrysomonade. Ann. Protistol. 2: 157-168. 1930.

————. Über die Verfestigung des Propoplasten im Gehäuse einer neuen Euglenine (Klebsiella). Arch. Prot. 73: 315-322. 1931.

————. Systematische Übersicht über die mit Flagellaten in Zusammenhang stehenden Algenreihen und Versuch einer Einrichtung dieser Algenstämme in die Stämme des Pflanzenreiches. Beih. bot. Cebtralbl. 48, Abt. 2: 317-332. 1931.

————. Über die Verbreitung endogener bezw. endoplasmatisch gebildeter Sporen bei den Algen. Beih. bot. Centralbl. 49, Abt. 1: 293-308. 1932.

————. Über drei auffallend konvergente zu verschiedenen Algenreihen gehörende epiphytische Gattungen. Beih. bot. Centralbl. 49, Abt. 1: 549-568. 1932.

Persoon, Christiaan Hendrik. Synopsis methodica Fungorum . . . Göttingen. 1801.

Perty, Maximilian. Zur Kenntniss kleinster Lebensformen . . . Bern. 1852.

Petersen, Hennig Eiler. Studien over Ferskvands-Phycomyceter. Bot. Tidsskr. 29: 345-440. 1909.

Petersen, Johs. Boye. Beiträge zur Kenntnis der Flagellatengeisseln. Bot. Tidsskr. 40: 373-389. 1929.

Poche, Franz. Das System der Protozoa. Arch. Prot. 30: 125-321. 1913.

Pratje, Andre. Noctiluca miliaris Suriray. Beiträge zur Morphologie, Physiologie und Cytologie. I. Morphologie und Physiologie (Beobachtungen an der lebenden Zelle). Arch. Prot. 42: 1-98. 1921.

Prévot, André-Romain. Études de systématique bactérienne. Ann. Sci. Nat. Bot. sér. 10, 15: 23-260. 1933.

Pribram, Ernst. A contribution to the classification of microorganisms. Jour. Bact. 18: 361-394. 1929.

Pringsheim, N. Über die Befruchtung der Algen. Monatsber. Akad. Wiss. Berlin 1855: 133-165. 1855.

Prowazek, Stanislas. Die Entwicklung von Herpetomonas. Arb. kaiserl. Gesundheitsamte 20: 440-452. 1903.

Rabenhorst, Ludwig. Deutschland's Kryptogamen-Flora . . . 2 vols. Leipzig. 1844, 1847.

————. Kryptogamen-Flora von Sachsen, der Ober Lausik, Thüringen und Nordböhmen . . . 2 vols. Leipzig. 1863, 1870.

———— and others. Dr. L. Rabenhorst's Kryptogamen-Flora von Deutschland, Oesterreich und der Schweiz . . . [construed as second edition of the work of 1844-1847.] 14 vols. Leipzig. 1879-1944.

Raper, John R. Tetrapolar sexuality. Quart. Rev. Biol. 28: 233-259. 1953.

Raper, Kenneth B. The communal nature of the fruiting process in the Acrasieae. American Jour. Bot. 27: 436-448. 1940.

Ratcliffe, H. L. Mitosis and cell division in Euglena spirogyra Ehrenberg. Biol. Bull. 53: 109-122. 1927.

Ritchie, Don. A fixation study of Russula emetica. American Jour. Bot. 28: 582-588. 1941.

Robertson, Muriel. Further notes on a trypanosome found in the alimentary tract of Pontobdella muricata. Quart. Jour. Micr. Sci. n. s. 54: 119-139. 1909.

Robinow, C. F. A study of the nuclear apparatus of bacteria. Proc. Roy. Soc. London B 130: 299-324. 1942.

————. Cytological observations on bacteria. Proc. 6th Internat. Congr. Exp. Cytol. 204-207. 1949.

Rosenberg, Marie. Die geschlechtliche Fortpflanzung von Botrydium granulatum Grev. Oesterreichische Bot. Zeit. 79: 289-296. 1930.

Rosebury, Theodor, Richard W. Linton, and Leon Buchbinder. A comparative study of dental aciduric organisms and Lactobacillus acidophilus. Jour. Bact. 18: 395-412. 1929.

Rosenvinge, L. Kolderup. The marine algae of Denmark. Vol. 1. Rhodophyceae. Mem. Acad. Roy. Sci. Lett. Danemark sér. 7, Sciences vol. 7. 1909-1931.

————. On mobility in the reproductive cells of the Rhodophyceae. Bot. Tidsskr. 40: 72-80. 1927.

von Rostafinski, Joseph Thomas. Versuch eines Systems der Mycetozoen. Strassburg. 1873.

————. Sluzowce (Mycetozoa) Monographia. Paris. 1875.

Rothmaler, Werner. Über das natürliche System der Organismen. Biol. Zentralbl. 67: 242-250. 1948.

Ruprecht, F. J. Tange des ochotskischen Meeres. in A. T. von Middendorf. Sibirische Reise, vol. 1, part 2: 193-435. 1851.

Ryan, Francis J., G. W. Beadle, and E. L. Tatum. The tube method of measuring the growth rate of Neurospora. American Jour. Bot. 30: 784-799. 1943.

Sabrosky, Curtis W., and Robert L. Usinger. Nomenclature of the human malaria parasites. Science 100: 190-192. 1944.

Saccardo, P. A. Sylloge Fungorum omnium hucusque cognitorum . . . 25 vols. Padua. 1882-1931.

Sachs, Julius. Lehrbuch der Botanik. 4th ed. Leipzig. 1874. English translation by Alfred W. Bennett and W. T. Thistleton-Dyer. Oxford. 1875.

de Saedeleer, Henri. Beitrag zur Kenntnis der Rhizopoden . . . Mém. Mus. Roy. Hist. Nat. Belgique 60: 1-112. 1934.

Salvin, S. B. Factors controlling sporangial type in Thraustotheca Primoachlya and Dictyuchus. American Jour. Bot. 29: 97-104. 1942.

Sartoris, George B. Studies in the life history and physiology of certain smuts. American Jour. Bot. 11: 617-647. 1924.

Sauvageau, M. C. Sur le développement et la biologie d'une laminaire (Saccorhiza bulbosa). Compt. Rend. 160: 445-448. 1915.

————. Sur la sexualité hétérogamique d'une laminaire (Saccorhiza bulbosa). Compt. Rend. 161: 796-799. 1915.

————. Sur un nouveau type d'alternance de générations chez les algues brunes; les Sporochnales. Compt. Rend. 182: 361-364. 1926.

Savile, D. B. O. Nuclear structure and behavior in species of the Uredinales. American Jour. Bot. 26: 585-609. 1939.

Schaeffer, Asa Arthur. Taxonomy of the amebas with descriptions of thirty-nine new marine and freshwater species. Carnegie Inst. Publ. 345: 1-116. 1926.

Schaffner, John H. The classification of plants. IV. Ohio Naturalist 9: 446-455. 1909.

Schaudinn, Fritz. Heliozoa. Thierreich Probelief.: 1-24. 1896.

————. Über den Zeugungskreis von Paramoeba eilhardi n. g. n. sp. Sitzber. Akad. Wiss. Berlin 1896: 31-41. 1896.

————. Über die Copulation von Actinophrys sol Ehrbg. Sitzber. Akad. Wiss. Berlin 1896: 83-89. 1896.

————. Ueber das Centralkorn der Heliozoen, ein Beitrag zur Centrosomenfrage. Verhandl. deutscher zool. Gess. 1896: 113-130. 1896.

————. Untersuchungen über den Generationswechsel bei Coccidien. Zool. Jahrb. Abt. Anat. 13: 197-292. 1900.

————. Untersuchungen über die Fortpflanzung einiger Rhizopoden. Arb. kaiserl. Gesundheitsamte 19: 547-576. 1902.

————. Generations- und Wirtswechsel bei Trypanosoma und Spirochaete. Arb. kaiserl. Gersundheitsamte 20: 387-439. 1903.

———— and Erich Hoffmann. Vorläufiger Bericht über das Vorkommen von Spirochaeten in syphilitischen Krankheitsprodukten und bei Papillomen. Arb. kaiserl. Gesundheitsamte 22: 527-534. 1905.

Scherffel, A. Phaeocystis globosa nov. spec. nebst einigen Beobachtungen über die Phylogenie niederer, insbesonderer brauner Organismen. Wiss. Meeresunters. n. f. 4, Abt. Heligoland: 1-29. 1900.

————. Kleiner Beitrag zur Phylogenie einiger Gruppen niederer Organismen. Bot. Zeit. 59: 143-158. 1901.

————. Endophytische Phycomyceten-Parasiten der Bacillariaceen und einige neue Monaden. Arch. Prot. 52: 1-141. 1925.

————. Zur Sexualität der Chytridineen. Arch. Prot. 53: 1-58. 1925.

Schiller, Jos. Die planktonischen Vegetationen des adriatischen Meeres. A. Die Coccolithophoriden-Vegetationen in den Jahren 1911-14. Arch. Prot. 51: 1-130. 1925.

————. Über Fortpflanzung, geissellose Gattungen und die Nomenklatur der Coccolithophoraceen nebst Mitteilung über Copulation bei Dinobryon. Arch. Prot. 53: 326-342. 1926.

Schmidt, Matthias. Makrochemische Untersuchungen über das Vorkommen von Chitin bei Mikroorganismen. Arch. Mikrobiol. 7: 241-260. 1936.

Schmitz, Fr. Untersuchungen über die Befruchtung der Florideen. Sitzber. kgl. preussischen Akad. Wiss. 1883: 215-258. 1883.

————. Systematische Übersicht der bisher bekannten Gattungen der Florideen. Flora 72: 435-456. 1889.

Schneider, Aimé. Sur les psorospermes oviformes ou coccidies. Arch. Zool. Exp. Gen. 9: 387-404. 1881.

Schultz, E. S. Nuclear division and spore-formation in the ascus of Peziza domiciliana. American Jour. Bot. 14: 307-322. 1927.

Schultze, Max. Die Bewegung der Diatomen. Arch. mikr. Anat. 1: 376-402. 1865.

Schulze, Franz Eilhard. Rhizopodenstudien V. Arch. mikr. Anat. 11: 583-596. 1875.

Schünemann, Erich. Untersuchungen über die Sexualität der Myxomyceten. Planta 9: 644-672. 1930.

Schuurmans Stekhoven, J. H. Jr. Die Teilung der Trypanosoma brucei Plimmer u. Bradford. Arch. Prot. 40: 158-180. 1919.

Schwartz, E. J. The Plasmodiophoraceae and their relationship to the Mycetozoa and Chytridieae. Ann. Bot. 28: 227-240. 1914.

Schwendener, S. Ueber die Beziehung zwischen Algen und Flechtengonidien. Bot. Zeit. 26: 289-292. 1868.

Setchell, William Albert. Parasitic Florideae, I. Univ. California Publ. Bot. 6: 1-34. 1914.

———— and Nathaniel Lyon Gardner. The marine algae of the Pacific Coast of North America. Univ. California Publ. Bot. 8, parts 1-3. 1919-1925.

———— and ————. Phycological contributions I. Univ. California Publ. Bot. 7: 279-324. 1920.

————. The coral reef problem in the Pacific. Proc. 3d Pan-Pacific Sci. Congr. Tokyo 323-329. 1926.

Sharp, Robert G. Diplodinium ecaudatum with an account of its neuromotor apparatus. Univ. California Publ. Zool. 13: 43-122. 1914.

Shear, C. L., and B. O. Dodge. Life histories and heterothallism of the red bread-mold fungi of the Monilia sitophila group. Jour. Agr. Res. 34: 1019- 1042. 1927.

Short, Robert B. Spindle twisting in the giant Amoeba. Science 102: 484. 1945.

von Siebold, C. Th., and H. Stannius. Lehrbuch der vergleichenden Anatomie. Berlin. vol. 1. 1848. vol. 2. 1846.

Simons, Etoile B. A morphological study of Sargassum filipendula. Bot. Gaz. 41: 161-182. 1906.

Sisler, Frederick D., and Claude E. ZoBell. Nitrogen fixation by sulfate-reducing bacteria indicated by nitrogen/argon ratios. Science 113: 511-512. 1951.

Sjöstedt, L. G. Floridean studies. Kgl. Fysiog. Sällsk. Handl. n. f. 37, no. 4: 1-95. 1926.

Skupienski, François-Xavier. Sur la sexualité chez les champignons myxomycètes. Compt. Rend. 165: 118-121. 1917.

————. Sur le cycle évolutif chez un espèce de myxomycète endosporé, Didymium difforme (Duby). Étude cytologique. Compt. Rend. 184: 1341-1344. 1927.

Smith, Arlo I. The comparative histology of some of the Laminariales. American Jour. Bot. 26: 571-585. 1939.

Smith, Ernest C. Some phases of spore germination of Myxomycetes. American Jour. Bot. 16: 645-650. 1929.

Smith, Gilbert Morgan. A preliminary list of algae found in Wisconsin lakes. Trans. Wisconsin Acad. 18: 531-565. 1916.

————. A second list of algae found in Wisconsin lakes. Trans. Wisconsin Acad. 19: 614-653. 1918.

————. Phytoplankton of the inland lakes of Wisconsin. Wisconsin Geol. and Nat. Hist. Survey Bull. 57, part 1: 1-243. 1920; part 2: 1-227. 1924.

————. The fresh-water algae of the United States. New York and London. 1933. Second edition. 1950.

—————. Cryptogamic botany. 2 vols. New York and London. 1938.

—————. Marine algae of the Monterey Peninsula, California. Stanford University. 1944.

————— (editor). Manual of phycology. Waltham, Massachusetts. 1951.

Smith, Theobald. Investigation of infectious diseases of domesticated animals. United States Dept. Agr. Bur. Anim. Ind. Ann. Reports 8 and 9: 45-70. 1893.

Snyder, William C., and H. N. Hansen. The species concept in Fusarium with reference to section Martiella. American Jour. Bot. 28: 738-742. 1941.

————— and —————. The species concept in Fusarium with reference to Discolor and other sections. American Jour. Bot. 32: 657-666. 1945.

Solms-Laubach, Herman. Note sur le Janczewskia, nouvelle floridée parasite du Chondria obtusa. Mém. Soc. Nat. Sci. Nat. Cherbourg 21: 209-224. 1877.

Sonneborn, T. M. Sex, sex inheritance and sex determination in Paramecium aurelia. Proc. Nat. Acad. Sci. 23: 378-395. 1937.

Sparrow, Frederick K., Jr. A classification of aquatic Phycomycetes. Mycologia 34: 113-116. 1942.

—————. Aquatic Phycomycetes. Ann Arbor. 1943.

—————. Observations on chytridiaceous parasites of phanerogams. I. Physoderma Menyanthis de Bary. American Jour. Bot. 33: 112-118. 1946.

—————. Observations on chytridiaceous parasites of phanerogams. II. A preliminary study of the occurrence of ephemeral sporangia in the Physoderma disease of maize. American Jour. Bot. 34: 94-97. 1947.

—————. Observations on chytridiaceous parasites of phanerogams. III. Physoderma Claytoniana and an associated parasite. American Jour. Bot. 34: 325-329. 1947.

Sprague, T. A. Standard-species. Kew Bull. 1926: 96-100. 1926.

Springer, Martha E. A morphologic study of the genus Monoblepharella. American Jour. Bot. 32: 259-269. 1945.

Stakman, Elvin C. The nature and importance of physiologic specialization in phytopathogenic fungi. Science 105: 627-632. 1947.

—————. Plant diseases are shifty enemies. American Scientist 35: 321-350. 1947.

Stanier, Roger Y., and C. B. van Niel. The main outlines of bacterial classification. Jour. Bact. 42: 437-466. 1941.

Stein, Friedrich. Ueber die Natur der Gregarinen. Arch. Anat. Phys. 1848: 182-223. 1848.

—————. Der Organismus der Infusionsthiere. 3 vols. Leipzig. 1859-1883.

Steinecke, Fritz. Die Phylogenie der Algophyten. Schr. königsberger Gelehrt. Gess. 8: 127-297. 1931.

Stevens, Edith. Cytological features of the life history of Gymnosporangium Juniperivirginianae. Bot. Gaz. 89: 394-401. 1930.

Stevens, Frank Lincoln. The compound oosphere of Albugo Bliti. Bot. Gaz. 28: 149-176, 225-245. 1899.

—————. Gametogenesis and fertilization in Albugo. Bot. Gaz. 32: 77-98, 157-169, 238-261. 1901.

—————. Studies in the fertilization of Phycomycetes. Bot. Gaz. 34: 420-425. 1902.

————— and A. C. Stevens. Mitosis of the primary nucleus in Synchytrium decipiens. Bot. Gaz. 35: 405-415. 1903.

Stokes, John H. Schaudinn: a biographical appreciation. Science 74: 502-506. 1931.

Strasburger, Eduard. Zur Entwicklungsgeschichte der Sporangien von Trichia fallax. Bot. Zeit. 42: 305-316, 321-326. 1884.

————. Kerntheilung und Befruchtung bei Fucus. Jahrb. wiss. Bot. 30: 351-374. 1897.

Strickland, Hugh Edwin, and others. Series of propositions for rendering the nomenclature of zoology uniform and permanent. Rept. 12th meeting British Assoc. 1842: 106-121. 1843.

Subrahmanyan, R. On somatic division, reduction division, auxospore-formation and sex differentiation in Navicula halophila (Grun.) Cl. Jour. Indian Bot. Soc. Iyengar Commemoration Volume: 239-266. 1947.

Svedelius, Nils. An evaluation of the structural evidences for genetic relationships in plants: algae. Proc. Internat. Congr. Pl. Sci. Ithaca 1: 457-471. 1929.

————. Zytologisch-Entwicklungsgeschichtliche Studien über Galaxaura, eine diplobiontische Nemalionales-Gattung. Nova Acta Reg. Soc. Sci. Upsaliensis ser. 4, 13, no. 4: 1-154. 1942.

Swellengrebel, N. H. Zur Kenntnis der Cytologie von Bacillus maximus buccalis (Miller). Centralbl. Bakt. Abt. 2, 16: 617-628, 673-681. 1906.

————. Note on the cytology of Calothrix fusca. Quart. Jour. Micr. Sci. n. s. 54: 623-629. 1910.

Swezy, Olive. Binary and multiple fission in Hexamitus. Univ. California Publ. Zool. 16: 71-88. 1915.

————. The genera Monocercomonas and Polymastix. Univ. California Publ. Zool. 16: 127-138. 1916.

Swingle, Deane B. Formation of the spores in the sporangia of Rhizopus nigricans and of Phycomyces nitens. United States Dept. Agr. Bur. Pl. Ind. Bull. 37: 1-40. 1903.

Swingle, Walter T. Zur Kenntniss der Kern- und Zelltheilung bei den Sphacelariaceen. Jahrb. wiss. Bot. 30: 297-350. 1897.

Tatum, E. L., and T. T. Bell. Neurospora. III. Biosynthesis of thiamin. American Jour. Bot. 33: 15-20. 1946.

———— and Joshua Lederberg. Gene recombination in the bacterium Escherichia coli. Jour. Bact. 53: 673-684. 1947.

————, R. W. Barratt, Nils Fries, and David Bonner. Biochemical mutant strains of Neurospora produced by physical and chemical treatment. American Jour. Bot. 37: 38-46. 1950.

Taylor, William Randolph. Recent studies of Phaeophyceae and their bearing on classification. Bot. Gaz. 74: 431-441. 1922.

Thaxter, Roland. On the Myxobacteriaceae, a new order of Schizomycetes. Bot. Gaz. 17: 389-406. 1892.

————. Contributions from the Cryptogamic Laboratory of Harvard University. XL. New or peculiar Zygomycetes. 2. Syncephalastrum and Syncephalis. Bot. Gaz. 24: 1-15. 1897.

————. Contributions toward a monograph of the Laboulbeniaceae. Mem. American Acad. 12: 187-429. 1896; 13: 217-469. 1908; 14: 311-426. 1924; 15: 427-580. 1926; 16: 1-435. 1931.

Theissen, F. Hemisphaeriales. Ann. Myc. 11: 468-469. 1913.

Thélohan, P. Observations sur les myxosporidies et essai de classification de ces organismes. Bull. Soc. Philomath. Paris sér. 8, 4: 165-178. 1892.

Thom, Charles, and Margaret B. Church. The Aspergilli. Baltimore. 1926.

——————— and Kenneth B. Raper. A manual of the Aspergilli. Baltimore. 1945.

Thomas, R. C. Composition of fungus hyphae I. The Fusaria. American Jour. Bot. 15: 537-547. 1928.

Thompson, R. H. Immobile Dinophyceae. I. New records and a new species. American Jour. Bot. 36: 301-308. 1949.

Thuret, Gustave.Recherches sur les zoospores des algues. Ann. Sci. Nat. Bot. sér. 3, 14: 214-260. 1850.

———————. Recherches sur la fécondation des fucacées suivies d'observations sur les anthéridies des algues. Ann. Sci. Nat. Bot. sér. 4, 2: 197-214. 1854; 3: 5-28. 1855.

———————. Essai de classification des nostochinées. Ann. Sci. Nat. Bot. sér. 6, 1: 372-382. 1875.

van Tieghem, Ph. Sur la classification des basidiomycètes. Jour. de Bot. 7: 77-87. 1893.

Tilden, Josephine E. A classification of algae based on evolutionary development, with special reference to pigmentation. Bot. Gaz. 95: 59-77. 1933.

———————. The algae and their life relations. Minneapolis. 1935.

Tippo, Oswald. A modern classification of the plant kingdom. Chron. Bot. 7: 203-206. 1942.

du Toit, P. J. Zur Systematik der Piroplasmen. Arch. Prot. 39: 84-104. 1918.

de Toni, J. B. Sylloge Algarum. 5 vols. Padua. 1889-1924.

Tseng, C. K. Utilization of seaweeds. Sci. Month. 59: 37-46. 1944.

———————. The terminology of seaweed colloids. Science 101: 597-602. 1945.

Tulasne, L.-R., and C. Tulasne. Mémoire sur les ustilaginées comparées aux urédinées. Ann. Sci. Nat. Bot. sér. 3, 7: 12-127. 1847.

Tulasne, R., and R. Vendrely. Demonstration of bacterial nuclei with ribonuclease. Nature 160: 225-226. 1947.

Turner, John P. Division and conjugation in Euplotes patella Ehrenberg with special reference to nuclear phenomena. Univ. California Publ. Zool. 33: 193-258. 1930.

Ülehla, Vladimir. Die Stellung der Gattung Cyathomonas From. im System der Flagellaten. Ber. deutschen bot. Gess. 29: 284-292. 1911.

Vanterpool, T. C., and G. A. Ledingham. Studies on "browning" root rot of cereals I. The association of Lagena radicicola n. gen.; n. sp., with root injury in wheat. Canadian Jour. Res. 2: 171-194. 1930.

Vlk, Vladimir. Über die Struktur der Heterokontengeisseln. Beih. bot. Centralbl. 49, Abt. 1: 214-220. 1931.

———————. Über die Geisselstrukutur der Saprolegnieenschwärmer. Arch. Prot. 92: 157-160. 1939.

Vokes, Margaret Martin. Nuclear division and development of sterigmata in Coprinus atramentarius. Bot. Gaz. 91: 194-205. 1931.

Vuillemin, Paul. Remarques sur les affinités des basidiomycètes. Jour. de Bot. 7: 164-174. 1893.

Wager, Harold. The life-story and cytology of Polyphagus Euglenae. Ann. Bot. 27: 173-202. 1913.

Walker, Leva B. Studies on Ascoidea rubescens—II. Cytological observations. Mycologia 27: 102-127. 1935.

Wallengren, Hans. Studier öfver ciliata infusorier II. Acta Univ. Lundensis 31, part 2, no. 7: 1-72. 1895.

Wehmeyer, Lewis E. A biologic and phylogenetic study of the stromatic Sphaeriales. American Jour. Bot. 13: 575-645. 1926.

Weimer, J. L. Some observations on the spore discharge of Pleurage curvicolla (Wint.) Kuntze. American Jour. Bot. 7: 75-77. 1920.

Weiser, Jaroslav. A new classification of the Schizogregarina. Jour. Protozool. 2: 6-12. 1955.

Weisz, Paul B. Morphogenesis in protozoa. Quart. Rev. Biol. 29: 207-229. 1954.

Wenyon, Charles Morley. Some observations on a flagellate of the genus Cercomonas. Quart. Jour. Micr. Sci. n. s. 55: 241-260. 1910.

——————. Protozoology. A manual for medical men, veterinarians, and zoologists. 2 vols. London. 1926.

West, G. S. A treatise on the British freshwater algae. Cambridge. 1904.

von Wettstein, Richard. Handbuch der systematischen Botanik. 2 vols. Leipzig and Wien. 1901-1908.

Whelden, Roy M. Cytological studies in the Tremellaceae II. Exidia. Mycologia 27: 41-57. 1935.

Wichterman, Ralph. Cytogamy: a new sexual process in joined pairs of Paramecium caudatum. Nature 144: 123-124. 1939.

——————. Cytogamy: a sexual process occurring in living joined pairs of Paramecium caudatum and its relation to other sexual phenomena. Jour. Morph. 66: 423-451. 1940.

Williams, A. E., and R. H. Burris. Nitrogen fixation by blue-green algae and their nitrogenous composition. American Jour. Bot. 39: 340-342. 1952.

Williams, J. Lloyd. Reproduction in Dictyota dichotoma. Ann. Bot. 12: 559-560. 1898.

Willstätter, Richard, and Arthur Stoll. Untersuchungen über Chlorophyll. Methoden und Ergebnisse. Berlin. 1913.

Wilson, C. M. Cytological study of the life history of Dictyostelium. American Jour. Bot. 40: 714-718. 1953.

—————— and Ian K. Ross. Meiosis in the myxomycetes. American Jour. Bot. 42: 743-749. 1955.

Wilson, Harriet L. Gracilariophila, a new parasite on Gracilaria confervoides. Univ. California Publ. Bot. 4: 75-84. 1910.

Wilson, H. V. Notes on a species of Pelomyxa. American Nat. 34: 535-550. 1900.

Wilson, Thomas B., and John Cassin. On a third kingdom of organized beings. Proc. Acad. Nat. Sci. Philadelphia 1863: 113-121. 1864.

Winogradsky, Sergius. Ueber Schwefelbacterien. Bot. Zeit. 45: 489-507, 513-523, 529-539, 545-559, 569-576, 585-594, 606-610. 1887.

——————. Recherches sur les organismes de la nitrification. Ann. Inst. Pasteur 4: 213-231, 257-275, 760-771. 1890.

——————. Clostridium Pastorianum, seine Morphologie und seine Eigenschaften als Buttersäureferment. Centralbl. Bakt. Abt. 2, 9: 43-54, 107-112. 1902.

Winslow, C.-E. A., Jean Broadhurst, R. E. Buchanan, Charles Krumweide Jr., L. A. Rogers, and George H. Smith. The families and genera of bacteria. Jour. Bact. 2: 505-566. 1917.

Winter, F. W. Zur Kenntnis der Thalamophoren. Arch. Prot. 10: 1-113. 1907.

Wolters, Max. Die Conjugation und Sporenbildung bei Gregarinen. Arch. mikr. Anat. 37: 99-138. 1891.

Woodcock, H. M. The haemoflagellates: a review of present knowledge relating to the trypanosomes and allied forms. Quart. Jour. Micr. Sci. n. s. 50: 151-331. 1906.

————. On the occurrence of nuclear dimorphism in a Halteridium parasitic in the chaffinch, and the probable connection of this parasite with a trypanosome. Quart. Jour. Micr. Sci. n. s. 53: 339-349. 1909.

————. Studies on avian haematoprotozoa. I. On certain parasites of the chaffinch (Fringilla coelebs) and the redpoll (Linota rufescens). Quart. Jour. Micr. Sci. n. s. 55: 641-740. 1910.

Woodruff, Lorande Loss, and Rhoda Erdmann. A normal periodic reorganization process without cell fusion in Paramaecium. Jour. Exp. Zool. 17: 425-520. 1914.

————. Eleven thousand generations of Paramoecium. Quart. Rev. Biol. 1: 436-438. 1926.

Woronin, M. Plasmodiophora Brassicae, Urheber der Kohlpflanzen-Hernie. Jahrb. wiss. Bot. 11: 548-574. 1878.

Yamanouchi, Shigeo. The life history of Polysiphonia violacea. Bot. Gaz. 41: 425-433. 1906.

————. The life history of Polysiphonia violacea. Bot. Gaz. 42: 401-449. 1906.

————. Mitosis in Fucus. Bot. Gaz. 47: 173-197. 1909.

————. The life history of Cutleria. Bot. Gaz. 54: 441-502. 1912.

Young, Edward Lorraine, III. Studies on Labyrinthula. The etiologic agent of the wasting disease of eel-grass. Amreican Jour. Bot. 30: 586-593. 1943.

Zederbauer, E. Geschlechtliche und ungeschlechtliche Fortpflanzung von Ceratium hirundinella. Ber. deutschen bot. Gess. 22: 1-8. 1904.

Ziegler, A. W. Meiosis in the Saprolegniaceae. American Jour. Bot. 40: 60-66. 1953.

ZoBell, Claude E. The term "pillbox" for describing diatoms. Chron. Bot. 6: 389. 1941.

Zopf, Wilhelm. Die Pilzthiere oder Schleimpilze. Breslau. 1885.

————. Zur Morphologie und Biologie der niederen Pilzthiere (Monadinen), zugleich ein Beitrag zur Phytopathologie. Leipzig. 1885.

Zuelzer, Margarethe. Bau und Entwicklung von Wagnerella borealis Mereschk. Arch. Prot. 17: 135-202. 1909.

INDEX

OF NAMES OF ORGANISMS AND GROUPS

Absidia, 123, 124
Acantharia, 189, 190, 195, 196, 197
Acanthochiasma, 197
Acanthocystida, 191, 193
Acanthocystidae, 193
Acanthocystis, 193
Acanthometra, 195, 197
Acanthometren, 197
Acanthometrida, 197
Acanthometron, 197
Acanthonida, 197
Acanthophracta, 195, 197
Acanthorhynchus, 218
Acanthospora, 218
Acanthosporida, 218
Acanthosporidae, 218
Acaulopage, 124
Acephalina, 215
Acervulina, 187
Acervulinida, 187
Acetobacter aceti, 24
Acetobacteriacea, 20, 24
Acetobacteriaceae, 24
Achlya, 70, 79
Achlya caroliniana, 78
Achlyogeton, 118
Achlyogetonacea, 115, 117
Achlyogetonaceae, 117
Achnanthea, 76
Achnantheae, 76
Achnanthes, 76
Achnanthaceae, 76
Achromatiacea, 33
Achromatiaceae, 33
Achromatium, 33
Achromatium oxaliferum, 32, 33
Achromobacter, 22
Achromobacteriacea, 19, 21
Achromobacteriaceae, 21
Acineta, 235
Acinetae, 235
Acinetaria, 235
Acinetida, 235
Acinetidae, 235
Acinetina, 235
Acinetoidea, 235
Acnidosporidea, 207
Acrasidae, 203
Acrasina, 203
Acrasis, 203
Acrita, 37
Acrochaetiacea, 47
Acrochaetiaceae, 47
Acrochaetium, 47
Actinelius, 197
Actinellida, 197

Actiniscea, 61, 62
Actinisceae, 62
Actinocephalida, 217, 218
Actinocephalidae, 218
Actinocephalus, 218
Actinolophus, 193
Actinomma, 195
Actinomma Asteracanthion, 196
Actinomonadida, 191
Actinomonadidae, 191
Actinomonas, 190, 193
Actinomyces Bovis, 25
Actinomycetaceae, 25
Actinomycetalea, 18, 24
Actinomycetales, 24
Actinomyxida, 219, 221
Actinophryida, 191, 193
Actinophryidae, 193
Actinophrys, 193
Actinophrys Sol, 193
Actinopoda, 189
Actinopodea, 189
Actinosphaerium, 193
Actinosphaerium Eichhornii, 192, 193
Actipylea, 195, 197
Actipyleen, 195
Actipylida, 197
Actipylina, 197
Acystosporidia, 190
Acyttaria, 179
Adelea, 211
Adeleida, 211, 212
Adeleidae, 211
Adeleidea, 211
Adeleina, 211
Adeleoidae, 211
Adelina, 211
Adinida, 98, 99
Adiniferidea, 96, 98
Aecidium, 147
Aerobacter aerogenes, 22
Agaricacea, 151, 153
Agaricaceae, 150, 151
Agaricales, 150
Agaricini, 151
Agaricus campestris, 119, 145, 152, 153
Agarics, 152
Aggregata, 209
Aggregata Eberthi, 210
Aggregatida, 210, 212
Aggregatidae, 210
Aglaozonia, 88
Agrobacterium, 23
Agrobacterium tumefaciens, 23
Agrostis, 148
Agyriales, 137

Agyrium, 137
Ahnfeldtia, 49
Akinetocystida, 216
Akinetocystidae, 216
Akinetocystis, 216
Akinetosporeae, 86
Albuginacea, 80, 81
Albuginaceae, 81
Albugo, 80, 81
Albugo Bliti, 80
Albugo Tragopogonis, 80
Alcaligenes fecalis, 22
Aleuria rutilans, 136
Algae, 9, 10, 69, 113, 118, 120, 177, 224
Algae, blue-green, 2, 3, 12, 13, 14, 17, 30, 37, 41, 117, 118
Algae, brown, 39, 69, 53, 179, 203
Algae, green, 38, 41, 53, 69, 82, 117, 118, 128, 203
Algae, red, 37, 39, 41, 82, 128, 140
Algae Zoosporeae, 86
Algen, 29, 120
Allantocystida, 216
Allantocystidae, 216
Allantocystis, 216
Allogromia, 183
Allogromiida, 183
Allogromiidae, 183
Allomorphina, 187
Allomyces, 111, 112, 113, 118
Allomyces anomalus, 116
Allomyces Arbuscula, 112, 115
Allomyces cystogenes, 112
Allomyces javanicus, 112, 114
Almond, 141
Alveolina, 185
Alveolinea, 185
Alveolinella, 185
Alveolinellidae, 185
Alveolinida, 185
Alveolinina, 185
Alternaria, 142
Alwisia, 175
Amanita, 152
Amanita muscaria, 152
Amaurochaetacea, 174, 175
Amaurochaetaceae, 175
Amaurochaete, 175
Amaurochaeteae, 171
Amaurochaetidae, 175
Amaurosporales, 171
Amiba diffluens, 37, 157, 201
Amiba divergens, 202
Ammodiscida, 185
Ammodiscidae, 185
Ammodiscus, 185
Ammodochidae, 62
Amoeba, 71, 118, 124, 157, 189, 201
Amoeba Proteus, 202
Amoebaea, 201
Amoebida, 10, 201
Amoebidae, 10, 201

Amoebina, 201
Amoebodiniaceae, 101
Amoebogeniae, 219
Amoebosporidia, 215
Amoebosporidies, 215
Amoebosporidiidae, 215
Amorphoctista, 37
Amorphozoa, 37
Amphibia, 220
Amphiacantha, 219
Amphiamblys, 219
Amphidinium, 101
Amphilonche, 197
Amphilothida, 103
Amphimonadaceae, 61, 158
Amphimonadidae, 61
Amphisolenia, 103
Amphisolenia laticincta, 104
Amphistegina, 187
Amphistomina, 191
Anabaena, 35
Anabaena circinnalis, 13
Anabaena inaequalis, 32
Ancylistales, 81
Ancylistes, 125
Ancylistinaeae, 81
Anemeae, 171
Angiococcus, 28
Angiogastres, 152
Angiospermeae, 82, 91
Animacule, 18
Animal kingdom, Animalia, Animals, 1, 2, 4, 6, 10, 68, 95, 111, 113, 159, 163, 167, 206, 214, 220, 223, 231, 233, 235
Anisochytridiales, 69
Anisochytrids, 57
Anisolpidiaceae, 69
Anisoplidium, 69
Anisolpidium Ectocarpii, 70
Anisonema, 109
Anisonema truncatun, 108
Anisonemida, 105, 108
Anisonemidae, 108
Anisonemina, 108
Anomalinidae, 187
Anopheles, 213
Anoplophryida, 230
Anoplophryinea, 229, 230
Anthophysis, 59
Anucleobionta, 6, 12
Ape, 213
Aphanizomenon, 35
Aphanomycopsis, 81
Aphrothoraca, 190, 193
Aphrothoracida, 190
Aplanosporeae, 86
Aplosporidies, 218
Apodachlya, 79
Apodachlyella, 79
Apodinidae, 102
Apodinium, 102
Appendiculatae, 73

Apple, 139, 148
Araceae, 67
Arachnula, 191
Araiospora, 79
Arcella, 205
Arcellidae, 205
Arcellina, 205
Archaelagena, 186
Archaias, 184, 185
Archangiacea, 28
Archangiaceae, 28
Archangium, 28
Archegregarina, 215
Archephyta, 17
Archezoa (of Haeckel), 17
Archezoa (of Perty), 223
Archi-Monothalamia, 183
Archimycetae, 110
Archimycetes, 110, 111
Archiplastidea, 18, 29
Archiplastideae, 30
Arcyria, 176
Arcyriacea, 174, 176
Arcyriaceae, 176
Arcyriidae, 176
Arthropods, 211, 212, 222
Arthrospira, 35
Asclepiadaceae, 161
Ascobolacea, 135
Ascochyta, 141
Ascocorticium, 137, 145
Ascocyclus, 88
Ascoidea, 130
Ascoidea rubescens, 127
Ascoideaceae, 130
Ascomycetae, 125
Ascomycetes, 120, 125, 140, 142, 145
Ascomyceten, 125
Ascosporeae, 125
Askeleta, 193
Aspergillus, 130, 131
Aspergilliales, 130
Aspidisca, 232
Aspirotricha, 229
Aspirotrichaceae, 229
Astasia, 96, 107
Astasiaceae, 107
Astasiaea, 96, 105, 107
Astasiidae, 107
Astasiina, 107
Asterigerina, 187
Asterigerinida, 187
Asterigerinidae, 187
Asterocyclina, 188
Asterocystis, 43
Asterophlyctis, 117
Astoma, 94, 96, 105
Astomaticae, 74
Astomina, 230
Astracanthida, 199
Astracanthidae, 199
Astrodisculus, 193

Astrolophida, 197
Astrolophus, 197
Astrorhiza, 183
Astrorhizida, 183
Astrorhizidaceae, 183
Astrorhizidae, 183
Astrorhizidea, 183
Astrorhizina, 183
Ataxophragmidae, 186
Ataxophragmidea, 186
Ataxophragmium, 186
Athene noctua, 162
Aulacantha, 199
Aulacanthida, 199
Aulacanthidae, 199
Aulactinium, 199
Aulosphaera, 199
Aulosphaerida, 199
Auricularia, 146
Auricularia Auricula, 146
Auriculariacea, 146, 148
Auriculariaceae, 146
Auriculariales, 146
Auriculariineae, 145, 146
Auricularineae, 146
Autobasidiomycetes, 146
Aves, 6
Axonoblasteae, 51
Azoosporidae, 191
Azoosporidea, 191
Azoosporidia, 190
Azotobacter, 14
Azotobacter Chroococcum, 23

Azotobacteriacea, 19, 23
Azotobacteriaceae, 23
Babesia, 214
Babesia bigemina, 206, 212, 214
Babesiida, 211, 214
Babesiidae, 214
Bacillacea, 19, 21
Bacillacei, 21
Bacillaria, 69, 75
Bacillariacea, 11, 55, 65, 69, 72
Bacillariaceae, 71
Bacillariales, 53, 71
Bacillarieae, 71
Bacillarioideae, 71
Bacillariophyceae, 71
Bacillariophyta, 71
Bacillus, 21
Bacillus alvei, 21
Bacillus Amylobacter, 21
Bacillus anthracis, 21
Bacillus, colon, 22
Bacillus, gas, 22
Bacillus Radicicola, 23
Bacillus, Shiga, 22
Bacillus subtilis, 18, 21
Bacteria, 2, 3, 4, 6, 7, 12, 13, 14, 17, 18,
  30, 38, 118, 119, 189, 222, 224, 231

Bacteriaceae, 21
Bacteriophyta, 17
Bacteroides, 22
Badhamia, 177
Balantidium, 230
Balantidium coli, 230
Bangia, 43
Bangia fuscopurpurea, 43
Bangiacea, 41
Bangiaceae, 41, 43
Bangialea, 40, 41, 52
Bangiales, 41
Bangieae, 41
Bangiineae, 41
Bangioideae, 41
Barbulanympha, 169
Barley, 6
Barrouxia, 210
Bartonella bacilliformis, 21, 214
Bartonellaceae, 20
Bartramia, 219
Bartramiidae, 218
Basidiobolacea, 125
Basidiobolaceae, 125
Basidiobolus, 119, 121
Basidiobolus ranarum, 125
Basidiomycetae, 142
Basidiomyceten, 142
Basidiomycetes, 121, 127, 128, 141, 142, 145
Basidiosporeae, 142
Bathysiphon, 183
Batrachospermaceae, 47
Batrachospermum, 47
Bdellospora, 124
Beetles, 177, 215, 217
Beggiatoa, 24, 30, 31, 32, 35
Beggiatoacea, 34, 35
Beggiatoaceae, 35
Bicoecaceae, 67
Bicoecidea, 67
Bicoekida, 67
Bicosoeca, 67
Biddulphia, 74
Biddulphiaceae, 74
Biddulphiea, 74
Biddulphieae, 74
Biflagellatae, 76
Bikoecidae, 67
Bikoecina, 67
Birds, 6, 210, 212, 213
Bitunicatae, 129
Blakesleea, 124
Blastocaulis, 26, 27
Blastocladia, 112, 113
Blastocladiacea, 110, 112
Blastocladiaceae, 112
Blastocladiales, 111
Blastocladiella, 112, 113
Blastocladiella cystogena, 115
Blastocladiineae, 111
Blastoderma, 130

Blastodinida, 100, 102
Blastodinidae, 102
Blastodinides, 102
Blastodinium, 102
Blastosporaceae, 44
Blepharisma, 230
Blue grass, 148
Blue-green algae, see Algae, Blue-green
Bodo, 159, 160, 199, 209, 212
Bodo edax, 161
Bodo Lacertae, 159
Bodonaceae, 159
Bodonidae, 159
Bodonidea, 158
Bodonina, 159
Boletus, 151
Bolivina, 188
Borelis, 185
Borrelia, 29
Borrelia recurrentis, 28, 29
Borrelia Vincenti, 29
Botrida, 198
Botrydiaceae, 67
Botrydiales, 63
Botrydiopsis, 66
Botrydium, 65, 67
Botryococcacea, 65, 66
Botryococcaceae, 66
Botryococcus, 66
Botryoglossum, 52
Botryoidea, 198
Botrytis, 140, 142
Bovista, 155
Braadrudosphaeridae, 60
Brachycystida, 217
Brefeldia, 175
Brefeldiaceae, 175
Brefeldiidae, 175
Brehmiella, 59
Brehmiella chrysohydra, 54
Brown algae, see Algae, Brown
Brucella, 22
Bulgariacea, 135
Bulimina, 188
Buliminida, 188
Buliminina, 188
Bumilleria, 66, 73
Bursaria, 230
Bursariidae, 230
Bursarina, 230

Cabbage, 178
Calcareae, 171
Calcarina, 187
Calcarinidae, 187
Calciconus, 60
Calciconus vitreus, 56
Calcisolenia, 60
Calcisolenidae, 60
Callocolax, 50
Callophyllis, 50

Calonectria, 142
Calonema, 177
Calonemeae, 171
Calonympha, 168
Calonymphida, 166, 167, 168
Calonymphidae, 168
Calothrix, 36
Calvatia, 155
Calyptosphaera, 60
Calyptosphaera insignis, 56
Camerina, 188
Camerinidae, 188
Camptonema, 193
Camptonematidae, 193
Campuscus, 191
Candida, 142
Cannobotryida, 198
Cannopilus, 63
Cannosphaerida, 199
Cannosphaeridae, 199
Cantharellales, 150
Carageen, 49
Carboxidomonas, 24
Carchesium, 233
Carchesium polypinum, 225
Carpomitra, 88
Carpomycetae, 119
Carpophyceae, 40
Carposporeen, 128
Caryococcus, 21
Caryospora, 210
Caryotropha, 211
Cassidulina, 188
Cassidulinida, 188
Cassidulinidae, 188
Castanellida, 200
Castanellidae, 200
Castanidium, 200
Cat, 6, 210
Catenariopsis, 69
Catenochytridium, 118
Cattle, 206, 214
Caulleryella, 215
Caulleryellidae, 215
Caulobacter, 26, 27
Caulobacter vibrioides, 26
Caulobacteriacea, 27
Caulobacteriaceae, 27
Caulobacterialea, 18, 25, 26
Caulobacteriales, 25
Cayeuxina, 186
Cellulomonas, 22
Cenolarcus, 195
Centipedes, 207, 210, 211
Centricae, 73, 74
Cepedia, 229
Cephalina, 217
Céphalopodes, 182
Cephalothamnium, 59
Cephalothamnium Cyclopum, 54
Cephalotrichinae, 18
Ceramiales, 51

Ceramiea, 51, 52
Ceramieae, 51
Ceratiidae, 103
Ceratiomyxa, 177, 221
Ceratiomyxa fruticulosa, 177, 178
Ceratiomyxacea, 177
Ceratiomyxaceae, 177
Ceratium (dinoflagellate), 103
Ceratium (myxomycete), 177
Ceratium Hirundinella, 103
Ceratomyxa, 221
Ceratomyxidae, 221
Ceratophyllus fasciatus, 160
Ceratospora, 216
Cercobodo, 159
Cercobodonidae, 159
Cercomonadida, 159
Cercomonadidae, 159
Cercomonadinea, 158
Cercomonas, 159, 161
Cercomonas Davainei, 165
Cercomonas Hominis, 165
Cercomonas longicauda, 160
Cercomonas obliqua, 165
Cercospora, 138, 139, 142
Chaenia, 229
Chaetangieae, 47
Chaetoceraceae, 74
Chaetoceros, 74
Chaetocladiaceae, 124
Chaetocladium, 123, 124
Chaetoproteida, 159, 163
Chaetoproteidae, 163
Chaetoproteus, 158, 160, 163, 202
Chaidae, 201
Chaidea, 201
Chalarothoraca, 190, 193
Chalarothoracida, 190
Challengerida, 200
Challengeridae, 200
Challengeron, 200
Chamaesiphon, 35, 36
Chamaesiphon incrustans, 32
Chamaesiphonacea, 34, 35
Chamaesiphonaceae, 33, 35
Champia, 51
Champiea, 51
Champieae, 51
Chantransia, 47
Chantransiaceae, 47
Chaos Protheus, 200, 201, 202
Chaosidae, 201
Chapmania, 187
Chapmaniida, 187
Chapmaniidae, 187
Characiopsis, 66
Characiopsis gibba, 64
Chestnut, 139
Chiastolida, 197
Chiastolus, 197
Chicken, 210
Chilodon, 230

Chilodon uncinatus, 225
Chilodontida, 230
Chilomastigidae, 165
Chilomastix, 165
Chilomastix davainei, 165
Chilomastix Hominis, 165, 237
Chilomastix Mesnili, 165
Chilomonadaceae, 98
Chilomonas, 94, 109
Chilomonas Paramaecium, 97
Chilostomella, 187
Chilostomellida, 187
Chilostomellidae, 187
Chlamydodon, 230
Chlamydodonta, 230
Chlamydodontida, 230
Chlamydodontidae, 230
Chlamydomonas, 61, 111
Chlamydomyxa, 191
Chlamydomyxidea, 190
Chlamydophora, 190, 193
Chlamydophorida, 190
Chlamydothrix ochracea, 32, 36
Chlamydotrichacea, 34
Chlamydotrichaceae, 36
Chlamydozoaceae, 20
Chloramoeba, 66
Chloramoeba heteromorpha, 64
Chloramoebacea, 65, 66
Chloramoebaceae, 66
Chloramoebidae, 66
Chlorarachnidae, 66
Chlorobacteriaceae, 31
Chlorobacterium, 33
Chlorobiacea, 31, 237
Chlorobium, 31
Chlorobotrydiaceae, 66
Chlorochromonas, 66
Chlorochytridion, 111
Chloromonadaceae, 109
Chloromonadales, 63, 105
Chloromonadida, 105
Chloromonadidae, 109
Chloromonadina, 63, 96, 105
Chloromonadinae, 94, 105
Chloromonadineae, 105
Chloromonads, 94
Chloromyxea, 221
Chloromyxées, 221
Chloromyxida, 221
Chloromyxidae, 221
Chloromyxum, 221
Chlorosaccacea, 65
Chlorosaccaceae, 65
Chlorosaccus, 55, 65, 66
Chlorosaccus fluidus, 64
Chlorotheciacea, 65, 66
Chlorotheciaceae, 66
Choanephoraceae, 124
Choano-Flagellata, 67
Choanocystidae, 194
Choanocystis, 194, 216

Choanoflagellata, 57, 61, 67, 68
Choanoflagellates, 57, 158
Choanosporidae, 216
Chondria, 52
Chondrieae, 51
Chondrioderma, 177
Chondrococcus, 28
Chondromyces, 28
Chondromyces aurantiacus, 26
Chondromyces crocatus, 26
Chondrus, 51
Chondrus crispus, 49
Chonotricha, 230, 231, 237
Chonotrichida, 230
Chordariacea, 88
Chordariaceae, 87
Chordariales, 87
Chordarieae, 87
Chromatiacea, 31, 237
Chromatiaceae, 31
Chromatium, 31
Chromobacterium, 22
Chromomonas, 98
Chromulina, 61, 62
Chromulina Pascheri, 56
Chromulinaceae, 62
Chromulinales, 61
Chromulinidae, 62
Chroococcacea, 33
Chroococcaceae, 33
Chroococcales, 33
Chroococcus, 32, 33
Chrysamoeba, 63
Chrysamoebida, 62, 63
Chrysamoebidae, 63
Chrysapsis, 62
Chrysarachniaceae, 63
Chrysarachnion, 63
Chrysidella, 98
Chrysocapsa, 59
Chrysocapsa paludosa, 54
Chrysocapsacea, 58, 59
Chrysocapsaceae, 59
Chrysocapsales, 61
Chrysocapsidae, 59
Chrysocapsina, 61
Chrysocapsinae, 61
Chrysocapsineae, 55, 61
Chrysochromulina, 58
Chrysococcus, 62
Chrysocrinus, 63
Chrysodendron, 59
Chrysomonadaceae, 59
Chrysomonadales, 61
Chrysomonadida, 61
Chrysomonadidae, 62
Chrysomonadina, 59, 61, 62
Chrysomonadinae, 61
Chrysomonadinea, 57
Chrysomonadineae, 55, 57, 61
Chrysomonads, 53, 83
Chrysomonas, 62

Chrysophaeum, 109
Chrysophyceae, 53, 55, 95
Chrysophycophyta, 53
Chrysophyta, 53
Chrysopyxis, 60
Chrysosphaera, 62
Chrysosphaeracea, 61, 62
Chrysosphaeraceae, 62
Chrysosphaerales, 61
Chrysosphaerella, 62
Chrysosphaerineae, 55, 61
Chrysospora, 62
Chrysothylakion, 63
Chrysotrichaceae, 60
Chrysotrichales, 61
Chrysotrichineae, 55, 61
Chytridiacea, 117, 118
Chytridiaceae, 110, 118
Chytridiales, 113
Chytridieae, 110
Chytridieen, 110, 118
Chytridiineae, 110
Chytridinae, 110
Chytridinea, 111, 113, 116
Chytridineae, 110, 113
Chytridium, 69, 110, 113, 118
Chytridium Olla, 110
Chytrids, 76, 110, 111, 119, 121, 125, 130,
    178
Chytriodinium, 102
Cienkowskiaceae, 177
Ciliata, 223, 228, 233
Ciliatea, 228
Cilio-flagellata, 94
Cilioflagellata, 96, 102
Ciliophora, 39, 223, 237
Ciliophryidae, 191
Ciliophrys, 193
Circoporida, 200
Circoporidae, 200
Circoporus, 200
Cladochytriacea, 115, 117
Cladochytriaceae, 117
Cladochytrium, 110, 117
Cladococcida, 195
Cladococcus, 195
Cladopyxida, 103
Cladosporium, 142
Cladothrix dichotoma, 33
Clastoderma, 175
Clathracea, 155
Clathraceae, 155
Clathrochloris, 31
Clathrulina, 194
Clathrulinida, 191, 194
Clathrulinidae, 194
Claudea, 51
Clavaria, 151
Clavariacea, 151
Clavariaceae, 151
Clavariei, 151
Clavati, 150

Claviceps purpurea, 139
Clonothrix fusca, 32, 36
Closterium, 125
Clostridium, 21
Clostridium botulinum, 21
Clostridium butyricum, 21
Clostridium Pastorianum, 21
Clostridium septicum, 21
Clostridium tetani, 21
Cnemidospora, 217
Cnidosporidea, 219
Cnidosporidia, 219, 220
Coccaceae, 20
Coccidia, 207, 210
Coccidians, 260, 209, 210, 212, 215
Coccididae, 210
Coccidiidea, 210
Coccidiomorpha, 207, 210
Coccidium, 210
Coccidium Schubergi, 207
Coccogonales, 33
Coccogonea, 31, 32, 33
Coccogoneae, 33
Coccolithaceae, 60
Coccolithidae, 60
Cocclithina, 60
Coccolithophora, 60
Coccolithophoridae, 55, 60
Coccolithus, 60
Coccomyces, 134
Coccomyxa, 221
Coccomyxida, 221
Coccomyxidae, 221
Cocconeidaceae, 76
Cocconeis, 72, 73, 76
Cocconemaceae, 75
Cocconemidae, 222
Coccosphaera, 60
Coccospora Slavinae, 222
Coccosporida, 222
Coccosporidae, 222
Coccus, 20
Cochliodinium, 101
Cochliopodiidae, 202
Cochliopodium, 202
Cochlonema, 124
Cockroach, 169, 217, 219
Codonoecina, 67
Codonosiga, 67
Codonosigidae, 67
Codosiga, 67
Coeloblastea, 46
Coeloblasteae, 51
Coelodendrida, 200
Coelodendrum, 199, 200
Coelomonadina, 105, 109
Coelosphaerium, 33
Coelosporidiidae, 218
Coelosporidium, 219
Coenenia, 203
Coffee, 148
Colaciacea, 105

Colaciaceae, 105
Colaciidae, 105
Colacium, 105
Colacium Arbuscula, 106
Coleosporiacea, 148
Coleosporiaceae, 148
Coleosporium, 143
Coleosporium Vernoniae, 143
Colepina, 229
Coleps, 229
Colletotrichum, 139, 140
Collida, 195
Collodaria, 194
Colloderma, 177
Collodermataceae, 177
Collosphaera, 195
Collosphaera Huxleyi, 196
Collosphaerida, 195
Collozoida, 195
Colpidium campylum, 227
Colpoda, 229
Colpodaea, 229
Colpodella, 189
Colpodidae, 229
Columniferae, 171
Comatricha, 175
Completoria, 125
Compsopogon, 44
Compsopogonacea, 41, 44
Compsopogonaceae, 44
Concharida, 200
Concharidae, 200
Conchulina, 205
Conferva, 66
Confervaceae, 66
Confervales, 63
Confervoidea, 63
Conger niger, 161
Conidiobolus, 125
Coniferinae, 9
Conifers, 148
Coniomycetes, 140
Conjugatae, 117
Conradiella, 62
Coprinus, 143, 152
Coprinus atramentarius, 153
Copromonas subtilis, 108
Cora, 151
Corallinacea, 50
Corallinea, 50
Corallineae, 50
Cordyceps, 139
Coreocolax, 50
Corethron, 74
Cormobionta, 6
Cornuspira, 185
Coronympha, 168
Corticium, 151
Corynebacteriacea, 19, 20
Corynebacteriaceae, 20
Corynebacteriidae, 20
Corynebacterium, 20, 21

Corynebacterium diphtheriae, 20
Coryneum, 141
Coryneum Beijerinckii, 141
Coscinodiscaceae, 74
Coscinodiscea, 74
Coscinodiscus, 74
Costia, 165
Costiidae, 165
Cothurnia, 233
Councilmania, 203
Crab, 218
Craigia, 163
Craspedomonadaceae, 67
Craspedomonadina, 67
Craspedotella, 102
Craterellus, 151
Craterium, 177
Crenothrix polyspora, 32, 36
Crenotrichacea, 35, 36
Crenotrichaceae, 36
Cribraria, 175
Cribrariacea, 173, 175
Cribrariaceae, 171, 175
Cribrariales, 171, 173
Cribrariidae, 175
Cribrospira, 186
Cristellaria, 187
Cristispira, 29
Cristispira Veneris, 26
Crithidia, 162
Cromodromys, 199
Cronartiacea, 148
Cronartiaceae, 148
Cronartium, 148
Cronartium ribicola, 148
Cryptobia, 160, 161, 209, 212
Cryptobiidae, 161
Cryptocalpis, 198
Cryptocapsales, 97
Cryptocapsineae, 95
Cryptocercus, 166, 169, 170
Cryptochrysis, 98
Cryptococcacea, 97, 98
Cryptococcaceae, 98
Cryptococcales, 96, 97
Cryptococcineae, 95
Cryptococcus, 98, 130
Cryptocystes, 219, 222
Cryptomonadaceae, 98
Cryptomonadalea, 96
Cryptomonadales, 96
Cryptomonadida, 97
Cryptomonadidae, 98
Cryptomonadina, 96, 97, 98
Cryptomonadinae, 96
Cryptomonadineae, 95, 96
Cryptomonads, 94, 194
Cryptomonas, 97, 98, 199
Cryptonemeae, 50
Cryptonemiales, 50
Cryptoneminae, 50
Cryptophyceae, 94, 96

Cryptospermea, 46, 47
Cryptospermeae, 47
Ctenomyces, 131
Ctenostomata, 230, 231
Ctenostomida, 231
Ctenostomina, 231
Cumagloia, 47
Cuneolina, 186
Cunninghamella, 124
Cup fungi, 134
Cupulata, 129, 134, 137, 145
Cupulati, 134
Currants, 148
Cutleria, 88
Cutleriacea, 88
Cutlerialea, 85, 88
Cutleriales, 88
Cyanomonas, 98
Cyanophyceae, 29
Cyanophyta, 17, 30
Cyathomonas, 97, 98
Cyathus, 155
Cyclammina, 186
Cyclidina, 230
Cyclidium, 230
Cycloclypeidae, 188
Cycloclypeina, 188
Cycloclypeus Carpenteri, 188
Cyclodinidae, 229
Cyclonexis, 59
Cyclonympha, 171
Cyclonymphidae, 169
Cycloposthium, 231
Cyclosiphon, 188
Cyclosporales, 91
Cyclosporeae, 82, 91
Cyclotella, 72, 73, 74
Cylindrospermum, 35
Cylindrospermum majus, 32
Cylindrosporium Pruni, 134
Cymbalopora, 180, 182, 187
Cymbella, 72, 73, 75
Cymbellea, 75
Cymbelleae, 75
Cyphoderia, 191
Cyrtellaria, 198
Cyrtida, 198
Cyrtoidea, 198
Cyrtophora, 62, 63
Cystidium, 198
Cystobasidium, 147
Cystobasidium sebaceum, 145
Cystochytrium, 69
Cystoflagellata, 94, 96, 99
Cytophaga Hutchinsonii, 26, 28
Cytophagacea, 28
Cytophagaceae, 28
Cytosporidia, 207
Cyttariacea, 135

Dacryomitra, 150
Dacryomyces, 150
Dacryomycetacea, 150
Dacryomycetaceae, 150
Dacryomycetalea, 146, 150
Dacryomycetales, 150
Dacryomycetineae, 150
Dactylophorida, 218
Dactylophoridae, 218
Dactylophorus, 218
Dactylosphaerium, 202
Daedalea, 151
Daldinia, 139
Dallingeria, 58
Dasyea, 51
Daucina, 188
Deer, 214
Delacroixia, 125
Delesseria, 51
Delesseria sinuosa, 49
Delesseriea, 51
Dematiaceae, 142
Dematiea, 142
Dematieae, 142
Dematiei, 141
Dendrocometes, 236
Dendrocometida, 236
Dendrocometidae, 236
Dendrocometina, 236
Dendromonadina, 59
Dendromonas, 59
Dendromonas virgaria, 54
Dendrosoma, 235
Dendrosomatidae, 235
Dendrosomida, 235
Dendrosomidae, 235
Dendrosomina, 235
Dentilina, 184
Derepyxis, 60
Dermateacea, 135
Dermatocarpa, 146, 152
Dermatocarpi, 152, 171
Dermocarpa, 36
Dermocarpa protea, 32
Dermocentor, 20
Desmarestia, 88, 89
Desmarestiacea, 88
Desmarestales, 87
Desmobacteriales, 33
Desmocapsa, 99
Desmocapsales, 98, 99
Desmocapsineae, 95, 99
Desmokontae, 94, 98, 99
Desmomastix, 99
Desmomonadales, 98, 99
Desmomonadineae, 95, 99
Desmothoraca, 190, 194
Desmothoracida, 190
Desmotrichum, 88
Deuteromycetes, 140
Deutschlandiaceae, 60
Devescovina, 167

Devescovinida, 167
Devescovinidae, 167
Devescovininae, 167
Diachea, 175
Dianema, 176
Dianemaceae, 176
Diaporthe, 139
Diatoma, 75
Diatomaceae, 69, 75
Diatomea, 53, 69, 71, 74
Diatomeae, 53, 69, 71, 74
Diatoms, 53, 71, 83, 117, 118
Diatrype, 139
Dictydiaethaliaceae, 175
Dictydiaethaliidae, 175
Dictydiaethalium, 175
Dictydium, 175
Dictyocha, 63
Dictyocha Fibula, 56
Dictyochaceae, 62
Dictyochidae, 62
Dictyoconoides, 187
Dictyoconus, 186, 198
Dictyophora, 155
Dictyosiphonales, 89, 91
Dictyosteliaceae, 203
Dictyosteliaceen, 203
Dictyostelidae, 203
Dictyostelium, 203
Dictyostelium discoideum, 204
Dictyostelium mucoroides, 204
Dictyota, 87
Dictyotacea, 87
Dictyotaceae, 86, 87
Dictyotales, 82, 86
Dictyotea, 85, 86
Dictyoteae, 82, 86
Dictyuchus, 78, 79
Didesmis, 231
Didiniidae, 229
Didinium, 229, 235
Didinium nasutum, 225
Didymiacea, 175, 177
Didymiaceae, 177
Didymidae, 177
Didymiidae, 177
Didymium, 177
Didymohelix ferruginea, 27
Didymophyes, 218
Didymophyida, 218
Didymophyidae, 218
Difflugia, 201, 205
Difflugiida, 205
Difflugiidae, 205
Dileptus, 230
Dimastigamoeba, 159
Dimorpha, 193
Dimychota, 17
Dinamoeba (dinoflagellate), 101
Dinamoeba (amoeba), 16, 202
Dinamoebidina, 100, 101, 237
Dinamoebidium varians, 101, 104

Dinastridium, 100
Dinenympha, 166
Dinenymphida, 165, 166
Dinenymphidae, 166
Dinifera, 102
Diniferidea, 103
Dinobryaceae, 60
Dinobryina, 58, 60
Dinobryon, 58, 60
Dinocapsaceae, 100
Dinocapsales, 99, 100
Dinocapsina, 99
Dinocapsineae, 95, 99
Dinococcales, 99, 100
Dinococcina, 99
Dinococcineae, 96, 100
Dinoclonium, 100
Dinocloniaceae, 100
Dinoflagellata, 94, 102
Dinoflagellatae, 94, 95
Dinoflagellates, 94, 199
Dinoflagellida, 103
Dinophyceae, 94, 103
Dinophysida, 103
Dinophysis, 103
Dinothrix, 100
Dinotrichales, 99, 100
Dinotrichineae, 96, 99
Dioxys, 66
Dioxys Incus, 64
Diplococcus, 20
Diplococcus pneumoniae, 20
Diploconida, 197
Diploconus, 197
Diplocystida, 216
Diplocystidae, 216
Diplocystis, 216
Diplodia, 141
Diplodinium, 224, 231
Diplomita, 60
Diplophlyctis, 117
Diplophysalis, 191
Diplophysalis stagnalis, 192
Dipodascus, 130
Dipodascus albidus, 132
Discellacea, 141
Discellaceae, 141
Dischizae, 215
Discida, 195
Disciformia, 73
Discoasteridae, 60
Discoidea, 195
Discolichenes, 134
Discomycetes, 133, 134
Discophrya, 235
Discophryida, 235
Discophryidae, 235
Discorbis, 180, 182
Discorbis mediterranensis, 182
Discorbis orbicularis, 182
Discosphaera, 60
Disporées, 209

Distephanus, 63
Distephanus Speculum, 56
Distigma, 107
Distomata, 163
Distomataceae, 166
Distomatinales, 163
Distomatineae, 163
Ditripodiidae, 62
Doassansia, 149
Dobellia binucleata, 210
Dobeliida, 210
Dobelliidae, 210
Dog, 210
Dolichocystida, 209, 214
Doliocystida, 216
Doliocystidae, 216
Doliocystis, 216
Dorataspida, 197
Dorataspis, 197
Dorataspis costata, 196
Dothideaceae, 137
Dothideales, 137, 138, 139, 140, 141
Drepanidium, 211
Duboscqia, 222
Dudresnaya purpurifera, 49
Dumontieae, 50

Earth star, 155
Earthworm, 215, 216
Eberthella, 22
Eberthella typhi, 22
Ebriaceae, 55, 62
Ebriidae, 62
Ebriopsidae, 62
Echinocystida, 189
Echinoderms, 216
Echinosteliaceae, 175
Echinostelium, 175
Ectocarpales, 86
Ectocarpea, 86
Ectocarpeae, 86
Ectocarpineae, 86
Ectocarpus, 70, 83, 86, 87
Ectocarpus Mitchelliae, 204
Ectocarpus siliculosus, 83
Ectosporeae, 177
Ectrogella, 81
Ectrogellacea, 81
Ectrogellaceae, 81
Eel, 161
Egregia Menziesii, 90, 91
Eimeria, 210
Eimerida, 210
Eimeridae, 210
Eimeridea, 210
Eimeriidea, 210
Eimeriinea, 210
Eimerioidae, 210
Elaeorhanis, 193
Elaphomyces, 131
Ellipsoidina, 188

Elphidium, 186, 187
Elphidium crispum, 181
Elvella, 135
Empusa, 125
Enchelia, 229
Enchelina, 229
Enchelis, 229
Enchelyidae, 229
Endamoeba, 202, 203
Endamoeba disparita, 202
Endamoeba histolytica, 202
Endamoebida, 201, 202
Endamoebidae, 202
Endocochlus, 124
Endogonacea, 123, 124
Endogonaceae, 124
Endogone, 123, 124
Endogonei, 124
Endolimax, 203
Endomyces, 130
Endomycetacea, 130
Endomycetaceae, 130
Endomycetalea, 130
Endomycetales, 129
Endosporea, 171
Endosporeae, 171
Endosporinei, 171
Endothia parasitica, 139
Endothyra, 186
Endothyridae, 186
Endothyrina, 186
Enerthenema, 175
Enerthenemaceae, 175
Enerthenemea, 174, 175
Entamoeba, 202, 203
Entamoeba coli, 202
Entamoeba dystenteriae, 202
Entamoeba gingivalis, 202
Enteridiea, 171
Enteridieae, 171
Enterobacteriaceae, 21
Entodiniomorpha, 230, 231
Entodiniomorphina, 231
Entodinium, 231
Entomophthora, 125
Entomophthoracea, 124
Entomophthoraceae, 124
Entomophthorales, 124
Entomophthorinea, 121, 124
Entomophthorineae, 124
Entophlycis, 113, 117
Entophysalidales, 33
Entosiphon sulcatum, 108
Eocronartium, 143, 147
Eocronartium muscicola, 145
Eouvigerina, 188
Ephelota, 236
Ephelotida, 236
Ephelotidae, 236
Ephelotina, 236
Ephemera vulgata, 222
Epiblasteae, 50

Epichrysis, 56, 62
Epiclintes, 233
Epidinium, 231
Epipyxis, 60
Epipyxis utriculus, 54
Epistylis, 233
Eremascus, 130
Eremascus albidus, 127
Eremospermeae, 77
Erica, 9
Ericae, 9
Erysiphe, 127, 132, 133
Erysiphe graminis, 132
Erysiphea, 133
Erysipheae, 133
Erythrocladia, 44
Erythropsis, 101
Erythrotrichia, 44
Erythrotrichia carnea, 44
Erythrotrichiaceae, 44
Erwina, 22
Erwinia amylovora, 22
Escherichia coli, 14, 15, 22
Ethmosphaerida, 195
Euactinomyxidae, 222
Euasci, 130
Eubacteria, 18, 25
Eubacteriales, 18
Eubasidii, 145
Euchrysomonadina, 61
Euchrysomonadinae, 61
Eucomonympha, 169
Eucyrtidina, 198
Eucyrtidium, 198
Eucyrtidium carinatum, 196
Eudesme, 88
Euflorideae, 44
Euglena, 38, 94, 107, 116, 117, 125
Euglena acus, 106
Euglena Spirogyra, 106, 107
Euglena viridis, 106
Euglenaceae, 105
Englenales, 105
Euglenamorpha, 105
Euglenida, 105
Euglenids, 94, 106
Euglenina, 105
Eugleninae, 94, 105
Euglenineae, 96, 105
Euglenocapsineae, 96
Euglenoidina, 96, 105
Euglenophycophyta, 94
Euglenophyta, 94
Euglypha, 191
Euglyphida, 191
Euglyphidae, 191
Eugregarinaria, 217
Eugregarinida, 217
Eumycetes, 119
Eumycetozoina, 171
Eumycophyta, 119
Eunotia, 75

Eunotiaceae, 75
Eunotiea, 75
Eunotieae, 75
Euphorbiaceae, 161
Euplotes, 227, 232, 233, 234
Euplotes Patella, 225, 226, 232
Eupodiscales, 73
Eurotium, 131
Eurychasma, 81
Eurychasmidium, 81
Eurysporea, 221
Eutreptia, 105
Excipula, 141
Excipulaceae, 141
Exidia, 143
Exoascalea, 129, 137
Exoascales, 137
Exoàscus, 137
Exobasidiacea, 151
Exobasidiaceae, 151
Exobasidiales, 150
Exobasidiineae, 150
Exobasidium, 151
Exosporea, 171, 177
Exosporeae, 177
Exosporinei, 177
Exuviaella, 99

Fasciolites, 185
Fauchea, 51
Faucheocolax, 51
Felis Catus, 6
Ferns, 125, 148
Filicineae, 1
Fisherinidae, 185
Fishes, 165, 210, 211, 219, 220, 222
Flabellina, 184, 187
Flagellata, 6, 55, 94, 96, 105
Flagellatae, 94
Flagellates, 10, 53, 55, 76, 94, 118
Flagellato-Eustomata, 105
Flagellato-Pantostomata, 158
Flatworms, 216
Flavobacterium, 22
Flea, 160
Flexostylida, 185
Floridea, 47, 50, 51
Florideae, 6, 40, 44, 51
Floridées, 40, 51
Floridineae, 44
Flowers of tan, 177
Fly, 213
Foaina, 167
Folliculina, 230
Fomes, 151
Foraminifera, 179, 182, 183, 185
Foraminifères, 179, 182
Foraminiferida, 179
Forficule, 217
Fragilaria, 75
Fragilariaceae, 75

Fragilariea, 75
Fragilarieae, 75
Frogs, 125, 210, 211
Frondicularia, 187
Fucaceae, 91
Fucales, 91
Fucea, 91
Fuceae, 91
Fucineae, 91
Fucacées, 82
Fucoidea, 83, 86, 91
Fucoideae, 53, 82
Fucus, 53, 91, 93
Fucus vesciculosus, 93
Fuligo septica, 177
Fungi, 39, 69, 76, 110, 119, 146, 150, 172
Fungi, bird's-nest, 155
Fungi imperfecti, 140
Fungilli, 39, 206
Furcellariea, 46, 50
Furcellarieae, 50
Fusarium, 142
Fusiformis, 29
Fusobacterium, 29
Fusulina, 188
Fusulinida, 188
Fusulinidae, 188

Galaxaura, 47
Galera tenera, 153
Gallionella, 27
Gallowaya, 148
Gammarus, 233
Gamocystis, 217
Ganymedes, 216
Ganymedida, 216
Ganymedidae, 216
Gasteromycetes, 152
Gastrobionta, 6
Gastrocarpeae, 50
Geaster, 155
Gelidiaceae, 49
Gelidialea, 46, 49, 50
Gelidiales, 49
Gelidieae, 49
Gelidium, 50, 51
Geophonus, 186, 187
Geoglossacea, 135
Giardia, 163, 166
Giardia enterica, 164, 166
Giardia Lamblia, 166
Gibberella, 142
Gigantomonas, 167
Gigartina mammilosa, 49
Gigartinales, 47
Gigartineae, 47
Gigartininae, 47
Glandulina, 187
Glaucocystis, 33
Glaucoma, 229
Glaucoma pyriformis, 227

Glenodinium, 94, 103
Globigerina, 184, 188
Globigerinida, 188
Globigerinidea, 183, 187
Globorotalia, 187
Globorotaliidae, 187
Gloeocapsa, 33
Gloeochaete, 33
Gloeochrysis, 62
Gloeodiniaceae, 100
Gloeodinium, 100
Gloeosporium, 139, 140, 141
Gloeotrichia, 36
Gloiophycea, 31, 32, 33
Gloiophyceae, 29, 33
Glomerella, 126, 127, 139, 140
Glomerella cingulata, 139
Glugea, 222
Glugeida, 222
Glugeidae, 222
Glugeidea, 222
Glugeidées, 222
Goat, 210
Gomphonema, 72, 75
Gomphonemaceae, 75
Gomphonemea, 75
Gomphonemeae, 75
Gomphosphaeria, 33
Gonapodiaceae, 112
Gonapodiineae, 112
Gonapodya, 112
Goniaulax, 103
Gonimophyllum, 52
Goniodoma, 103
Goniostomum, 109
Goniotrichaceae, 43
Goniotrichopsis, 43
Goniotrichum, 43
Gonococcus, 20
Gonospora, 216
Gooseberries, 146
Goussia, 209, 210
Goussia Schubergi, 207, 208, 237
Gracilaria, 49
Grains, 149
Granuloreticulosa, 179
Graphidiacea, 134
Graphidiaceae, 134
Graphidiales, 133
Grasses, 149
Green algae, see Algae, Green
Gregarina, 206, 217
Gregarina conica, 217
Gregarina cuneata, 217
Gregarina ovata, 217
Gregarinae, 206, 216
Gregarinarien, 217
Gregarines 206, 209, 215, 219
Gregarinida, 207, 217
Gregarinidae, 217
Gregarinidia, 207
Gregarininea, 217

Gregarinoidae, 217
Gregarinoidea, 217
Gregarinomorpha, 207
Gromia, 179, 191
Gromida, 191
Guepinia, 150
Guepinia apathularia, 145
Gurleya, 222
Guttulina, 203
Guttulina sessilis, 204
Guttulinacea, 201, 203
Guttulinaceae, 203
Guttulineae, 203
Guttulineen, 203
Guttulinidae, 203
Guttulinopsis, 203
Gymnamoebae, 201
Gymnamoebida, 201
Gymnascales, 130
Gymnoascaceae, 130
Gymnoascus, 131
Gymnocraspedidae, 67
Gymnodiniacea, 99, 100
Gymnodiniaceae, 100
Gymnodiniales, 99
Gymnodinida, 100
Gymnodinidae, 100
Gymnodiniidae, 100
Gymnodinina, 99
Gymnodinioidae, 99
Gymnodinium, 100
Gymnodinium Lunula, 101, 104
Gymnodinium striatum, 104
Gymnogongrus, 49
Gymnosporangium, 143, 148
Gymnosporidae, 211
Gymnosporidiida, 209, 211
Gymnostomata, 229
Gymnostomataceae, 229
Gymnostomina, 229
Gyrodinium, 101
Gyromonas, 166
Gyrophragmium, 152
Gyrosigma, 75

Haemamoeba, 213
Haemamoeba malariae, 213
Haemamoeba vivax, 213
Haemogregarina, 211
Haemogregarinida, 211, 212
Haemogregarinidae, 211
Haemogregarinina, 211
Haemoproteidae, 212
Haemoproteus, 213
Haemoproteus Columbae, 212, 213
Haemosporidae, 211
Haemosporidia, 207, 211, 212
Haemosporidiida, 211
Haliarchnion, 49
Halicryptina, 198
Haliomma, 195

Haliomma capillaris, 196
Haliommatina, 195
Halkyardia, 187
Halopappaceae, 60
Halopappus, 60
Halosphaeraceae, 66
Halteria, 231
Halteridiida, 212
Halteridiidae, 212
Halteridium, 212
Halteriidae, 231
Halterina, 231
Hantkenina, 187
Hantkeninidae, 187
Hantschia, 75
Haploactinomyxidae, 222
Haplobacteriacei, 18
Haplocyta, 215
Haplodinium, 99
Haplospora, 87
Haplosporangium, 124
Haplosporangium lignicola, 122
Haplosporidia, 218
Haplosporidies, 218
Haplosporidiida, 218
Haplosporidiidae, 218
Haplosporidiidea, 209, 218
Haplosporidium, 218
Haplostichinae, 82
Haplozoonidae, 102
Hauerinina, 185
Hedriocystis, 194
Helicosorina, 185
Heliodiscus, 195
Heliodiscus Phacodiscus, 196
Heliolithae, 58
Helioflagellida, 189
Helioflagellidae, 191
Heliozoa, 63, 157, 189, 190, 205
Heliozoariae, 189, 190
Heliozoida, 189
Helminthocladeae, 47
Helminthosporium, 142
Helotiacea, 135
Helvellacea, 135
Helvellales, 134
Helvellineae, 134
Hemiascales, 130
Hemiasceae, 130
Hemiasci, 129
Hemiascineae, 130
Hemibasidii, 145, 149
Hemicristellaria, 187
Hemicyclomorpha, 18
Hemidinium, 101
Hemileia vastatrix, 148
Hemisphaeriaceae, 134
Hemisphaeriales, 133
Hemitrichia, 177
Hemitrichia intorta, 176
Hemophilus, 22
Henneguya, 221

Hepatozoon, 211
Herpetomonas, 161, 162
Heterocapsaceae, 65
Heterocapsales, 63
Heterocapsineae, 55, 63
Heterocarpea, 41, 44, 52
Heterocarpeae, 40, 44
Heterochlorida, 63
Heterochloridaceae, 66
Heterochloridae, 66
Heterochloridales, 63
Heterochloridea, 63
Heterochloridineae, 55, 63
Heterochromonas, 59
Heterococcales, 63
Heterococcineae, 55, 63
Heterodermaceae, 175
Heterodermeae, 171
Heterogeneratae, 82
Heterohelicida, 188
Heterohelicidae, 188
Heterohelix, 188
Heterokonta, 11, 55, 83
Heterokontae, 53, 55, 63
Heteromastigoda, 158
Heteromonadina, 59
Heteronema, 109
Heteronemidae, 108
Heterophryida, 191, 193
Heterophryidae, 193
Heterophrys, 193
Heterosiphonales, 63
Heterosiphoneae, 55, 63
Heterostegina, 188
Heterotricha, 228, 230
Heterotrichaceae, 230
Heterotrichales, 63
Heterotrichida, 230
Heterotrichina, 230
Heterotrichineae, 55, 63
Hexacontium, 195
Hexaconus, 197
Hexactinomyxon, 222
Hexamastix, 167
Hexamastix Termopsidis, 164
Hexamita, 163, 166
Hexamitidae, 166
Hirmocystis, 217
Hodotermitidae, 167
Hoferellus, 221
Holocyclomorpha, 18
Holomastigotoides, 169
Holomastigotoidida, 169
Holomastigotoididae, 169
Holophrya, 229
Holophryidae, 229
Holotricha, 228, 229
Holotrichida, 229
Homalogonata, 69
Homo sapiens, 6
Honey bees, 222
Hoplonympha, 169

Hoplonympha natator, 170
Hoplonymphida, 169
Hoplonymphidae, 169
Hoplorhynchus, 218
Hordeum vulgare, 6
Hormogonales, 34
Hormogoneae, 34
Horse, 231
Hyalobryon, 60
Hyalodiscida, 201, 202
Hyalodiscidae, 202
Hyalodiscus, 202
Hyaloklossia, 211
Hyaloria, 149
Hyalospora, 217
Hydnacea, 151
Hydnaceae, 151
Hydnangiacea, 155
Hydnangiaceae, 155
Hydnei, 151
Hydnum, 151
Hydra, 203, 233
Hydramoeba, 203
Hydrocoleum, 35
Hydrogenomonas, 24
Hydruracea, 61, 62
Hydruraceae, 62
Hydruridae, 62
Hydrurina, 62
Hydrurus, 61
Hydrurus foetidus, 56, 62
Hyella, 36
Hymenogastraceae, 155
Hymenogastrales, 152
Hymenogastrea, 155
Hymenogastrei, 155
Hymenogastrineae, 152
Hymenomonadacea, 58, 60
Hymenomonadaceae, 60
Hymenomonadidae, 60
Hymenomonas, 60
Hymenomycetales, 150
Hymenomycetes, 150
Hymenomycetineae, 150
Hymenothecii, 150
Hymenostomata, 229
Hyperammina, 183
Hyperamminidae, 183
Hypermastigida, 168
Hypermastigina, 158, 166, 168
Hyphochytriacea, 69
Hyphochytriaceae, 69, 117
Hyphochytrialea, 57, 61, 69, 70, 111
Hyphochytriales, 69
Hyphochytrium, 69, 117
Hyphochytrium catenoides, 70
Hyphomycetes, 121, 140, 141
Hypnodiniaceae, 100
Hypocreaceae, 137
Hypocreales, 137, 138, 139, 142
Hypodermia, 146, 147
Hypodermii, 147

Hypomyces, 142
Hypomyces Solani var. Cucurbitae, 126, 127
Hypotricha, 228, 232, 233
Hypotrichaceae, 233
Hypotrichida, 233
Hypotrichina, 233
Hysterangiacea, 155
Hysterangiaceae, 155
Hysteriacea, 134
Hysteriaceae, 133, 134
Hysteriales, 133, 141
Hysteriineae, 133, 134
Hysterophyta, 119

Ichthyophthirius, 229
Ichthyosporidium, 219
Imperforida, 183
Infusoires, 223
Infusoires suceurs, 235
Infusoria, 2, 37, 95, 118, 223, 228, 232, 235
Inoperculata, 135
Inophyta, 39, 119
Insects, 69, 113, 117, 118, 124, 125, 155, 159, 161, 165, 167, 216, 217, 220
Invertebrates, 161, 210, 211, 215, 216
Iodamoeba, 203
Irish moss, 49
Irpex, 151
Isoachlya, 79
Isocarpeae, 69
Isochrysidaceae, 59
Isochrysidae, 59
Isochrysidales, 57
Isogeneratae, 82
Isospora, 210

Janczewskia, 52
Jarrina, 210
Joenia, 169
Joeniidae, 169
Joeniidea, 168
Joenina, 169
Joenopsis, 169
Jola, 147
Junipers, 148

Kalotermes, 169
Kalotermitidae, 167
Kalotermitinae, 166, 168
Karyamoebina, 203
Karyolysus, 211
Kelps, 82, 83, 89, 90
Keramosphaera, 185
Keramosphaeridae, 185
Keramosphaerina, 185
Kerona, 233
Klebsiella (bacterium), 7, 22

Klebsiella pneumoniae, 22
Klebsiella (flagellate), 7
Klebsiella alligata, 106
Klossia, 211
Klossiella, 211
Kofoidia, 168, 169
Kofoidiida, 169
Kofoidiidae, 169
Kurthia, 21
Kurthiacea, 19, 21, 237

Laboulbenia, 140
Laboulbenia Guerinii, 140
Laboulbenia Rougetii, 140
Laboulbeniaceae, 140
Laboulbenialea, 129, 140
Laboulbeniales, 140
Laboulbenieae, 140
Laboulbeniineae, 140
Laboulbeniomycetes, 140
Labyrinthula, 203, 204
Labyrinthula macrocystis, 203
Labyrinthulida, 201, 203
Labyrinthulidae, 203
Lachnea scutellata, 127, 136
Lachnobolus, 176
Lacrymaria, 229
Lactobacillaceae, 20
Lactobacillus, 20
Lactobacteriaceae, 20
Lagena (oomycete), 82
Lagena (rhizopod), 82, 184, 187
Lagenaceae, 187
Lagenidae, 186
Lagenidea, 185
Lagenidiacea, 81, 82
Lagenidiaceae, 82
Lagenidialea, 76, 81, 111, 118
Lagenidiales, 81
Lagenidium, 82
Lagenina, 186
Lagenocystis, 82, 237
Lagenocystis radicicola, 82, 237
Lagynida, 191
Lagynion, 63
Lagynis, 191
Laminaria, 91
Laminaria yezoensis, 92
Laminariaceae, 89
Laminariales, 89
Laminariea, 85, 89
Laminarieae, 89
Lampoxanthium, 195
Lampramoebae, 205
Lamproderma, 175
Lamprodermaceae, 175
Lamprospora leiocarpa, 136
Lamprosporales, 171
Lankesteria, 216
Larcarida, 195
Larcoidea, 195

Latrostium, 69
Laurencia, 52
Leangium, 177
Leathesia, 88
Lecudina, 216
Lecudinidae, 216
Leeches, 161, 211
Legerella, 211
Leidyopsis, 169
Leishmania, 162
Leishmania brasiliensis, 162
Leishmania Donovani, 162
Leishmania tropica, 162
Lemanea, 47
Lemna, 69
Lenticulina, 187
Lenticulites, 187
Lentospora, 221
Lenzites, 151
Leocarpus, 177
Leocarpus fragilis, 176
Lepidoderma, 177
Lepidoderma Chailletii, 176
Lepochromulina, 62
Leptodiscida, 100, 102
Leptodiscidae, 102
Leptodiscus, 102
Leptolegnia, 79
Leptomitaceae, 79
Leptomitales, 77
Leptomitea, 77, 79
Leptomiteae, 79
Leptomitus, 79
Leptomonas, 162
Leptospira, 29
Leptospira icteroides, 29
Leptospira icterohaemorrhagiae, 29
Leptospironympha, 169
Leptostromatacea, 141
Leptostromataceae, 141
Leptotheca, 221
Leptothrix, 27
Leptothrix ochracea, 27, 36
Leptotrichacea, 27
Leptotrichaceae, 20
Leptotrichacei, 20, 27
Leptotrichia, 20
Leucocytozoidae, 212
Leucocytozoon, 213
Leuvenia, 66
Liagora tetrasporifera, 47, 49
Licea, 175
Liceacea, 173, 175
Liceaceae, 175
Liceales, 171
Liceidae, 175
Lichenes, 119
Lichens, 119, 120
Ligniera, 179
Lindbladia, 175
Lionotus, 230
Listeria, 21

Lithocampe, 198
Lithochytridina, 198
Lithocircus, 198
Lithocircus productus, 196
Lithocolla, 193
Lithocollidae, 193
Lithocyclia, 195
Lithocyclidina, 195
Lithocystis, 216
Litholophida, 197
Litholophus, 197
Lituola, 186
Lituolidaceae, 186
Lituolidae, 186
Lituolidea, 185, 186
Lituolina, 186
Liverworts, 10
Lizards, 211
Lobosa, 201
Loborhiza, 117
Lobster, 211
Loftusia, 186
Loftusiidae, 186
Loftusiina, 186
Lophomonadidae, 169
Lophomonadida, 168, 169
Lophomonadina, 168
Lophomonas, 168, 169
Loxodes, 230
Lychnaspis, 197
Lycogala, 172, 175
Lycogala epidendrum, 176
Lycogalaceae, 171, 175
Lycogactida, 174, 175
Lycogalactidae, 175
Lycogalales, 171
Lycogalopsis, 155
Lycogalopsis Solmsii, 145
Lycoperdacea, 155
Lycoperdaceae, 155
Lycoperdales, 152
Lycoperdineae, 152
Lycoperdon, 155
Lyngbya, 13, 35
Lytothecii, 152

Macrocystis pyrifera, 90, 91
Macromastix, 58
Macrotrichomonas, 167
Macrotrichomonas pulchra, 164
Maize, 6
Mallomonadidae, 62
Mallomonadinea, 61, 62
Mallomonas, 61, 62
Mallomonas roseola, 56
Mammals, 166, 210
Man, Mankind, Men, 6, 159, 165, 210, 213, 230
Margarita, 176
Margaritaceae, 176
Margaritida, 174, 176

Margaritidae, 176
Massospora, 124, 125
Mastigamoeba, 158, 163
Mastigamoeba aspera, 160
Mastigamoebidae, 163
Mastigella, 163
Mastigophora, 6, 55, 94, 95
Mastotermitidae, 167
Matthewina, 186
Mayorella, 202
Mayorellida, 201, 202
Mayorellidae, 202
Medusetta, 200
Medusettida, 200
Megachytriaceae, 118
Megachytrium, 118
Melampsora, 143
Melampsoracea, 148
Melampsoraceae, 148
Melanconiacea, 141
Melanconiaceae, 141
Melanconialea, 141
Melanconiales, 141
Melanophycea, 11, 55, 82
Melanophyceae, 82
Melanospermeae, 82
Melitangium, 28
Melosira, 72, 73, 74
Melosiraceae, 74
Melosireae, 74
Meningococcus, 20
Menoidium, 103, 107, 109
Menoidium incurvum, 108
Menospora, 218
Menosporida, 218
Menosporidae, 218
Meridiea, 75
Meridieae, 75
Meridion, 75
Meridionaceae, 75
Merismopedia, 33
Merocystis, 211
Merogregarina, 215
Merogregarinida, 215
Merogregarinidae, 215
Merolpidiaceae, 117
Meroselenidium, 215
Mesocaena, 63
Mesogloia, 88
Mesogloiacea, 88
Metachaos, 202
Metadevescovina, 167
Metaphyta, 6
Metasporeae, 117
Metazoa, 6
Metchnikovella, 219
Metchnikovellida, 219
Metchnikovellidae, 219
Methanomonas, 24
Micrococcacea, 19, 20
Micrococcaceae, 20
Micrococcus, 20

Microcoleus, 35
Microglena, 62
Micromycopsis, 117
Micropeltidacea, 134
Micropeltidaceae, 134
Microrhopalodina, 166
Microsphaera, 132, 133
Microsphaera alni, 133
Microsporidia, 222
Microsporidies, 222
Microthyriacea, 134, 141
Microthyriaceae, 134
Microthytriales, 133
Miescher's tubes, 206
Mieschersche Schläuche, 206, 214
Mikrogromia, 183
Miliola, 182, 185, 201
Milioles, 179
Miliolida, 185
Miliolidae, 185
Miliolidea, 183, 185
Miliolina, 185
Mindeniella, 79
Mischococcacea, 65, 66
Mischococcaceae, 66
Mischococcus, 66
Mites, 211
Mitraspora, 221
Mitrati, 134
Molds, 142
Molds, water, 77
Mollisiacea, 135
Monaden, 189
Monades, 59
Monadidae, 59, 60
Monadidea, 158
Monadina, 57, 58, 59, 158
Monadineae Tetraplasteae, 191
Monadineae Zoosporeae, 191
Monads, collared, 38
Monas, 38, 54, 59, 60, 158
Monas amyli, 189
Monas Okenii, 31
Monascus, 131
Monera, 6, 12
Moneres, 12, 189
Monilia, 135, 140, 142
Monilia sitophila, 139
Moniliacea, 142
Moniliaceae, 142
Moniliales, 141
Monkeys, 213
Monoblepharella, 112
Monoblepharella Taylori, 114
Monoblepharidacea, 112
Monoblepharidaceae, 112
Monoblepharidalea, 111, 114
Monoblepharidales, 111
Monoblepharideae, 110
Monoblepharidineae, 110, 111
Monoblepharis, 111, 112
Monocercomonadida, 167

Monocercomonadidae, 167
Monocercomonas, 167
Monocercomonoides, 165
Monocilia, 66
Monociliaceae, 66
Monocystid gregarines, 209
Monocystida, 216
Monocystidae, 216
Monocystidea, 209, 215
Monocystiden, 216
Monolpidiaceae, 118
Monomychota, 17
Monopylaria, 190, 196, 198
Monopylea, 198
Monopyleen, 198
Monopylida, 198
Monopylina, 198
Monoschizae, 215
Monosiga, 67, 68
Monosomatia, 179, 183
Monosporea, 210
Monosporées, 209
Monostomina, 191
Morchella, 135
Morchella conica, 136
Mortierella, 124
Mortierellacea, 123, 124
Mortierellaceae, 124
Mosquitoes, 162, 213
Moss, Irish, 49
Mosses, 10
Mouse, Mice, 211, 214
Mrazekia, 222
Mrazekiida, 222
Mrazekiidae, 222
Mucedinaceae, 142
Mucedineae, 142
Mucedines, 129, 130, 135
Mucor, 121, 123
Mucor Mucedo, 121, 123
Mucoracea, 123
Mucoraceae, 123
Mucorales, 121
Mucorina, 121, 128
Mucorineae, 121
Mucorini, 121
Mucronina, 188
Mushrooms, 145, 151
Mussels, 211, 218
Mutinus, 155
Mycetalia, 119
Mycetoideum, Regnum, 119
Mycetosporidium, 179
Mycetozoa, 119, 157, 171, 176, 203
Mycetozoen, 171, 172
Mycetozoida, 171
Mychota, 1, 4, 6, 8, 10, 12
Mycobacteriacea, 25
Mycobacteriaceae, 25
Mycobacterium, 25
Mycobacterium leprae, 25
Mycobacterium tuberculosis, 25

Mycochytridinae, 113
Mycoderma mesentericum, 24
Mycophyceae, 77
Mycophyta, 119
Mycoporacea, 139
Mycosphaerella, 139
Mycosphaerella personata, 138
Myrioblepharis, 112
Myriogloiacea, 88
Myrionema, 89
Myrionematacea, 88
Myriospora, 211
Myxidiea, 221
Myxidiées, 221
Myxidiida, 221
Myxidiidae, 221
Myxidium, 221
Myxobacter, 28
Myxobacteria, 12, 14
Myxobacteriacea, 28
Myxobacteriaceae, 27, 28
Myxobacteriales, 27
Myxobactralea, 26, 27
Myxobactrales, 27
Myxobolea, 221
Myxobolées, 221
Myxobolida, 221
Myxobolidae, 221
Myxobolus, 221
Myxoceratida, 221, 237
Myxoceros, 221, 237
Myxoceros Blennius, 220, 221, 237
Myxoceros sphaerulosa, 221, 237
Myxochloridae, 66
Myxochrysidaceae, 63
Myxochrysidae, 63
Myxochrysis, 63
Myxochytridinae, 113
Myxococcacea, 28
Myxococcaceae, 28
Myxococcus, 28
Myxococcus coralloides, 26
Myxocystoda, 99
Myxogastres, 171
Myxomycetes, 10, 157, 171, 172, 178
Myxomyceten, 172
Myxomycidium flavum, 143
Myxomycophyta, 171
Myxophyceae, 17, 29, 30
Myxophykea, 29
Myxophyta, 171
Myxoproteus, 221
Myxoschizomycetae, 27
Myxoschizomycetes, 18, 27
Myxosoma, 221
Myxosomatida, 221
Myxosomatidae, 221
Myxosporidia, 206, 219, 220
Myxothallophyta, 171
Myzocytium, 82

Naegelliella, 62
Naegelliellaceae, 62
Naegelliellidae, 62
Naegleria, 159
Najadea, 60
Nassellaria, 198
Nassellida, 198
Nassoidea, 198
Nassula, 230
Nassulidae, 230
Nautilus, 182, 186, 187
Navicula, 72, 73, 75
Naviculaceae, 75
Naviculales, 74
Naviculea, 75
Naviculeae, 75
Neactinomyxon, 222
Nebela, 205
Nebelida, 205
Nebelidae, 205
Nectria, 141, 142
Nectria cinnabarina, 139
Nectrioidaceae, 141
Nectrioideae, 141
Neisseria gonorrhoeae, 20
Neisseria intracellularis, 20
Neiseria meningitidis, 20
Neisseria Weichselbaumii, 20
Neisseriacea, 19, 20
Neisseriaceae, 20
Neisseriacées, 20
Nemalion, 47
Nemalion multifidum, 49
Nemalionales, 47
Nemalioninae, 47
Nemastomatales, 47
Nematochrysidaceae, 60
Nematochrysis, 61
Nematocystida, 219
Nematodes, 113, 118, 124
Nematothecia, 141
Nematothecii, 141
Neogregarina, 215
Neosporidia, 206, 207, 219
Nephroselmidacea, 98
Nephroselmidaceae, 98
Nephroselmidae, 98
Nephroselmis, 98
Nereocystis, 89
Nereocystis Luetkeana, 90, 91
Neurospora, 139, 140
Neurospora crassa, 127
Neusinidae, 186
Neusina, 186
Nevskia, 27
Nidularia, 155
Nidulariaceae, 155
Nidulariales, 152
Nidulariea, 155
Nidulariei, 155
Nidulariineae, 152
Nina, 217, 218

Nitrobacter, 24
Nitrobacter Winogradskyi, 24
Nitrobacteriacea, 20, 24
Nitrobacteriaceae, 24
Nitromonas, 24
Nitrosococcus, 24
Nitrosococcus nitrosus, 24
Nitrosomonas europaea, 24
Nitrosomonas javanensis, 24
Nitzschia, 75
Nitzschiacea, 75
Nitzschiaceae, 75
Noctiluca, 95, 99, 102
Noctiluca miliaris, 102
Nectiluca scintillans, 102, 104
Noctilucae, 94, 99
Noctilucida, 100, 102
Noctilucidae, 102
Nodosalida, 186
Nodosarella, 188
Nodosaria, 184, 187
Nodosarida, 186
Nodosaridae, 186
Nodosarina, 186, 188
Nodosaroum, 186
Nodosinella, 186
Nodosinellida, 186
Nodosinellidae, 186
Nonion, 184, 187
Nonionidea, 187
Nonionideae, 187
Nonionina, 187
Nosema, 222
Nosema bombycis, 206, 222
Nosematidae, 222
Nostoc, 35
Nostocacea, 34, 35
Nostocaceae, 35
Nostochineae, 33
Nowakowskiella, 118
Nowakowskiellacea, 117, 118
Nowakowskiellaceae, 118
Nubecularina, 185
Nucleophaga, 118
Nuda, 201
Nummulitaceae, 188
Nummulites, 188
Nummulitida, 188
Nummulinidae, 188
Nummulitina, 188
Nummulitinidea, 183, 185, 188

Oats, 148
Ochromonadaceae, 59, 60
Ochromonadalea, 54, 56, 57, 61, 64, 67,
    85, 165
Ochromonadales, 57
Ochromonadidae, 59
Ochromonas, 58, 59, 60
Ochromonas granularis, 54
Octomitus, 166

Octomyxa, 179
Octospora, 222
Oicomonadacea, 159, 161
Oicomonadaceae, 161
Oicomonadidae, 161
Oidium, 142
Oikomonas, 161
Oligochaet worms, 222
Oligonema, 177
Oligosporea, 209, 210
Oligotricha, 230
Oligotrichaceae, 230
Oligotrichida, 230
Oligotrichina, 230
Olpidiacea, 115, 118
Olpidiaceae, 118
Olpidopsidacea, 81
Olpidopsidaceae, 81
Olpidiopsis, 81
Olpidium, 113, 118
Olpidium Allomycetos, 116
Ommatida, 195
Onygena, 131
Oodinidae, 102
Oomycetes, 11, 53, 55, 65, 76, 78, 111,
    118, 119, 121, 125, 127, 177, 178, 179
Oosporeae, 77
Opalina, 225, 227, 229
Opalinalea, 228, 237
Opalinida, 228
Opalinidae, 228, 229
Opalinina, 229
Opalininea, 228
Opalinoea, 225, 229
Operculata, 135
Operculina, 188
Ophiocytiaceae, 66
Ophiocytium, 66
Ophiocytium parvulum, 66
Ophiotheca, 176
Ophrydium, 233
Ophryocystis, 215
Ophryocystida, 215
Ophryocystidae, 215
Ophryodendrida, 236
Ophryodendridae, 236
Ophryodendrina, 236
Ophryodendron, 236
Ophryoglena, 229
Ophryoglenidae, 229
Ophryoscolecidae, 231
Ophryoscolecids, 225
Ophryoscolecina, 231
Ophryoscolex, 231
Ophthalmidium, 184, 185
Opisthokonta, 39, 110, 121, 237
Opistokonten, 111
Orbitoides, 188
Orbitoidida, 188
Orbitoididae, 188
Orbitolina, 186
Orbitolinida, 186

Orbitolinidae, 186
Orbitolites, 185
Orbulina, 188
Orbulinida, 188
Orcadella, 175
Orcadellaceae, 175
Orcadellidae, 175
Orcheobius, 211
Orobias, 188
Ortholithinae, 58
Orthopteran, 217
Orthosporeae, 117
Oscillaria malariae, 213
Oscillatoria, 30, 35, 36
Oscillatoria Princeps, 13
Oscillatoria splendida, 32
Oscillatoriacea, 34, 35
Oscillatoriaceae, 35
Owl, 162
Ox, Oxen, 162, 231
Oxymonadida, 165, 166
Oxymonadidae, 166
Oxymonadina, 163
Oxymonas, 163, 166
Oxyphysis, 103
Oxyrrhis, 101
Oxyrrhis marina, 101
Oxytocum, 103
Oxytricha, 233
Oxytrichidae, 233
Oxytrichina, 233

Pacinia, 23
Pacinia cholerae-asiaticae, 23
Padina, 87
Palatinella, 62
Pantostomatales, 158
Pantostomatida, 158
Pantostomatineae, 158
Paradinida, 98
Pardinidae, 98
Paradinium Pouchetii, 97, 98
Paraisotricha, 231
Parajoenia, 167
Paramaecium, 223, 224, 225, 226, 227,
    229
Paramaecium Aurelia, 226, 227
Paramaecium Bursaria, 226
Paramaecium caudatum, 226
Paramaecium multimicronucleatum, 226
Parameciina, 229
Paramoeba Eilhardi, 98
Paramoebida, 98
Paramoebidae, 98
Paramoecidae, 229
Parasitella, 123
Parvobacteriaceae, 22
Pasteurella avicida, 22
Pasteurella pestis, 22
Pasteurellacea, 19, 22, 23, 237
Pasteuria, 26, 27

Patellariacea, 135
Patellina, 181, 182, 185
Patouillardina, 149
Patouillardina cinerea, 145
Pavonina, 188
Peach, 137
Pectobacterium, 23
Pectobacterium carotovorum, 23
Pedangia, 186
Pedilomonas, 111
Pedinella, 62, 63
Pegidia, 188
Pegidiida, 188
Pegidiidae, 188
Pelodictyon, 31
Pelomyxa, 202
Pelomyxa carolinensis, 200, 201
Pelomyxa palustris, 202
Peneroplidae, 185
Peneroplidea, 185
Peneroplidina, 185
Peneroplis, 181, 184, 185
Penicillium, 130, 131
Penicillium notatum, 25, 131
Pennatae, 74
Pentatrichomonas, 165
Pentatrichomonas obliqua, 164, 167, 237
Peranema, 108, 109
Peranema trichophorum, 108
Peranemaceae, 108
Peranemina, 108
Perforida, 186
Periblasteae, 47
Perichaena, 176
Perichaenacea, 174, 176
Perichaenaceae, 176
Peridinaea, 102, 103
Peridinea, 96
Peridineae, 94, 96, 103
Peridiniaceae, 103
Peridiniales, 102
Peridinidae, 103
Peridinina, 103
Peridinioidae, 103
Peridinium, 94, 103
Peridinium cinctum, 104
Perionella, 66
Peripylaria, 194
Peripylea, 194
Peripyleen, 194
Peripylida, 194
Peripylina, 194
Perisporia, 131
Perisporiacea, 129, 131
Perisporiaceae, 131
Perisporiales, 131
Peritricha, 233
Peritrichaceae, 233
Peritrichida, 233
Peritrichinae, 18
Peritromidae, 233
Peritromina, 233

Peritromus, 233
Peronospora, 81
Peronosporacea, 80, 81
Peronosporaceae, 81
Peronosporales, 80
Peronosporina, 76, 80
Peronosporinae, 80
Peronosporineae, 80
Peziza, 127, 135
Peziza domiciliana, 127
Pezizacea, 135
Pezizales, 134
Pezizineae, 134
Pestallozia, 141
Pfeifferella mallei, 22
Phacidiaceae, 133, 134
Phacidiacei, 133
Phacidialea, 129, 133, 135
Phacidiales, 133
Phacidiea, 134, 141
Phacidieae, 134
Phacidiineae, 133, 134
Phacus, 94, 106, 107
Phaenocystes, 219
Phaenocystida, 219
Phaeocapsa, 98
Phaeocapsaceae, 98
Phaeocapsales, 96
Phaeococcus, 98
Phaeoconchia, 198, 199
Phaeocystia, 198
Phaeocystina, 199
Phaeocystis, 58
Phaeocystis globosa, 54
Phaeodaria, 199
Phaeodariae, 198
Phaeodermatium, 63
Phaeogromia, 198, 199
Phaeophyceae, 53, 82, 95
Phaeophycophyta, 53
Phaeophyta, 39, 53
Phaeoplakaceae, 98
Phaeoplax, 98
Phaeosphaera, 59
Phaeosphaeria, 190, 196, 198, 199
Phaeosporales, 86
Phaeosporeae, 82, 86
Phaeothamnion, 61
Phaeothamnionacea, 58, 60
Phaeothamnionaceae, 60
Phaeozoosporea, 85, 86, 87
Phaeozoosporeae, 86
Phagomyxa, 179
Phalanasteriaceae, 67
Phalanasteriidae, 67
Phalanasterium, 67
Phalanasterium digitatum, 68
Phallaceae, 155
Phallales, 152
Phallineae, 152
Phalloidea, 155
Phalloidei, 155

Phallus, 155
Phlebotomus, 21
Phleospora, 139
Phlyctidiacea, 115, 117
Phlyctidiaceae, 117
Phlyctidium, 117
Phlyctorhiza, 117
Phoma, 141
Phomaceae, 141
Phomales, 141
Phomatacea, 141
Phomataceae, 141
Phomatalea, 141
Phomatales, 141
Phormidium, 32, 35
Phragmidium, 147, 148
Phragmidium violaceum, 147
Phycochromaceae, 29
Phycomyces, 123, 124
Phycomyces nitens, 122
Phycomyceten, 76
Phycomycetes, 76
Phycomycophyta, 76
Phyllactinia, 132, 133
Phyllactinia corylea, 127
Phyllophora, 49
Phyllosiphon, 67
Phyllosiphonacea, 67
Phyllosiphinaceae, 67
Physaraceae, 171, 177
Physarales, 171, 174
Physarea, 174, 177
Physaridae, 177
Physarum, 177
Physarum notabile, 176
Physarum polycephalum, 176
Physematium, 189, 195
Physoderma, 115, 117
Physodermataceae, 117
Physomonas, 59
Phytodiniacea, 99, 100
Phytodiniaceae, 100
Phytodinidae, 100
Phytodinium, 100
Phytomastigophorea, 55
Phytomonas (bacterium), 7, 23
Phytomonas (flagellate), 7, 161
Phytomonas Donovani, 160
Phytomyxida, 111, 171, 177
Phytomyxidae, 179
Phytomyxinae, 177
Phytomyxini, 177
Phytophthora, 80
Phytophthora infestans, 81
Phytosarcodina, 171
Phytozoidea, 94, 105
Pigeon, 212
Pileati, 150
Pileocephalus, 218
Pilobolus, 121, 124
Pinaciophora, 193
Pinacocystis, 193

Pines, 148
Pinnularia, 72, 75
Pipetta, 195
Piptocephalidacea, 123, 124
Piptocephalidaceae, 124
Piptocephalis, 123, 124
Piroplasma, 214
Pisces, 1
Plagiotomidae, 230
Plagiotomina, 230
Planopulvinulina, 187
Planorbulina, 187
Planorbulinidae, 187
Plant kingdom, Plantae, Plants, 1, 2, 4, 6,
    8, 10, 24, 38, 61, 67, 95, 113, 117, 118,
    130, 137, 148, 151, 161, 177, 179, 202
Plasmodida, 213
Plasmodidae, 213
Plasmodiida, 211
Plasmodiophora, 179
Plasmodiophora Brassicae, 178
Plasmodiophoraceae, 179
Plasmodiophorales, 177
Plasmodiophorea, 179
Plasmodiophoreae, 179
Plasmodiophoreen, 179
Plasmodiophorina, 177
Plasmodium, 212, 213
Plasmodium falciparum, 214
Plasmodium malariae, 213
Plasmodium vivax, 213
Plasmodroma, 157
Plasmopara viticola, 81
Platychrysis, 58
Platygloea, 147
Platynoblasteae, 51
Platysporea, 221
Plectascales, 130
Plectascineae, 130
Plectellaria, 198
Plectida, 198
Plectobasidiales, 152
Plectobasidiineae, 152
Plectofrondicularia, 188
Plectoidea, 198
Plectonema, 36
Plectonida, 198
Pleurage curvicolla, 128
Pleurocapsa, 36
Pleurocapsacea, 35, 36
Pleurocapsaceae, 36
Pleuromonas (dinoflagellate), 99
Pleuromonas (zoomastigote), 159
Pleuronemidae, 230
Pleurosigma, 75
Pleurostomella, 188
Pleurostomellida, 188
Pleurostomellidae, 188
Pleurotricha, 233
Pleurotrichidae, 233
Pleurotus, 152
Pleurotus ostreatus, 152

Plistophora, 222
Plistophoridae, 222
Plocapsilina, 186
Plocapsilinidae, 186
Plowrightia morbosa, 140
Pneumobacillus, 22
Podangium, 28
Podaxacea, 152
Podaxaceae, 152
Podaxon, 152
Podocyathus, 236
Podophrya, 235
Podophyridae, 235
Podophryina, 235
Podosphaera, 132, 133
Polyangiaceae, 28
Polyangidae, 27
Polyangium, 28
Polychaos, 202
Polychytrium, 117
Polycystidea, 209, 216
Polycystina (of Ehrenberg), 189, 198
Polycystina (of Delage and Hérouard), 217
Polydinida, 101
Polygastrica, 223
Polykrikida, 100, 101
Polykrikos, 101
Polymastigida, 158, 163, 164
Polymastigidae, 165
Polymastigina, 158, 163, 165
Polymastix, 163, 165
Polymastix melolonthae, 164
Polymorphina, 187
Polymorphinida, 187
Polymorphinidae, 187
Polymorphinina, 187
Polymyxa, 178, 179
Polyphagaceae, 117
Polyphagus, 111, 117
Polyphagus Euglenae, 116, 117
Polyporacea, 151
Polyporaceae, 151
Polyporales, 150
Polyporei, 151
Polyporus, 151
Polysiphonia nigrescens, 49
Polysiphonia violacea, 45, 46
Polysiphonieae, 51
Polysomatia, 179, 185
Polysphondylium, 203
Polysphondylium violaceum, 204
Polysporea, 209, 211
Polystichinae, 82
Polystictus, 151
Polystomella, 186, 187
Polystomella crispa, 181
Polystomellina, 187
Polythalamia, 179, 185
Polytoma, 61
Pontifex, 202
Pontisma, 81

Pontosphaera, 60
Pontosphaeraceae, 60
Porospora, 218
Porosporida, 218
Porosporidae, 218
Porphyra, 43
Porphyra laciniata, 42
Porphyra tenera, 42, 43
Porphyra umbilicaris, 42, 43
Porphyraceae, 43
Porphyrea, 41, 43
Porphyreae, 43
Porphyridiacea, 41
Porphyridiaceae, 41
Porphyridiales, 41
Porphyridium, 3, 40
Porphyridium cruentum, 41
Postelsia palmaeformis, 90, 91
Poteriochromonas, 60
Poteriodendron, 67
Poteriodendron petiolatum, 68
Pouchetia, 101
Pouchetiida, 100, 101
Pouchetiidae, 101
Prasiola, 3, 40, 44
Prasiolaceae, 44
Primalia, 37
Primigenium, Regnum, 37
Proboscoidella, 166
Progastréades, 94, 95
Pronoctiluca, 101
Prorocentraceae, 99
Prorocentrales, 99
Prorocentridae, 99
Prorocentrina, 99
Prorocentrinea, 98
Prorocentrinen, 99
Prorocentrum, 99
Prorodon, 229
Protamoeba, 189
Proteomyxa, 189, 190
Proteomyxiae, 189, 190
Proteomyxida, 189
Proteromonadidae, 159
Proteromonadina, 158
Proteromonas, 159
Proterospongia Haeckeli, 68
Proteus diffluens, 201
Proteus vulgaris, 22
Protista, 4, 6, 37, 189
Protistes trichocystifères, 94, 95
Protoascineae, 130
Protobasidiomycetes, 145, 146, 150
Protobionta, 6, 37
Protochrysis, 98
Protociliata, 228
Protoctista, 1, 4, 6, 8, 10, 37
Protodermieae, 171
Protodinifer, 101
Protodiniferida, 100, 101
Protodiniferidae, 101
Protodiscineae, 137

Protodontia Uda, 145
Protoflorideae, 41
Protogenes, 189
Protomastigales, 158
Protomastigida, 158
Protomastigina, 158
Protomastigineae, 158
Protomonas, 189, 191
Protomonadina, 158
Protomyces, 130
Protoopalina, 229
Protoopalinidae, 229
Protophyta, 6, 12, 18
Protoplasta, 39, 111, 157
Protoplasta filosa, 190
Protopsis, 101
Protozoa, 6, 12, 29, 37, 39, 223
Prowazekia, 159
Prunoidea, 195
Prunophracta, 197
Prymnesiidae, 58
Prymnesium, 58
Pseudomonas, 23
Pseudomonas aeruginosa, 23
Pseudospora, 159, 189, 191
Pseudosporea, 191
Pseudosporeae, 191
Pseudosporeen, 191
Pseudosporidae, 191
Pseudotetraedron, 66
Pseudotetraedron neglectum, 64
Psorosperms, 206
Psychodière, Regne, 37
Psychodiés, 37
Pteridomonas, 193
Pterocephalus, 218
Pterospora, 216
Ptychodiscida, 103
Puccinia, 143, 147, 148
Puccinia graminis, 147, 148
Puccinia Malvacearum, 148
Pucciniaceae, 148
Pucciniales, 147
Puffballs, 155, 172
Punctariales, 89
Pycnospermeae, 82, 89
Pylaiella, 86
Pyrenomycetales, 138
Pyrenomycetes, 137
Pyrenomycetineae, 137
Pyrgo, 185
Pyrocystis, 100
Pyronema, 127, 134, 135, 137
Pyronema confluens var. igneum, 127
Pyronemacea, 135
Pyrrhophycophyta, 94
Pyrrhophyta, 39, 94, 182
Pyrsonympha, 166
Pyrsonymphina, 163
Pythiacea, 80
Pythiaceae, 80

Quadrula, 205

Rabbit, 210
Raciborskya, 100
Radaisia, 36
Radioflagellata, 190
Radiolaria, 189, 190, 194, 196
Radiolariae, 189
Radiolarida, 189
Ralfsia, 87, 89
Ralfsiacea, 88
Ramularia, 139
Ramulinina, 187
Raphidophrys, 193
Raphidozoum, 195
Rat, 160
Ravenelia, 148
Red algae, see Algae, Red
Regne Psychodière, 37
Regnum Mycetoideum, 119
Regnum Primigenium, 37
Reophacida, 186
Reophacidae, 186
Reophax, 186
Reptiles, 212, 220
Reticularia, 175, 179
Reticulariacea, 174, 175
Reticulariaceae, 175
Reticularieae, 171
Reticulitermes, 171
Reticulosa, 179
Retortomonadidae, 165
Retortomonadina, 163
Retortomonas, 163, 165
Rhabdogeniae, 207
Rhabdosphaera, 60
Rhipidiacea, 77, 79
Rhipidiaceae, 79
Rhipidium, 79
Rhizammina, 183
Rhizamminidae, 183
Rhizaster, 63
Rhizidiacea, 115, 117
Rhizidiaceae, 117
Rhizidiomyces, 69
Rhizidiomyces apophysatus, 70
Rhizidiomycetaceae, 69
Rhizidium, 113, 117
Rhizinacea, 135
Rhizo-Flagellata, 158
Rhizobiacea, 19, 22, 23
Rhizobiaceae, 22
Rhizobium, 23
Rhizobium Leguminosarum, 23
Rhizochloridaceae, 66
Rhizochloridae, 66
Rhizochloridales, 63
Rhizochloridea, 63
Rhizochloridineae, 55, 63
Rhizochloris, 66
Rhizochrysidaceae, 63

Rhizochrysidae, 63
Rhizochrysidina, 61
Rhizochrysidinae, 61
Rhizochrysidineae, 55
Rhizochrysis, 61, 63
Rhizochrysis Scherffeli, 56
Rhizocryptineae, 95
Rhizoctonia, 142
Rhizodiniales, 99, 101
Rhizodininae, 95, 99
Rhizoflagellata, 157, 158, 160, 178, 192
Rhizomastigaceae, 163
Rhizomastigida, 158
Rhizomastigina, 158, 163
Rhizomastix, 163
Rhizopoda, 6, 63, 157, 172, 179, 184, 200, 205
Rhizopoda radiaria, 189, 194
Rhizopods, 179
Rhizopodes, 179
Rhizopogonacea, 155
Rhizopogonaceae, 155
Rhizopus, 121
Rhizopus nigricans, 122, 124
Rhizosolenia, 74
Rhizosoleniacea, 74
Rhizosoleniaceae, 74
Rhodobacillacea, 31, 237
Rhodobacillus, 31
Rhodobacteria, 30, 31
Rhodobacteriaceae, 31
Rhodochaetacea, 41, 43
Rhodochaetaceae, 43
Rhodochaete, 43
Rhodochorton, 47
Rhodomelaceae, 51
Rhodomeleae, 51
Rhodomonas, 98
Rhodomonas baltica, 97
Rhodophyceae, 6, 40
Rhodophycophyta, 40
Rhodophyllis, 49
Rhodophyta, 39, 40, 44
Rhodopseudomonas, 31
Rhodospermeae, 40
Rhodospirillum, 31
Rhodymeniacea, 51
Rhodymeniaceae, 51
Rhodymeniales, 51
Rhodymenieae, 51
Rhodymeninae, 51
Rhoicosphenia, 76
Rhoicosphenia curvata, 72
Rhopalodia, 75
Rhynchocystida, 216
Rhynchocystidae, 216
Rhynchocystis, 216
Rhynchomonas, 159
Rickettsia Melophagi, 21
Rickettsia Prowazekii, 21
Rickettsia Rickettsii, 21
Rickettsiacea, 19, 20, 118

Rickettsiaceae, 20
Rivularia, 36
Rivulariacea, 34, 36
Rivulariaceae, 36
Roach, 166, 168, 170
Rodents, 211
Roesia, 69
Rosaceae, 148
Rotalia, 184, 187
Rotaliaceae, 187
Rotalida, 187
Rotalidae, 187
Rotalidea, 187
Rotalina, 187
Rotifers, 113, 118, 219
Rozella, 118
Rugipes, 202
Rupertia, 187
Rupertiidae, 187
Russula, 143
Russula emetica, 145
Rusts, 145, 147
Rye, 148

Saccamminidae, 183
Saccharomyces cerevisiae, 130
Saccharomycetacea, 130
Saccharomycetaceae, 130
Saccharomycetes, 130
Saccinobaculus, 163, 166
Sagosphaerida, 199
Sagrina, 188
Salmonella, 22
Salpingoeca, 67
Salpingoeca ampullacea, 68
Salpingoeca Clarkii, 68
Salpingoecidae, 67
Sappinia, 203
Sappinia diploidea, 203
Sappinia pedata, 204
Sappiniaceae, 203
Sappiniidae, 203
Saprolegnia, 76, 79
Saprolegnia ferax, 78
Saprolegnia mixta, 78
Saprolegniaceae, 77
Saprolegniales, 77
Saprolegniea, 77
Saprolegnieae, 77
Saprolegniineae, 77
Saprolegnina, 77
Saprolegninae, 77
Sapromyces, 79
Saprospira, 29
Sarcina, 20
Sarcocystida, 214
Sarcocystidae, 214
Sarcocystidea, 214
Sarcocystis, 214
Sarcocystis Miescheriana, 214
Sarcocystis Muris, 214

Sarcodina, 6, 172, 200
Sarcosporidia, 207, 214
Sargassaceae, 91
Sargassea, 92
Sargasseae, 92
Sargassum, 93
Sargassum Horneri, 93
Sarkodina, 63, 157, 200
Schaudinella, 216
Schaudinellida, 216
Schaudinellidae, 216
Schinzia Leguminosarum, 23
Schizocystida, 215
Schizocystidae, 215
Schizocystinea, 215
Schizocystis, 215
Schizodinium, 102
Schizogoniacea, 41, 44
Schizogoniaceae, 44
Schizogonium, 44
Schizogregarinaria, 215
Schizogregarinida, 209, 215
Schizomycetae, 17, 18
Schizomycetes, 18, 206
Schizomycophyta, 17
Schizophyta, 12, 18
Schizophytae, 12
Schizosporea, 18
Schläuche, Mieschersche, 206, 214
Sciadiaceae, 66
Sciadophora, 218
Sclerocarpa, 129, 133, 135, 137, 145
Sclerocarpi, 137
Scleroderma, 143
Sclerodermataceae, 155
Sclerodermatales, 152
Sclerodermea, 155
Sclerodermei, 155
Sclerotinia, 140
Sclerotinia cinerea, 135, 136
Scytomonas pusilla, 108
Scytonema, 36
Scytonematacea, 34, 35
Scytonemataceae, 35
Sebacina, 149
Sebacina sublilacina, 145
Sebdenia, 49
Selenidium, 215
Seleniida, 215
Seleniidae, 215
Selenococcidiida, 211
Selenococcidiidae, 211
Selenococcidinea, 210
Selenococcidium intermedium, 211
Sennia, 97, 98
Sepedonei, 141
Septata, 217
Septobasidium, 147
Septoria, 139, 141
Sheep, 210, 214
Shigella, 22
Shigella dysenteriae, 22

Serratia, 22
Siderocapsa, 27
Sideromonas, 27
Siedleckia, 215
Silicina, 185
Silicoflagellata, 55, 56, 57, 61, 62, 64, 67, 69
Silicoflagellatae, 55, 62
Silicoflagellidae, 62
Silicoflagellina, 61
Silkworms, 206, 222
Sinuolinea, 221
Siphonaria, 117
Siphonogenerina, 188
Siphonomycetae, 77
Siphonophyceae, 55
Siphonotestales, 62
Sirolpidiacea, 81
Sirolpidiaceae, 81
Sirolpidium, 81
Sirosiphon, 36
Sirosiphonacea, 34, 36
Sirosiphonaceae, 36
Slavina, 222
Smuts, 145, 149
Snails, 161, 211
Snakes, 210
Snyderella, 168
Snyderella Tabogae, 164
Solenodinium, 199
Sorangiacea, 28
Sorangiaceae, 28
Sorangium, 28
Soranthera, 89
Sorites, 185
Soritidae, 185
Soritina, 185
Sorodiscus, 179
Sorophoreen, 203
Sorosphaera, 179
Sphacelaria, 86
Sphacelarialea, 85, 86
Sphacelariales, 86
Sphacelariea, 86
Sphacelarieae, 86
Sphaeractinomyxon, 222
Sphaerastrum, 193
Sphaerellaria, 194
Sphaeria, 138, 141
Sphaeria Scirpi, 128
Sphaeriaceae, 137
Sphaeriales, 137, 138, 139, 141
Sphaerida, 195
Sphaeridea, 194
Sphaerioidaceae, 141
Sphaerioideae, 141
Sphaerita, 118
Sphaerobolacea, 155
Sphaerobolaceae, 155
Sphaerobolus, 155
Sphaerocapsa, 197
Sphaerocapsida, 197

Sphaerocladia, 112, 113
Sphaerococcales, 47
Sphaerococcoidea, 46, 47, 50
Sphaerococcoideae, 47
Sphaeroeca, 67
Sphaeroidea, 195
Sphaeroidina (genus of Rhizopoda), 187
Sphaeroidina (family of Radiolaria), 195
Sphaeromyxa, 221
Sphaerophracta, 197
Sphaerophrya, 235
Sphaeropsidales, 141
Sphaeropsideae, 141
Sphaerospora, 221
Sphaerosporida, 221
Sphaerosporidae, 221
Sphaerosporea, 221
Sphaerotheca, 127, 133
Sphaerotheca pannosa, 133
Sphaerotilacea, 33
Sphaerotilaceae, 33
Sphaerotilalea, 30, 33, 237
Sphaerotilus, 30
Sphaerotilus natans, 33
Sphaerozoen, 194
Sphaerozoida, 195
Sphaerozoum, 189, 195
Spirillacea, 19, 23
Spirillaceae, 23
Spirillina, 181, 182, 185
Spirillinidea, 185
Spirillinina, 185
Spirillum, 24
Spirochaeta, 29
Spirochaeta cytophaga, 26, 27
Spirochaeta plicatilis, 28, 29
Spirochaetacea, 29
Spirochaetaceae, 29
Spirochaetae, 27
Spirochaetalea, 28
Spirochaetales, 28
Spirochaets, 12, 14, 166, 167
Spirochona, 233, 235
Spirochonidae, 231
Spirochonina, 230, 231
Spirocystida, 215
Spirocystidae, 215
Spirocystidées, 215
Spirocystis, 215
Spirodinium, 231
Spirodiscus, 66
Spirodiscus fulvus, 64, 66
Spirogyrales, 121
Spirolina, 185
Spironema, 222
Spirophyllum, 27
Spirostomum, 230
Spirotricha, 230
Spirotrichida, 230
Spirotrichonympha, 168, 169
Spirotrichonymphidae, 169
Spirotrichonymphina, 168

Spirulina, 35
Sponges, 37, 67
Spongocarpeae, 50
Spongospora, 179
Spongurida, 195
Spongurus, 195
Sporobolomyces, 145
Sporochnales, 87
Sporochnea, 88
Sporochnoidea, 85, 87, 89
Sporochnoideae, 87
Sporochnus, 93
Sporodinia, 124
Sporochytriaceae, 117
Sporomyxa, 179
Sporozoa, 111, 206, 207, 219
Sporozoans, 21, 162
Sporozoaires, 207
Sporozoaria, 206, 207
Spumaria, 177
Spumariaceae, 177
Spumellaria, 194, 195
Spyrida, 198
Spyridieae, 51
Spyridina, 198
Spyroidea, 198
Squamarieae, 50
Squids, 210
Staphylococcus, 20
Staurocyclia, 195
Staurojoenina, 169
Staurojoenina assimilis, 170
Staurojoeninida, 169
Staurojoeninidae, 169
Stelangium, 28
Stemonitaceae, 171, 175
Stemonitales, 171, 174
Stemonitea, 174, 175
Stemonitidae, 175
Stemonitis, 175
Stemonitis splendens, 176
Stenophora, 217
Stenophorida, 217
Stenophoridae, 217
Stentor, 227, 230
Stentor coeruleus, 225
Stentoridae, 230
Stentorina, 230
Stephanida, 198
Stephanonympha, 168
Stephida, 198
Stephoidea, 198
Stereotestales, 62
Stereum, 151
Stictaceae, 134
Stictea, 134
Sticteae, 134
Stictidaceae, 134
Stictideae, 133
Stigonema, 36
Stigonemataceae, 36
Stilbaceae, 142

Stilbeae, 142
Stilbellacea, 142
Stilbellaceae, 142
Stilbosporei, 141
Stilbum, 142
Stilophora, 88
Stilotricha, 233
Stipitochloridae, 66
Stipitococcacea, 65, 66
Stipitococcaceae, 66
Stipitococcus, 66
Stokesiella, 60
Stomaticae, 74
Stomatoda, 223, 228, 233
Stomatophora, 216
Stomatophorida, 216
Stomatophoridae, 216
Streblomastigida, 165, 166
Streblomastigidae, 166
Streblomastix, 163, 168
Streblomastix Strix, 164, 166
Streblonema, 86
Streptococcus, 20
Streptomyces, 25
Streptomycetaceae, 25
Streptothrix, 25
Striatae, 74
Stylobryon, 60
Stylocephalida, 218
Stylocephalidae, 218
Stylocephalus, 218
Stylochrysalis, 59
Stylocometes, 236
Stylodinium, 100
Stylonychia, 227, 232, 233
Stylopage, 124
Stylopyxis, 60
Stylorhynchidae, 218
Stypocaulon, 83, 84, 86
Suctorea, 235
Suctoria, 235
Surirella, 71, 73, 75
Surirella saxonica, 72, 73
Surirellaceae, 75
Surirellea, 75
Surirelleae, 75
Swine, 210, 214
Symbelaria, 194
Symploca Muscorum, 13
Synactinomyxida, 222
Synactinomyxidae, 222
Synactinomyxon, 222
Synchytriacea, 115, 117
Synchytriaceae, 117
Synchtrium, 117
Syncephalastrum, 124
Syncephalastrum racemosum, 122
Syncephalis, 123, 124
Syncephalis nodosa, 122
Syncephalis pycnosperma, 122
Syncollaria, 194
Syncrypta, 59

Syncryptaceae, 59
Syncryptida, 58, 59
Syncryptidae, 59
Syncystida, 216
Syncystidae, 216
Syncystis, 216
Syndinidae, 102
Synedra, 72, 75
Syntamiidae, 86
Synura, 55, 59
Synura Uvella, 54
Synuraceae, 59
Syracosphaera, 60
Syracosphaera Quadricornu, 56
Syracosphaeraceae, 60
Syracosphaeridae, 60
Syracosphaerinae, 57, 60

Tabellaria, 75
Tabellariaceae, 75
Tabellariea, 75
Tabellarieae, 75
Taphrina, 127, 137
Taphrina aurea, 137
Taphrina deformans, 127, 136, 137
Teliosporeae, 142
Telomyxa, 222
Telomyxa glugeiformis, 222
Telomyxida, 222
Telomyxidae, 222
Telosporidea, 207
Telosporidia, 207
Tentaculifera, 224, 228, 235
Teratonympha, 171
Teratonympha mirabilis, 170
Teratonymphida, 169
Teratonymphidae, 169
Termites, 166, 167, 168, 169
Termitidae, 168
Termopsis, 166, 168
Testacea, 205
Testacida, 205
Testaceolobosa, 205
Tetractinomyxida, 222
Tetractinomyxidae, 222
Tetractinomyxon, 222
Tetradinium, 100
Tetradinium javanicum, 104
Tetrahymena, 229
Tetrahymena Geleii, 227
Tetramitaceae, 165
Tetramitida, 165
Tetramitidae, 165
Tetramitina, 165
Tetramitus, 165
Tetramyxa, 179
Tetrasporeae, 82, 86
Tetrasporées, 209
Tetrataxis, 186
Textularia, 182, 186
Textulariaceae, 186

Textularidae, 186
Textularidea, 185
Textularina, 186
Textulinida, 186
Thalamophora, 179
Thalassicolla, 189, 194, 195
Thalassicollen, 194, 195
Thalassicollida, 195, 199
Thallochrysidacea, 62, 63
Thallochrysidaceae, 63
Thallochrysis, 63
Thamnidium, 124
Thaumatomastix, 109
Thaumatomonadidae, 109
Thaumatonema, 109
Thaumatonemidae, 109
Thecamoeba, 202
Thecamoebae, 205
Thecamoebida, 201, 202
Thecamoebidae, 202
Theileria, 214
Theileridae, 214
Thelephora, 151
Thelephoracea, 151
Thelephoraceae, 151
Thelephorei, 151
Thelohania, 222
Theoconus, 198
Thiere, 172
Thiobacillus, 24
Thiobacteria, 30, 31, 35
Thiorhodaceae, 31
Thioploca, 35
Thiospira, 24, 31
Thiospirillum, 31
Thiothrix, 35
Thoracosphaeraceae, 60
Thoracosphaeridae, 60
Thorea, 47
Thraustochytriacea, 81, 82
Thraustochytriaceae, 82
Thraustochytrium proliferum, 82
Thraustotheca, 79
Ticks, 161, 206
Tilletia, 149
Tilletia Tritici, 145
Tilletiacea, 149
Tilletiaceae, 149
Tilopteridales, 86
Tilopteridea, 87
Tilopterideae, 87
Tilopteris, 87
Timothy, 148
Tinoporidea, 187
Tinoporus, 187
Tintinnidae, 231
Tintinnids, 224
Tintinnina, 231
Tintinnodea, 231
Tintinnoinea, 231
Tipulocystis, 215
Toads, 125

Toadstools, 151
Tokophrya, 235
Tokophrya Lemnarum, 234
Tolypothrix, 35, 36
Torula, 130
Torulopsis, 130
Toxonema, 222
Tracheliidae, 230
Trachelina, 230
Trachelius, 230
Trachelomonas, 94, 106, 107
Transchelia, 143
Tremella, 149
Tremella Auricula, 146
Tremellacea, 149
Tremellaceae, 149
Tremellales, 149
Tremellina, 146, 149, 150
Tremellineae, 146, 149
Tremellinei, 149
Tremellini, 149
Tremellodendron, 149
Trepomonadida, 165, 166
Trepomonadidae, 166
Trepomonas, 166
Treponema, 29
Treponema macrodentium, 29
Treponema microdentium, 29
Treponema pallidum, 28, 29
Treponema pertenue, 29
Treponematacea, 29
Treponemataceae, 29
Tretomphalus, 180
Triactinomyxon, 222
Triactinomyxidae, 222
Tribonema, 65, 66, 73, 95
Tribonema bombycina, 64
Tribonematacea, 65, 66
Tribonemataceae, 66
Triceratium, 74
Tricercomitus, 167
Tricercomitus Termopsidis, 164
Trichamoeba, 202
Trichia, 176, 177
Trichiacea, 174, 176
Trichiaceae, 171, 176
Trichiales, 171, 174
Trichiidae, 177
Trichina, 177
Trichinaceae, 176
Trichoblasteae, 51
Trichocystifères, Protistes, 94, 95
Trichodina, 235
Trichodinidae, 235
Trichomitus, 166
Trichomonadida, 166, 167
Trichomonadidae, 166, 167
Trichomonadina, 158, 164, 166
Trichomonads, 165
Trichomonas, 166, 167
Trichomonas hominis, 165
Trichomonas tenax, 164, 167

Trichomonas Termopsidis, 168
Trichomonas vaginalis, 167
Trichonympha, 168, 169, 170
Trichonympha Campanula, 168, 170
Trichonympha sphaerica, 168
Trichonymphida, 169
Trichonymphidae, 169
Trichonymphidea, 168
Trichonymphina, 168
Trichophyton, 142
Trichospermi, 152, 171
Trichostomata, 229
Tridictyopus elegans, 196
Trigonomonas, 166
Triloculina, 184, 185
Trimastigaceae, 58
Trimastigida, 58, 165
Trimastigidae, 58
Trimastix, 58
Trinema, 191
Triplagia, 198
Triposolenia, 103
Triposolenia Ambulatrix, 104
Tripylaria, 199
Tripylea, 199
Tripyleen, 199
Tripylina, 199
Triticina, 188
Trochammina, 186
Trochamminida, 186
Trochamminidae, 186
Trochamminina, 186
Truffles, 135
Tryblidacea, 134
Tryblidaceae, 134
Tryblidieae, 133
Trypanophidae, 161
Trypanophis, 161
Trypanoplasma, 161
Trypanoplasmida, 159, 161
Trypanoplasmidae, 161
Trypanosoma, 162
Trypanosoma Brucii, 160, 162
Trypanosoma Cruzi, 162
Trypanosoma equinum, 162
Trypanosoma equiperdum, 162
Trypanosoma Evansi, 162
Trypanosoma gambiense, 162
Trypanosoma Lewisi, 160
Trypanosomata, 158
Trypanosomatidae, 161
Trypanosomes, 161, 212
Trypanosomidae, 161
Trypanosomidea, 158
Tuberacea, 135
Tuberaceae, 134
Tuberales, 134
Tuberculariaceae, 141
Tuberculariea, 141
Tubercularieae, 141
Tubercularini, 141
Tuberineae, 134

Tubifer, 175
Tubiferaceae, 175
Tubiferida, 174, 175
Tubiferidae, 175
Tubinella, 185
Tubulina, 175
Tubulinaceae, 175
Tubulinidae, 175
Tuburcinia, 149
Tulasnella, 149, 150
Tulasnella sphaerospora, 145
Tulasnellales, 149
Tulostoma, 155
Tulostomataceae, 155
Tulostomea, 155
Tulostomei, 155
Tunicates, 216
Turillina, 188
Turkeys, 210
Turtles, 211
Tuscarilla, 200
Tuscarora, 200
Tuscarorida, 200

Ulvina aceti, 24
Umbina aceti, 24
Uncinula, 132, 133
Uniflagellatae, 110
Urceolaria, 235
Urceolaridae, 235
Urceolarina, 235
Urceolus, 109
Uredinacea, 148
Uredinaceae, 148
Uredinales, 145, 147
Uredineae, 147
Urédinées, 147
Uredo, 147
Uredo linearis, 147
Urnula, 235
Urocentridae, 230
Urocentrina, 230
Urocentrum, 230
Uroglena, 59
Uroglenopsis, 59
Uroleptus, 233
Uromyces, 143
Urophagus, 166
Urophlyctis, 117
Urospora, 216
Urosporida, 216
Urosporidae, 216
Urosporidium, 218
Urostyla, 233
Urostylida, 233
Urostylidae, 233
Ustilaginacea, 149
Ustilaginaceae, 149
Ustilaginales, 149
Ustilaginea, 146, 149
Ustilagineae, 149

Ustilago, 149
Ustilago Heufleri, 145
Ustilago Hordei, 145
Uterini, 134, 137
Uvella, 59
Uvellina, 188
Uvigerina, 188
Uvigerinida, 188
Uvigerinidae, 188

Vacuolaria, 65, 109
Vacuolaria viridis, 108
Vacuolariaceae, 109
Vaginicola, 233
Vaginifera, 233
Vaginulina, 187
Vahlkampfia, 202, 203
Valsa, 139
Valvulina, 186
Valvulinidae, 186
Vampyrella, 118, 189, 191, 192
Vampyrellacea, 191
Vampyrellaceae, 191
Vampyrelleae, 191
Vampyrellidae, 191
Vampyrellidea, 190
Vaucheria, 67, 76
Vaucheria Gardneri, 64
Vaucheria sessilis, 64
Vaucheriacea, 57, 63, 64
Vaucheriaceae, 63, 67
Vaucheriales, 63
Vaucherioideae, 55
Venturia, 139
Venturia inaequalis, 139
Verbeekina, 188
Vermes, 9
Verneulina, 186
Verneulinidae, 186
Veronica, 69
Verrucariacea, 139
Vertebralina, 184, 185
Vertebrates, 161, 165, 166, 167, 210, 211
Vibrio, 23
Vibrio Protheus, 201
Virgulina, 188
Volvox Chaos, 201
Vorticella, 223, 226, 233, 235
Vorticellidae, 233
Vorticellina, 233
Vorticiales, 179
Vorticialis, 186, 187
Vulvulina, 186

Wagnerella, 193, 194
Wardia, 221

Water molds, 77
Whales, 71
Wheat, 148
Wood roach, 166, 169
Worms, 215, 217, 220
Worms, annelid, 216, 219, 221
Worms, oligochaet, 222
Worms, polychaet, 211
Worms, siphunculid, 210
Woronina, 179
Woroninaceae, 179
Woroninidae, 179
Wrangelieae, 47

Xanthomonadina, 63
Xanthomonas, 23
Xenococcus, 36
Xiphacantha, 197
Xylaria, 139

Zanardinia, 88
Zea Mays, 6
Zelleriella, 229
Zonaria, 87
Zooflagellata, 157
Zoomastigina, 157
Zoomastigoda, 157, 178
Zoomastigophorea, 157
Zoopagacea, 123, 124
Zoopagaceae, 124
Zoopagales, 121
Zoopage, 124
Zoophagus, 81
Zoosporidae, 191
Zoosporidea, 191
Zoosporidia, 190
Zoothamnium, 233
Zooxanthellae, 194
Zostera, 203
Zschokkella, 221
Zygochytrium, 118
Zygocystis, 216
Zygocystida, 216
Zygocystidae, 216
Zygomyceteae, 121
Zygomyceten, 121
Zygomycetes, 76, 118, 120, 121, 122, 127, 141
Zygophyceae, 53
Zygophyta, 53
Zygorhynchus, 124
Zygostephanus, 198
Zythiacea, 141
Zythiaceae, 141